Models of the
Structural-Functional
Organization
of Certain
Biological Systems

The MIT Press
Cambridge, Massachusetts and
London, England

Models of the
Structural-Functional
Organization
of Certain
Biological Systems

Edited by
I. M. Gelfand with
V. S. Gurfinkel
S. V. Fomin
M. L. Tsetlin

Translated from
the Russian by
Carol R. Beard

Foreword by
Peter H. Greene

Translation reviewed by
John S. Barlow, M.D.

Originally published under the auspices of the Academy of Science of the USSR (Institute of Biological Physics) under the title "Modeli Strukturno-Funktionalnoy Organisatsii Nyekotoyh Biologichyeskikh Sistem," Moscow, 1966.

ISBN 262 07 042 1 (hardcover)

Library of Congress catalog card number: 68–20046.

Set in Monotype Times New Roman
by William Clowes & Sons, Limited
printed on Glatfelter Old Forge
by The Maple Press Company
and bound
by The Maple Press Company
in the United States of America

N. A. Bernstein (1896–1966)

CONTENTS

Introduction
xi

Dedication
xxxiii

Preface
xxxvii

1.
Mathematical Modeling of   I. M. Gelfand and
Mechanisms of the Central   M. L. Tsetlin
Nervous System
1

I

The Spreading of
Excitation in Continuous Media
and the Electrical Properties of
Branching Structures

2.
An Analysis of the Functional   Yu. I. Arshavskiy,
Properties of Dendrites in   M. B. Berkinblit, S. A. Kovalev,
Relation to Their Structure   V. V. Smolyaninov, and
25   L. M. Chaylakhyan

3.
The Electrical Behavior of the   M. B. Berkinblit, S. A. Kovalev,
Myocardium as a System and the   V. V. Smolyaninov, and
Characteristics of the Cellular   L. M. Chaylakhyan
Membrane of the Heart
78

4.
The Problem of the Electrical   V. V. Smolyaninov
Properties of Syncytia
132

5.
The Periodic Blocking of   M. B. Berkinblit
Impulses in Excitable Tissues
155

II

The Organization of Certain
Parts of the Central Nervous
System and the Modeling of
Their Operation

6.
Characteristics of the
Respiratory Neurons of
Different Levels of the Central
Nervous System
193

I. A. Keder-Stepanova

7.
A Model of a System of Neurons
with Periodic Volley Activity,
Resistant to Random Afferent
Influences
234

I. A. Keder-Stepanova and
N. N. Rikko

8.
Some Special Features of
Organization of the Cerebellar
Cortex
251

V. V. Smolyaninov

III

The Regulation of Movements

9.
Some Problems in the Analysis
of Movements
329

I. M. Gelfand, V. S. Gurfinkel,
M. L. Tsetlin, and M. L. Shik

10.
An Analysis of Physiological
Tremor by Means of a
General-Purpose Computer
346

V. S. Gurfinkel, L. E. Sotnikova,
O. D. Tereshkov, S. V. Fomin,
and M. L. Shik

11.
Concerning Tuning before
Movement
361

V. S. Gurfinkel, Ya. M. Kots,
V. I. Krinskiy, E. I. Paltsev,
A. G. Feldman, M. L. Tsetlin,
and M. L. Shik

12.
The Control of Movements of
the Joints and Kinesthetic
Afferentation
373

Ya. M. Kots, V. I. Krinskiy,
V. L. Naydin, and M. L. Shik

13.
The Compensation of Respiratory
Disturbances of the Erect Posture
of Man as an Example of the
Organization of Interarticular
Interaction
382

V. S. Gurfinkel, Ya. M. Kots,
E. I. Paltsev, and
A. G. Feldman

Index
396

This book surveys the ongoing activity of several theoretical and experimental groups at the Institute of Problems of Information Transmission (formerly called the Institute of Biological Physics). The several chapters introduce, and explain the reasons behind, programs of research that collectively represent a remarkably broad and important effort toward understanding the nervous system and other excitable systems at the levels of cells, tissues, organs, and tasks. Since these chapters were written, their authors have remained leaders in the areas they survey, and the further fruits of their research have added to the value of the present chapters as unified introductions to the subjects that enable the reader to appreciate the fundamental questions and coherent points of view that have guided the authors' research. Such guides (particularly when the facts continue to affirm their reliability) are badly needed when studying a large-scale system like the nervous system, whose meaningful elements of information are largely unknown, so that, without a guide, one may be encouraged to amass reams of unanalyzable and infinitely analyzable data.

It is easy for the English-speaking scientist to keep up with the subsequent research of these authors: simply look in *Biophysics* (English translation of *Biofizika*) every two months, particularly under the headings "Biophysics of complex systems. Mathematical models" and "Letters to the Editor." Therefore, this introduction will mainly discuss how their investigations of various levels of biological systems fit together with each other and with work in other fields.[1]

The first chapter discusses mathematical modeling that may become applicable to the study of large-scale systems, including biological systems. The authors begin with a method they had developed for minimizing a function of many variables. If the value of the function is regarded as the elevation of a surface, then their procedure is designed to work efficiently

[1] I thank M. A. Aizerman, M. H. Cohen, S. V. Fomin, G. L. Gottlieb, V. S. Gurfinkel, and R. Llinás, for helpful discussions of some of the ideas reviewed here; and T. A. Easton for making available to me his translation of this book, which first brought to my attention work of these authors that enabled me to crystallize ideas that had long tantalized me. My research concerning the implications of Part III has been supported by the Information Systems Branch of the Office of Naval Research under Contracts Nonr 2121(17) NR 049–148 and N00014–67–A–0285–0012.

in case the minimum lies somewhere along the bottom of a long, narrow ravine that has a relatively level bottom but steep sides. The special difficulty of such a search is that, if the precise direction of the ravine is unknown, a step large enough with respect to the gentle slope of the bottom usually climbs up the steep wall, thus arriving at a higher, rather than a lower point. The authors have developed an efficient way to escape this difficulty.

Since the reader can easily compare this solution with the literature on optimization, I shall only mention the authors' use of this example to discuss research on large-scale systems such as the nervous system. In Chapter 1 [and at greater length in 19] the authors explain how special features of a situation (here, the steep walls and almost level bottom) bring into existence special problems and methods that have no significance in the most general situations (e.g., in optimization of general functions, not known to be organized in any special way). Typically, universes of restricted generality allow the successful operation of simple mechanisms that will not work in general universes. In the present example, what is important for the use of the method is not the a priori knowledge of the direction of the ravine, but only the fact that there is some organization of the variables that can be described as a ravine. In [19], the authors emphasize the importance of utilizing such crude a priori knowledge of organization, analogous to a hypothesis about the universe of possibilities, in controlling the myriad degrees of freedom of the muscular system. Remarking upon the extraordinary precision required to strike a ball properly in billiards, of a fineness exceeding the resolution of the eye, they draw the following analogy:

Perhaps the paradoxical increase in precision occurs owing to the use of a suitable system of hypotheses. Thus, for example throwing a [circle] of large radius onto a millimetre mesh, one can determine its diameter with a precision considerably better than 1 mm. Here the use of the hypothesis that the figure in question is a [circle] accounts for the "precision paradox." This hypothesis organizes the measurements and allows one to work up their results so that great precision is attained . . .

It is possible to suppose that the precision of a complex system of proprioceptors considerably exceeds the resolving power of the eye or the precision of a single proprioceptor. We have such a situation typically in the case of physiological measuring systems. Their precision considerably exceeds the precision of the separate elements on account of the use of a suitable system of organizing hypotheses.

The authors go on, in Chapter 1, to cite examples of searching movements of the joints in minimizing deviations from a desired posture, in which they suppose that the mechanism of their "method of ravines" may be found. Although the deviations do fluctuate in a fashion qualitatively resembling the movements of ravine search, it is not clear that the precise ravine algorithm is being utilized. Nonetheless, the idea of seeking ways in which special features of organization allow novel techniques to emerge seems to me an essential step in coming to understand how the brain's tasks emerge from its tissues. Such phenomena cannot be apprehended by studying subjects like automaton theory, decision theory, or theories of changes in synaptic conductivities, without also taking into account special structural features analyzed at their appropriate levels of organization.[2] In an age of brain measurements that fail to provide models for what the brain does, and of unstructured cybernetic network models that are expected, when properly adjusted, to perform wonders, we must urgently look to the work of scientists such as the authors, who seek hypotheses at an appropriate level of organization.

Chapter 1 continues with other examples of systems that make use of hypotheses about their universe, notably a sequence of automata engaged in maximizing payoffs obtainable from a collection of environments. The authors' "asymptotically optimal" sequence of automata consists of automata whose competence eventually increases with lateness of position within the sequence, so that their index within the sequence becomes a measure of the complexity of their state space or memory, as they take into account the features of their environments. I mention these automata merely to say that a colleague has almost convinced me that the word "party" in the discussion—meaning a chess-playing party, together with discussion and analysis—is the very best translation of what the authors had in mind.

Chapter 1 concludes with speculations about the existence of some minimal principle in the functioning of the nervous system, specifically a

[2] Analogously, the placement of colored tiles in a mosaic is determined, in part, by structural features that must be stated in terms—such as parallel lines and vanishing points—that cannot be expressed in terms of tiles, even though the mosaic consists of nothing but tiles.

principle of least "interaction." It seems to me, however, that the authors point to examples in which something is minimized, and then call that something—whatever it is—"interaction." Now biology abounds with instances of structure that is optimal for some purpose [53, 62]. However, if we consider all these examples of optimization, we may ask whether the whole collection possesses the kind of unity expressed by minimum principles of physics, which stem somehow from the presence of analytical dynamics at the confluence of several mainstreams of mathematics (analysis, differential geometry, topology). Although only time will tell whether a generalized "minimum principle" is fruitful, all the authors' examples are certainly important to study. Can these examples be related to the mathematical theory of Mesarovic, Macko, and Takahara [46] concerning the forms of interaction among subsystems needed in order to optimize a multilevel system?

We turn now to the chapters on the spread of excitation in branching structures and in continuous media, and applications of these ideas to modeling parts of the nervous system. Chapter 2 concerns dendritic excitation of neurons. Contrary to much of the literature, the authors suppose that all-or-none excitation of the dendrites plays an important role in the electrical activity of neurons. Thus, they adopt the hypothesis that the dendritic membrane is not essentially unlike the membrane of the body of the neuron. From this hypothesis and data on the electronic properties of the dendrites, the authors derive interesting possibilities for influences of dendritic synapses upon the soma, for influences of somatic excitation upon the excitability of the dendrites, for the role of recurrent axon collaterals synapsing upon the dendrites, and for logical operations upon signals through local modifications of dendritic excitability, particularly at the nodes of branching. All agree that such possibilities are attractive (and would allow tuning operations, as in Part III), but not everyone would agree that these possibilities actually exist. Analysis of electrical properties of dendrites is relatively advanced (e.g., in the authors' references and [49, 50, 51], as well as numerous papers in *Biophysics*),[3] and evidence for all-or-none conduction within dendrites has

[3] Analogous studies [e.g., 55 and Chapter 4] make use of Green's function methods to analyze spread of excitation in branching structures and networks and other excitable media.

been forthcoming [39, 40, 41], but, as the authors point out, experimental conclusions about dendritic activity have always been uncertain and controversial, largely because the interpretation of data has rested upon plausible, yet not firmly established, assumptions about the nature of the electrical fields surrounding branched neurons. Now, however, intracellular recordings of dendrites and somata in alligator Purkinje cells [42] have directly demonstrated that the dendrites are able to generate spikes. By introducing procion yellow stain through their electrodes, Llinás and Nicholson were able to mark the precise places where the spikes were recorded.

The authors of Chapter 2 suppose only that the dendrites, as well as the soma and axon hillock, are excitable, and do not insist that the membrane is thoroughly homogeneous. Indeed, the axon hillock is more excitable than the soma, and Llinás and Nicholson found particular locations on the dendrites that were more sensitive than other dendritic locations. In fact, they found evidence that a sensitive spot near a bifurcation of the dendrite generates spikes that spread electrotonically until they reach the next bifurcation, where a sensitive spot regenerates new spikes. From their analysis of current density measurements, they conclude that the sources of the spikes are at the nodes of branching, although Y. Y. Zeevi (unpublished) believes the sources are slighly proximal to the nodes. Through controlled excitation of specific regions of the neuron, Llinás and Nicholson found that the effect of a dendritic spike will be transmitted up some dendrites, but not others, depending upon the pattern of excitation. In their observations of alligator Purkinje cells, they were unable to reset or modulate dendritic activity through somatic excitation, although they believe the situation may be different in certain motor neurons. Their results lead them, in accord with the views of Chapter 2, to believe in the prevalence and importance of dendritic switching and local modification.

Thus, there appears, at last, to be indisputable evidence for the existence of neurons satisfying the hypotheses underlying the analysis of Chapter 2. It will be particularly valuable if these experimental techniques will be come usable to refine the necessarily crude ideas of those who have tried to extract and synthesize meaning from our sketchy knowledge of activation and tuning in neural circuitry [9, 10, 16, 22, 31, 43, 44, 45, 61].

I shall discuss Chapters 3 through 7 as a unit, in regard to their placement in even broader contexts than those explicitly mentioned by the authors. These chapters concern the wavelike spread of excitation in excitable media. The elements of these systems, in idealized form, have three states: excited, refractory, and unexcited but responsive. An excited element triggers its neighbors after some phase delay. This mechanism leads to propagated waves, which can maintain themselves through excitation of new elements and elements that have recovered from refractoriness. Waves in excitable media are very different from the wave motion ordinarily encountered in physics.[4]

Physicists have long studied the propagation of waves, such as sound waves and electromagnetic waves, that are superposable: the deviations coming from two different waves are added. In contrast to classical wave motion, wave amplitudes in excitable media are discrete valued, for, in regard to signaling, a cell is, at a particular time, either on or off. Because the signal from a neighbor's response is sufficient to initiate the response of a cell, the cell's response is the same regardless of the number of neighbors signaling at a given time: it either becomes excited or it does not. Therefore, the waves are not additive in the sense of linear superposition, and other waves do not exert any influence on an excited point in a given wave. Classical waves freely penetrate through each other, and, coming to the boundary of the medium, they can be partly reflected and partly transmitted. In contrast, owing to the refractory period, when two waves in an excitable medium collide, they are both annihilated, and waves cannot be reflected from the boundary of the medium; transmission often cannot occur, because the boundary coincides with the boundary of the whole biological system or with a boundary of a tissue across which there is no communication. Thus, one expects important features to be uniquely characteristic of wave propagation in excitable media. At the center of the exploration of these features are the investigations represented in this book and numerous subsequent publications in *Biophysics*. Although the studies of boundary conditions and particular effects of

[4] I am following summaries given in [20, 33].

shape and internal imperfections lead to predictions of behavior peculiar to these models, one can still use boundary conditions and analogues of the Huyghens' construction [e.g., 20, 28] to rebuild an entire theory of wave propagation in excitable media.

The origin of much of this work was the Wiener-Rosenblueth study of heart muscle [63], which led, through the contributions of others, to models for cardiac fibrillation, as reviewed in this book and related studies [e.g., 29, 33]. However, as pointed out by the authors, such theories do not apply just to the heart. Wave conduction also occurs in peristaltic waves of the gut, and rudiments of the present ideas may be seen in analysis of the central nervous system [8]. Still more general than these instances is, for example, the Zhabotinsky reaction [67], in which a concentration wave propagating in a two-dimensional liquid phase system of bromomalonic acid and ferroin shows (in dramatic motion pictures) the characteristic features of wave interaction in excitable media.

B. C. Goodwin and M. H. Cohen [20] proposed that some embryonic fields may be considered as excitable media, and that control of development takes place by propagation of waves that they have called "organizing waves." They proposed that position within an embryonic field may be known by the phase-shift between two waves, one of which is propagated with more delay than the other, and they have shown how such a mechanism might account for many features of the spatial and temporal organization of developing systems.

M. H. Cohen and A. D. J. Robertson [14, 52] have analyzed the development of the amoeboid slime mold *Dictyostelium discoideum* from the point of view of an organizing wave propagating in an excitable medium, and have developed a quantitative theory of the early aggregation of these and other slime molds; and these waves are related to various models for the control of development in [13].

Still more examples of activity in active media are explored in Chapters 5, 6, 7, which examine the production of stable modes of oscillation in heart and respiratory system, and cite experimental investigations of similar mechanisms in insects. Further studies may be found in *Biophysics* [e.g., 6, 30]. For additional discussions of rhythm generation in insect flight, see also [65, 66]. A mechanism controlling respiratory rhythms is analyzed, more from the point of view of Part III of this book, by means of experiments conducted in M. A. Aizerman's laboratory [59, 60, 68, 69, 70]. An entirely different approach to rhythms of neural activity, through a statistical mechanics of coupled oscillators, has been pursued by J. D. Cowan [15].

In summary, the biophysical, chemical, and electrical phenomena found in excitable media are not special and unusual occurrences, but play an important role in myocardium, in respiration, in other neuro-muscular systems, in the central nervous system, in development, and even in relatively simple chemical systems. The authors represented here themselves provide an organizing wave of activity in this important subject!

Chapter 8 presents information about the geometric organization of tissues, cells, and connections in the cerebellar cortex, and complements related studies such as [16, 17, 38, 39, 57, 58], which include some discussions of possible relations between structure and function.

The motivation for the ongoing work introduced in Part III, on the regulation of movements, is the idea that the higher motor centers do not control each motor neuron or each mechanical degree of freedom individually; rather, they transmit signals that change the states and the mode of interaction of lower centers, which then, relatively autonomously, produce combinations of movements. The first stage of a motor task, according to this view, is the coming together of joints and muscles into a small number of closely knit groups, with a resultant reduction in the number of degrees of freedom of the motor apparatus, since the higher centers have only to select, and possibly tune, these groups [47]. This research program was inspired by the ideas of the physiologist N. A. Bernstein [7, 12]. The masses of undigested details, the lacks of agreement, and the inconclusiveness that mark the long history of investigations of motor mechanisms, arise from our limited ability to recognize the significant informational units of movement, so that any observation probably overlooks something relevant. In this situation, we can once more be thankful to the present group for bringing to this subject an organizing point of view. This point of view places special emphasis upon the rigorous analysis of mechanisms that have long been classical concerns of neurophysiologists, the programmed combinations, or *synergies*, of muscle movements that play a primary role in voluntary movement. In Part III, the authors tell what they are trying to do, and why they are trying to do it. Their coherent program has yielded striking results, concerning which the last five years of *Biophysics* form a systematic treatise. I shall name only a few of the general areas of this later research and discuss, instead, some important related research in other laboratories and what I believe to be still wider implications of this work than the authors had in mind.

The general areas of research include search for correct postural, configurations and other examples of minimization [e.g., 5, 35, 36], tuning before movement [e.g., 18, 32]; and mechanisms of local, spinal, and supraspinal control of limbs in walking and running dogs and cats, including mesencephalic cats [e.g., 34, 48, 54].

Closely related mechanisms of tuning interactions between voluntary and postural mechanisms were studied by G. L. Gottlieb, G. C. Agarwal, and L. Stark [21], who tested the excitability of the synapse from the Ia spindle afferents to the alpha motor neurons of the human soleus and gastrocnemius by means of an H-reflex elicited at different times during target-tracking movements of the ankle. They concluded that there is a mechanism for facilitating and inhibiting this reflex, which operates separately from the mechanism that actively drives the alpha motor neuron pool in voluntary activity; and that the stretch reflex in the antagonist muscle is turned off at the spinal cord, rather than at the spindle. Thus, they view their data as indicating the existence of a mechanism within the spinal cord for the independent regulation of the gain of the monosynaptic reflex arcs just prior to the initiation of a movement: increasing the gain of the agonist myotatic loop, decreasing that of the antagonist loop. G. L. Gottlieb has informed me that work in progress shows that inhibition of reflexes can be totally independent of whether the antagonistic muscle is activated. Specifically, reflex contraction of the human soleus can be inhibited without requiring the generation of alpha activity, by relaxing the soleus from a previously active state, in the absence of anterior tibial contraction. It will be important to integrate these findings with the Russian results, and with the mass of information in [22] relating the activity of muscles, alpha and gamma motor neurons, and their control systems.

In the laboratory of M. A. Aizerman and coworkers at the Institute of Automation and Remote Control, an extensive series of experiments [1, 2, 3, 4, 11] showed that to hold the arm in a particular position (under the conditions of their experiments), opposing muscles alternately pull the arm one way and then the other, producing a tremor of about 10 Hz. When the subject moves his arm slowly in one direction, under these conditions, half of each cycle of the tremor is biased to have a slightly larger amplitude than the other half. It is remarkable that the pull to either side during a half-cycle is the resultant of pulls by *both* opposing muscles; each pulls ten times per second, but first one, and next the other, pulls harder. The investigators hypothesize that, whereas the brain sends the same signals to each muscle in the limb ten times per second, the inter-neuronal pools have been set up ahead of time so that each muscle responds in the proper amount. (Is this mechanism of pretuning related to that of Gottlieb et al?) This is a further example of the beneficial use of low-level mechanisms: other low-level systems and the brain pretune motor automatisms, in such a way that the signals from the brain may be very simple. In this example, these signals do not have to address each muscle separately. (In addition, control by the rapid alternation of opposing influences has a well-known linearizing effect, that allows graded control to be exerted by highly nonlinear and discontinuous systems, such as relays.)

This mechanism for the adjustment of body position is a particular case of a fundamental mechanism for adjustment of muscular tension or velocity shown by a series of experiments in Aizerman's laboratory [1, 2, 3, 4, 11] to take part in other muscular search activities of very different kinds (cf. the discussion by Gelfand and Tsetlin in Chapter 1), such as maintaining a rapid oscillation of a limb, minimizing the magni-tude of a painful stimulus that is some function of muscle activity, and regulating the electrical activity of the diaphragm and the duration of inspiration and expiration. In each of these cases, a "splash" of muscular activity occurs wherever some significant quantity (such as an error or reference signal or painful stimulus) exceeds a threshold. It is remarkable that splashes occur simultaneously in *all* muscles—not just the muscle producing the error, etc. Of course, these splashes, or surges of contraction, occur at times systematically related to the error (etc.) of the relevant muscle, but uncorrelated with the states of all the irrelevant muscles.

For example [2, 3, 37, 68, 69] the electromyographic signal from a muscle (of rat, cat, or man) was fed to a nonlinear function generator whose output voltage was applied to the body as far as possible from the muscle producing the EMG. Thus, the output voltage was a painful stimulus that could be made to depend, in various desired fashions, upon the EMG. Surges of contraction in the muscle produces the EMG, *and in all other measured muscles too*, agreed with the statement above, by occurring whenever the state of the muscle that produced the EMG was such that the given function exceeded a threshold. By varying the shape of the function, various unusual distributions of surge amplitude were obtained, in complete accord with the hypothesis. Movements of the diaphragm of a rabbit under various experimental conditions of normal and altered CO_2 concentration and physical resistance to air flow were explained by a simple model, whereby surges occur whenever a function of diaphragm tension exceeds a variable threshold that depends mainly upon lung volume, and slightly upon CO_2 concentration [2, 3, 59, 60, 70]. Is there a connection between this model and the model proposed by authors of the present book and their coworkers [34, 54] for the transition, in walking, from the support phase to the transfer phase of a dog's leg when a function of its joint angles reaches a threshold value influenced by the state of the other legs? Is this model related to the questions about searching behavior asked by Gelfand and Tsetlin in Chapter 1? It would be extremely valuable to perform some of the experiments from the two laboratories under conditions that might allow tentative contact between Aizerman's results and the results in this book.

The nervous system shows the great extent to which complex behavior can be governed by simple commands, if one starts with a repertoire of subsystems that can be activated by simple inputs to do useful things that they can do relatively autonomously. Part III shows how further adaptability results if a subsystem may be tuned in advance, so that a standardized activating signal produces an appropriate variant of the subsystem's response. The remainder of this introduction will concern what I believe to be far reaching implications of these ideas of Part III, as they apply to hybrid (analogue-digital) control systems in general, that must monitor and coordinate many degrees of freedom, whether biological, as in this book, or artificial, as in process control and unmanned spacecraft or lunar explorers. In all these systems, much local control must take place, so that an executive controller can make decisions without being overwhelmed by having to regulate all the degrees of freedom. For this reason, according to Part III, the nervous system uses, and machines of the future will need to use, a hierarchical form of control, in which actions are composed by selecting and activating small low-level analogue systems. These low-level systems are tuned by other low-level systems in such a way that a high-level executive system can direct them by means of relatively simple signals, without having to regulate too many degrees of freedom. The basic procedures of such a system are to select low-level devices (e.g., microcircuits of an integrated circuit chip) that crudely do some part of the desired action, to tune these partial actions into closer approximations to the desired action, and to fit the partial actions together into a realization of the desired action as a whole. In computer terminology, this could be called a *compiler* that turns task-oriented commands into programs for neural and electronic microcircuit relization of multidimensional hybrid control tasks.

What organization would be required for such a compiler, patterned after the ideas of Section III, to work? In what form could the commands and memories be stated, when the executive seldom knows just which circuit elements are operating, or how these are tuned? Can there be units of information that behave deterministically, even though the executive can rarely specify control functions more narrowly than to place them within broad classes of possible realizations? A general theory of this style of operation would help us know what are the significant observables to measure in the nervous system, and would be expressly fitted to the types of control systems that will be made possible by currently emerging large-scale integrated circuit technology [25].

For example, in the generation of a function[5] that represents some coordination between system variables, it is common in the nervous system, and desirable in artificial systems, for the generator to be activated and tuned by independent subsystems. Just as resetting the clock to Daylight Saving Time advances all acts, while leaving all decisions unchanged, so do pretunings of motor neurons by low-level postural reflexes allow the appropriate variant of a movement to result from a standardized command by the executive level, in this case, the higher centers of the brain. In the fullest exploitation of this idea, the executive only needs to remember which generator worked well in a given situation, and can remain ignorant even of the existence of tuning. As a result, the smallest elements of information available to the executive may not be functions, but families of functions parametrized by possible tunings, i.e., families akin to *homotopy classes of functions.*

Another characteristic procedure is the piecewise realization of functions by the responses of several low-level neural or electronic microcircuit function generators. Of course, this requires low-level detectors that signal when the state of the system lies in a domain wherein the output of a particular function generator matches the required function. In the topology generated by the detectable subsets of the state space, one can argue [24] that the continuous function generators are just the ones that can be operated without high-level supervision (since there is a detector for some input domain that will produce outputs satisfying any specifiable tolerance). Two low-level generators are interchangeable if their outputs agree throughout some detectable neighborhood of the current input, since a detector for that neighborhood could equally well become conditioned to activating either generator without affecting the required function; hence, the functions are regarded as equivalent. Therefore, these equivalence classes, or *germs of functions,* rather than individual functions, become units of information.

Another characteristic operation involved in piecewise low-level

---

[5] The domains of these functions can be varied and complicated spaces that represent the structure of tasks, as well as the physical structure of the systems. A few examples are mentioned in [24]. The executive controller must have uniform methods that continue to work in the face of this variety and the controller's probable ignorance of the explicit structure of these state spaces.

realization is adjusting signals to compensate for the differing characteristics of two effectors when switching from one to the other during performance of a task. An example might be low-level computation of the transformation between the states of activation of two different groups of muscle fibers (or rocket engines) that could be used to perform the same movement. Further arguments about interchangeability of these transitional adjustments lead to the recognition of another equivalence class (of cohomologous transition functions) as a significant unit of information. In addition, two functions are interchangeable without executive notice, and can thus be considered equivalent, if they differ by a transformation that can be realized by some low-level adjustment.

Finally, the definition of an action consists not only in coordination of states that actually exist, but in a coherent pattern of coordination of all the states that could exist in a family of potential circumstances. (For instance, to transfer an object from one hand to another, the nervous system must be prepared to generate any of the coordinations that would be required for a family of possible transfer locations, postures, weights of the object, and methods of low-level realization.) It has seemed useful to view all these coordinations at once, by representing these families in the format of fiber spaces of one sort or another [26, 27, 56], in which a point of the base space is a potentially given condition, and its fiber a set of potential states associated with the given state in ways too varied to explain here. One now considers the general methods distilled from the ideas of Part III. Each, if totally exploited, leads to a situation that can be described in a cumbersome paragraph that translates into a simple (and standard) definition for structures or mappings of fiber spaces. Basic standard theorems [26, 27, 56] then can be interpreted as possibilities for the systematic control of hierarchical systems, despite the tremendous uncertainties generated when the ambiguities of the previously described equivalence relations are propagated through the systems. Roughly speaking, the equivalence classes serve as "ballparks," into which it is sufficient for the executive to transfer the state; once the state enters the ballpark, it will be automatically brought to the correct position without further attention. The theorems may be interpreted to show that, although the ambiguities inevitably lead to erroneous signals, these signals will never be moved outside their correct ballparks, or equivalence classes. Hence, the equivalence classes mentioned above seem to be systematically behaving units of information in situations in which the individual elements themselves will behave in haphazard fashion. We need to know such units of information before we can conclude, for example (as two recent symposia attempted to do) whether the nervous system is stochastic or deterministic and what might be the coding it uses.

It is probably easier to apply these principles to artificial devices than to understanding the nervous system, and in our laboratory, we are constructing models of coordination organized according to the ideas of Section III and the possibilities shown by the mathematics. In roughest terms, we try to regulate the models not through solution of equations, but through coordination of crude reflexes and rules of thumb with finer adjustments, as when one does something "by feel" (such as tossing a piece of paper into the wastebasket from across the room). The price of using the simpler mode of operation by feel is that the data and commands within the systems must assume a more complex form (involving the equivalence classes and fiber space structures) than others have considered.

Thus, the ideas of Part III seem ideally suited to large-scale integrated circuit technology for controllers, as well as to understanding the nervous system, yet they are never abstracted in *Referativniy Zhurnal Kibernetika* (*Cybernetics Abstracts*), and cyberneticians here and abroad generally start with other points of view. This is why I have discussed at length the wider implications of Part III, in addition to its place in the remarkable program of research surveyed in this book.

Peter H. Greene
Department of Theoretical Biology
and Committee on Information
Sciences
The University of Chicago

REFERENCES

1.
M. A. Aizerman and E. A. Andreeva, *Automation and Remote Control* 1968:29, 452 (*Avtomatika y Telemekhanika* 1968: No. 3, 103).
2.
——, and E. A. Andreeva, *On Some Control Mechanisms of Skeletal Muscles* (English and Russian versions), Moscow: Institute of Automation and Remote Control, 1968.
3.
——, and V. S. Gurfinkel, eds. (*Issledovaniye protsessov upravleniya myshechnoy aktivnostyu* (*Investigations of Processes of Control of Muscular Activity*), Moscow: Nauka, 1970.
4.
E. A. Andreeva, Kh. A. Turakhanov, and V. I. Chernov, *Automation and Remote Control* 1969:30, 1988 (*Avtomatika y Telemekhanika* 1969: No. 12, 111).
5.
G. A. Arutyunyan, V. S. Gurfinkel, and M. L. Mirskiy, *Biophysics* 1969:14, 1162 (*Biofizika* 1969:14, 1103).
6.
I. S. Balakhovskiy, *Biophysics*, 1966:11, 143 (*Biofizika* 1966:11, 129).
7.
N. A. Bernstein, *O postroyenii dvizheniy* (*On the Construction of Movements*), Moscow: Medgiz, 1947.
8.
R. L. Beurle, *Phil. Trans. Roy. Soc.* 1956:B240, 55.
9.
J. M. Brookhart and K. Kubota, in *Progress in Brain Research, Vol. 1: Brain Mechanisms*, ed. G. Moruzzi, A. Fessard, and H. H. Jasper, Amsterdam: Elsevier, 1963, p. 38.
10.
V. B. Brooks, in *Information Processing in the Nervous System*, ed. K. N. Leibovic, New York: Springer-Verlag, 1969, p. 231.
11.
V. I. Chernov, *Automation and Remote Control* 1968:29, 1090 (*Avtomatika y Telemekhanika* 1968: No. 7, 86).
12.
L. V. Chkaidze, *Biophysics* 1967:12, 203 (*Biofizika* 1967:12, 174).

13.

M. H. Cohen, Models for the control of development, *Soc. Exp. Biol. Symp.* 1971: in press.

14.

————, and A. D. J. Robertson, Wave propagation in early aggregation of *Dictyostelium discoideum*, and Chemotaxis and the early stages of aggregation in the slime molds, *J. Theor. Biol.* 1971: 31, in press.

15.

J. D. Cowan, in *Lectures on Mathematics in the Life Sciences, Vol. 2*, ed. M. Gerstenhaber, Providence, R.I.: American Mathematical Society, 1970.

16.

J. C. Eccles, in *Information Processing in the Nervous System*, ed. K. N. Leibovic, New York: Springer-Verlag, 1969, p. 245.

17.

————, M. Ito, and J. Szentágothai, *The Cerebellum as a Neuronal Machine*, New York: Springer-Verlag, 1967.

18.

A. G. Feldman, *Biophysics* 1966, 766 (*Biofizika* 1966 : 11, 667).

19.

I. M. Gelfand and M. L. Tsetlin, *Russian Math. Surveys* 1962: 17, 95 (*Uspekhi Matem. Nauk* 1962 : 17, 3).

20.

B. C. Goodwin and M. H. Cohen, *J. Theor. Biol.* 1969: 25, 49.

21.

G. L. Gottlieb, G. C. Agarwal, and L. Stark, *J. Neurophysiol.* 1970: 33, 365.

22.

R. Granit, *The Basis of Motor Control*, London and New York: Academic Press, 1970.

23.

P. H. Greene, Problems of organization of the motor system, *Institute for Computer Research Quarterly Report*, No. 29, May 1, 1971, University of Chicago and *J. Theor. Biol.*, to appear.

24.

————, Organization of hierarchical hybrid control systems, *Institute for Computer Research Quarterly Report*, University of Chicago, and *Int. J. Man-Machine Systems*, to appear.

25.

————, Information Structure and Theory of Tasks for Hierarchical Hybrid Control Systems, a Proposal for Research, 1971, unpublished.

26.
S-T. Hu, *Homotopy Theory*, New York: Academic Press, 1959.
27.
D. Husemoller, *Fibre Bundles*, New York: McGraw-Hill, 1966.
28.
A. V. Kholopov, *Biophysics* 1968:13, 1231 (*Biofizika* 1968:13, 1058).
29.
————, *Biophysics* 1969:14, 726 (*Biofizika* 1969:14, 688).
30.
A. B. Kogan, V. N. Yefimov, A. A. Chumachenko, and V. A. Safonov, *Biophysics* 1969:14, 758 (*Biofizika* 1969:14, 718).
31.
P. G. Kostyuk and D. A. Vasilenko, *Proc. IEEE* 1968:56, 1049.
32.
Ya. M. Kots, *Biophysics* 1969:14, 1146 (*Biofizika* 1969:14, 1087).
33.
V. I. Krinskiy, in *Systems Theory Research*, Vol. 20, ed. A. A. Lyapunov (translation of *Problemy Kibernetiki*), New York: Plenum, 1971.
34.
A. S. Kulagin and M. L. Shik, *Biophysics* 1970:15, 171 (*Biofizika* 1970:15, 164).
35.
O. C. J. Lippold, *J. Physiol.* 1970:206, 359.
36.
————, *Sci. Amer.* 1971:224, No. 3, 65.
37.
A. I. Litvintsev. *Automation and Remote Control* 1968:29, 464 (*Avtomatika y Telemekhanika* 1968: No. 3, 119).
38.
R. Llinás, ed. *Neurobiology of Cerebellar Evolution and Development*, Chicago: American Medical Association, Education and Research Foundation, 1969.
39.
————, in *The Neurosciences, Second Study Program*, ed.-in-chief, F. O. Schmitt, New York: Rockefeller University Press, 1970, p. 409.
40.
————, and D. E. Hillman, in *Neurobiology of Cerebellar Evolution and Development*, ed. R. Llinás, Chicago: American Medical Association, Education and Research Foundation, 1969, pp. 69, 71.

41.

————, and C. Nicholson, in *Neurobiology of Cerebellar Evolution and Development*, ed. R. Llinás, Chicago: American Medical Association, Education and Research Foundation, 1969, p. 431.

43.

A. Lundberg, in *Muscular Afferents and Motor Control*, ed. R. Granit, New York: Wiley and Stockholm: Almqvist and Wiksell, 1966, p. 275.

44.

A. Lundberg, *Electroenceph. clin. Neurophysiol.* 1967: Supplement 25, 35.

45.

L. A. Maksimenko, *Biophysics* 1968: 13, 599 (*Biofizika* 1968:13, 502).

46.

M. D. Mesarovic, D. Macko, and Y. Takahara, *Theory of Multi-Level Hierarchical Systems*, New York: Academic Press, 1970.

47.

Obituary: M. L. Tsetlin, *Biophysics* 1966: 11, 1080 (*Biofizika* 1966: 11, 939).

48.

G. N. Orlovskiy, *Biophysics* 1970: 15, 178 (*Biofizika* 1970: 15, 171).

49.

W. Rall, *J. Neurophysiol.* 1967: 30, 1138; 1968: 31, 884.

50.

————, in *The Neurosciences, Second Study Program*, ed.-in-chief, F. O. Schmitt, New York: Rockefeller University Press, 1970, p. 552.

51.

————, R. E. Burke, T. G. Smith, P. G. Nelson, and K. Frank, *J. Neurophysiol.* 1967: 30, 1169.

52.

A. D. J. Robertson, The control of development in *Dictyostelium discoideum*, to appear in *Lectures on Mathematics in the Life Sciences, Vol. 4*, ed. M. Gerstenhaber. Providence, R.I.: American Mathematical Society, in press.

53.

R. Rosen, *Optimality Principles in Biology*, London: Butterworth, 1967.

54.

M. L. Shik and G. N. Orlovskiy, *Biophysics* 1965: 10, 1148 (*Biofizika* 1966: 10, 1037).

55.

V. V. Smolyaninov, *Biophysics* 1970: 15, 133 (*Biofizika* 1970: 15, 130).

56.
N. Steenrod, *The Topology of Fibre Bundles*, Princeton: Princeton University Press, 1951.

57.
J. Szentágothai, *Proc. IEEE* 1968:56, 960.

58.
————, in *The Neurosciences, Second Study Program*, ed.-in-chief, F. O. Schmitt, New York: Rockefeller University Press, 1970, p. 427.

59.
L. A. Tenenbaum, *Automation and Remote Control* 1966:27, 1951 (*Avtomatika y Telemekhanika* 1966:27, no. 11, 127).

60.
————, in *Issledovaniye protsessov upravleniya myshechnoy aktivnostyu* (*Investigations of Processes of Control of Muscular Activity*), ed. M. A. Aizerman and V. S. Gurfinkel, Moscow: Nauka, 1970, p. 148.

61.
C. A. Terzuolo and R. Llinás, in *Muscular Afferents and Motor Control*, ed. R. Granit, New York: Wiley and Stockholm: Almqvist and Wiksell, 1966, p. 373.

62.
D. W. Thompson, *On Growth and Form*, revised edition, New York: Macmillan, 1945.

63.
N. Wiener and A. Rosenblueth, *Arch. Inst. Cardiologia de Mexico* 1946:16, 205.

64.
D. M. Wilson, *Ann. Rev. of Entomology* 1966:11, 103.

65.
————, in *The Neurosciences, Second Study Program*, ed.-in-chief, F. O. Schmitt, New York: Rockefeller University Press, 1970, p. 397.

66.
————, and I. Waldron, *Proc. IEEE* 1968:56, 1058.

67.
A. N. Zaikin and A. M. Zhabotinsky, *Nature* 1970:225, 535.

68.
L. M. Zakharova and A. I. Litvintsev, *Automation and Remote Control* 1966:27, 1942 and 1968:29, 835 (*Avtomatika y Telemekhanika* 1966:27, no. 11, 115 and 1968, No. 5, 166).

69.
————, and A. I Litvintsev, in *Issledovaniye protsessov upravleniya myshechnoy aktivnostyu* (*Investigations of Processes of Control of Muscular Activity*), ed. M. A. Aizerman and V. S. Gurfinkel, Moscow: Nauka, 1970, p. 74.
70.
N. V. Zavalishin and L. A. Tenenbaum, *Automation and Remote Control* 1968:29, 1456 (*Avtomatika y Telemekhanika* 1968:No. 9, 106).

In memory of N. A. Bernstein

Nikolai Aleksandrovich Bernstein is dead. He was an outstanding scholar, and many ideas of contemporary physiology are connected with his name. His creative work is notable for its remarkable unity and its profound world outlook in natural science.

Born in Moscow in 1896, Bernstein studied at Moscow University. Besides subjects in general biology and medicine, he also attended courses in physical mathematics. A great mathematical culture appeared in his astute papers in biomechanics and communicated the clearness of his overall ideas. After completing the university, he worked for several years as a physician, specializing in neuropathology and psychiatry. In 1922 his scientific work began in the field of biomechanics and physiology of movement. The primitive reflexological concepts that prevailed at the time restricted the development of this important area of physiology. Even in his earliest works a new, original approach to its problems began to show. He understood the importance of the study of the mechanics of movement and the peculiarity of structure of the motor apparatus of higher animals having a large number of degrees of freedom. A great service by Bernstein was the introduction of new precise methods of research—cyclogrammetry and kymocyclography—which made it possible to record the trajectories and times of the motor elements.

If before him the majority of physiologists underestimated the role of the kinematic and dynamic characteristics of movements and their functional expediency, then Bernstein for the first time formulated the problem of the study of movements of higher animals and man as a problem of control of a complex kinematic system. Turning his attention to the fact that together with active muscle forces, inertia and reactive forces and initial posture play an important role, he conceived the idea of the ambiguous dependence of the motor effect on the motor command. The impossibility of prior calculation of the influence of these numerous factors during either the duration or the results of movement led him to an orderly study of sensory corrections as an essential mechanism of the construction of movements. Clearly formulated even in 1928, his concept of sensory corrections was, essentially, one of the first attempts of study of feedback not only in biological systems but in other systems as well.

After Bernstein's work, the inadequacy of the concept of the reflex arc for an understanding of the process of movement became clear. He formulated a remarkably simple conclusion that the basic problem of coordination of movements is the overcoming of a large number of degrees of freedom, in other words, the lowering of the number of the independent variables controlling movement. For different motor tasks this goal is realized, according to his concepts, at different levels of the nervous system. Moreover, the controlling level for a given class of movements is that at which there is already possible an acceptance of solutions of correction of movement.

Solutions of this sort, naturally, can be formulated only on the basis of a calculation of all essential information about the course of movement of a given class. Control of walking on flat terrain, for example, can be realized at a lower level than control of such movements as throwing a ball at a target or the working movements of a skilled machine operator. Depending on the task, motor acts of similar structure can, according to Bernstein, be controlled at different levels. Bernstein studied with enthusiasm problems connected with a study of the work and play movements of a human being. He is the author of fundamental studies of walking, running, and jumping.

The theory of the control of movements written up in a monograph by Bernstein, "Concerning the Construction of Movements" (1947), was awarded a state prize. Bernstein's works constitute an epoch in the physiology of movements and in many respects determined the paths of development of this area of science. The works on the physiology of movements in this collection were likewise realized under the influence of Bernstein's ideas.

Over a span of many years, Bernstein developed ideas which form a study of the physiology of activity. His basic thought was that expediency is predominant in the physiology of animals. Any act of voluntary activity of an animal is aimed at the achievement of a certain goal—a predictable future. This goal determines the choice of action and the way of calculating the conditions in which it is realized. Moreover, mechanisms of control and stabilization, among them both reflexes and feedbacks (sensory corrections), are used. They make it possible to follow the approach to the goal in the course of the action itself by means of the comparison of the attained result with the model of the future one. Bernstein sharply criticized the attempts to build animal physiology as merely a physiology of reflexes and responses to the actions of the environment, i.e., of a stimulus-response type. He assumed that the reflex is not an element of action but just elementary action. The physiology of activity excludes comprehension of the active work of an animal as an equilibration with the environment.

Bernstein well understood the inadequacy of the concept of feedback, and the importance of the active search activity of an animal. An unusually gifted man, he was not afraid to return again and again to facts considered firmly established and find contradictions however protected by words and quotations. These views of Bernstein are stated in his last works.

Bernstein was not only one of the first propagandists of cybernetics in our country, he was, essentially, one of its founders. The fundamental ideas of Bernstein exert with every year an ever greater influence on physiological thinking.

A scholar of deep principle, Bernstein courageously defended his opinions, sharply contradictory to the then prevailing primitive dogmatic concepts. His scientific opinions were not acknowledged for a long time, and for several years he was deprived of the opportunity to carry out experimental work. Until the last days of his life, he did not cease intensive work, and he was the initiator and leader of a series of investigations. He always was surrounded by able young people who generously gave of their time, learning, and experience. Those who knew N. A. Bernstein deeply respected and loved him.

THE PRINCIPAL WORKS OF N. A. BERNSTEIN

1.
Obshchaya biomekhanike (*General Biomechanics*), Moscow, 1926.
2.
Die Kymocyclographische Methode der Bewegungsuntersuchungen, *Handb. d: biol. Arbeitsmeth.*, ed. Abderhalden, 1927:5, no. 5a.
3.
Klinicheskie puti sovremennoy biomekhaniki—Sb trudov Instituta usovershenstvovaniya vrachey (*Clinical Methods of Contemporary Biomechanics—Collected Works of the Advanced Training Institute for Surgeons*), Kazan, 1929, vol. 1.
4.
Tekhnika izucheniya dvizheniy (*The Technology of the Study of Movements*), Leningrad-Moscow, 1934.
5.
Problema vzaimootnosheniya koordinatsii i lokalizatsii (Problems of the Interrelation of Coordination and Localization), *Arkhiv, Biol. Nauk* 1935:38, no. 1.
6.
Issledovaniya po biodinamike lokomotsiy (*Research in the Biodynamics of Locomotion*), Moscow-Leningrad, 1930, Book 1.
7.
Issledovaniya po biodinamike khody, bega, pryzhka (*Research in the Biodynamics of Walking, Running and Jumping*), Moscow, 1940.
8.
O postroenii dvizheniy (*Concerning the Construction of Movements*), Moscow, 1947.
9.
Ocherednye problemy fiziologii aktivnosti (Routine Problems of the Physiology of Activity), *Problemy kibernetiki* no. 6 (1961), pp. 101–160.
10.
Puti razvitiya fiziologii i svyazannye s nimi zadachi kibernetiki—sb Biologicheskie aspekty kibernetiki (*Paths of Development of Physiology and of Related Problems of Cybernetics.* Collection *Biological Aspects of Cybernetics*), Moscow: Idz-vo AN SSSR, 1962, pp. 52–65.
11.
Ocherki po fiziologii dvizheniy i fiziologii aktivnosti (*An Outline of the Physiology of Movements and the Physiology of Activity*), Moscow, 1966.

At the present time it is becoming increasingly clear that the complexity of physiological systems as objects of contemporary experimental research is so great that even a complete description of the elements and their interrelationship is in itself insufficient for an understanding of a system's principles of operation. This dilemma leads to the necessity of using new methods for the study of complex systems. One of these methods can be, it seems to us, a model decription of a system's functioning. In this case the essential demands of the structure and properties of the model are in the first place that it should provide a correct description of the phenomenology of the functioning system; and in the second place that the postulates used in the construction of the model should correspond to the real properties of the elements of the modeled system and to the interrelationship of the elements. Furthermore, inasmuch as one and the same phenomenology usually allows several model descriptions, it is desirable to be concerned about finishing up the model to the level where it becomes possible to make conclusions which allow experimental verification. In other words, one of the most important results of modeling should be a prediction of certain properties of the system not obvious beforehand. Then the corresponding model can be useful in understanding the functioning of the system investigated.

However, the naturalist's experience with the empirical method tells him that even a successful model description of a physiological system should be considered not as the result but rather as only the beginning of the investigation.

In this collection there is included only that type of model in which some real physiological system is modeled. In any case we honestly tried to keep to a minimum the number of papers containing constructions not having well-grounded roots in biology and not of interest to mathematicians.

The reader will easily observe that part of the papers included in the book have a specifically model and others, on the other hand, a traditionally experimental character. It was difficult, maybe impossible, to avoid this discrepancy, since a model description and a physiological experiment, from our point of view, are consecutive stages of research.

It may seem to the reader that the individual articles of the collection are not closely connected with one another. However, the authors themselves were well aware of their true interrelationship. Questions of physiology touched upon in the collection were repeatedly discussed at a seminar which assembled the authors of these papers and enabled the working out of general viewpoints.

For the reader-mathematician it would probably be helpful during the reading of the series of articles of the collection to have on hand such books as *Conduction of the Nervous Impulse* by Alan L. Hodgkin (1964); *Nervous Transmission* by I. Tasaki (1953); *Electrophysiology of the Heart* by B. Hoffman and P. Cranefield (1960); *Physiology of Nerve Cells* by J. C. Eccles (1957); *Receptors and Sensory Perception: A Discussion of Aims, Means, and Results of Electrophysiological Research*, by R. Granit (1955); and *O postroenii dvizneniy* (*Concerning the Construction of Movements*) by N. A. Bernstein (1947).

On May 30, 1966, when this book was already complete, Mikhail Lvovich Tsetlin, one of its authors and editors—a brilliant scientist and remarkable man—died.

He was not just a versatile, talented, and broadly educated scholar. His distinguishing characteristic was that in his views he proceeded from a simple, sensible, and very important idea that the universe is organized as a single whole. Consequently, he always looked for general principles having a broad application. This trait, in conjuction with a remarkable ability to talk about the most abstract ideas simply and interestingly, and at the same time his cordial charm the basis of which was his sincerity, kindness, and straightforwardness in dealing with people, were the reason why Tsetlin played such a large role in working out general viewpoints and in assembling our collection.

This loss is a painful and irretrievable one for all of us.

Models of the
Structural-Functional
Organization
of Certain
Biological Systems

1

Mathematical Modeling of
Mechanisms of the Central
Nervous System

I. M. Gelfand and
M. L. Tsetlin

This paper is not a survey of the methods of mathematical modeling of mechanisms of the central nervous system; it is devoted to certain mathematical models connected with the physiology of the central nervous system. We limit ourselves to a statement of the fundamental ideas; readers interested in the components of the mathematical apparatus and details of the physiological applications can find supplementary information in the Bibliography at the end of this chapter.

Three models will be described in this paper. The first of them is connected with the modeling of expedient activity in the example of search methods for the extremum of a nonstationary function of many variables. The second model is devoted to a modeling of the behavior of collectives consisting of automata having expedient behavior. Both these models result from a study of behavior which we understand as a single process of study of environment, acceptance, and realization of a decision; we are not examining here questions about concrete physiological mechanisms realizing behavior. The third model is an attempt to describe mathematically the work of the simplest excitable tissues. These models are rather heterogeneous, but we would like to see in them examples of the realization of some overall principle. The concluding section of the paper is devoted to an attempt to formulate a sort of principle: the principle of the least interaction.

1. We will examine the working of a specific complex system directed to the attainment of a goal in a changing environment. We assume that the system has an opportunity to determine the degree of proximity to the goal; information being used by the system is obtained by it in the very process of goal-directed activity. By system complexity we mean here the number of parameters defining its state.

The problem of the "intelligent" functioning of such a complex system is in many respects similar to the problem of finding the minimum of a function of many variables.

Let the function $F(x_1, \ldots, x_n, y_1, \ldots, y_m)$ characterize our system's property of work—the goal distance, so to say. We will call the parameters $x_1, \ldots, x_n$ the *working parameters* of the system. Let us assume that the system can change the values of the working parameters and measure the values of the function F. The independent variables $y_1, \ldots, y_m$ are the system's intrinsic parameters which depend on time and, perhaps, on the variables $x_1, \ldots, x_n$; the system is not able to influence the intrinsic parameters.

We will write the function $F(x_1, \ldots, x_n, y_1, \ldots, y_m)$ as $\Phi(x_1, \ldots, x_n, t)$, we will call the function Φ the *estimator* of the system.

When the estimator is assigned analytically, a search for the minimum is, of course, possible by means of a way that is trivial for classical mathematics: we must differentiate the function Φ with respect to each of the independent variables and equate the obtained derivations to zero. After this the problem is reduced to the solution of a system of equations the number of which equals the number of working parameters of the system. In practical computing problems it turns out, however, that the solution of such a system is by no means simpler than directly finding the extremum. Moreover, for some considerable number of independent variables the solution of any of these problems goes far beyond the range of possibilities of contemporary computing technology.

However, we will by no means assume that the estimator is assigned analytically. Consequently, the selection of the necessary values of the parameters must by necessity be carried out experimentally. The time dependence of Φ (this dependence is by no means assumed certain) means that the search for the necessary values of the independent variables becomes continuous. Moreover, the search consists of a succession of experiments, each of which includes choosing some point of space of the change of the system's working parameters and the measuring of the corresponding values of Φ. The various search tactics differ in the ways of using the information about the values of Φ, measured in a series of previous experiments, for the choice of the following point of space of the working parameters. The tactics are successful if they manage thus to select these experiments so that the values are small. Consequently, finding the absolute minimum becomes absurd, and a successful continuous search of the minimum means only that the system is almost all the time in those domains of change of the working parameters in which the values of the estimator are relatively small.

By an organization we mean the sort of singularities of problems (or situations) that distinguish these problems from the mass of all problems and that can facilitate and hasten acceptance of a solution. These singularities are not, as a rule, known exactly beforehand, and are only more or less probable. Therefore, for use of the organization there is essential the promotion of those or other hypotheses and the construction of solution methods (tactics) based on these hypotheses. It does not work, as a rule, to check the hypotheses directly, for a criterion of the accuracy of the hypothesis is the successfulness of the application of the tactics based on it. The search tactics used by a specific system can be produced by the characteristics of that very system. In this instance the successfulness of the search serves as a criterion of the fitness of the problem organization and of the system's structure, which solves this problem. The use of the hypotheses and the tactics based on them involve the intentional rejection of examination of all possible cases, among them rejection of random situations seemingly most probable.

A very important circumstance here is the limitation of time that can be spent in the solution of the problem. Since the environment changes in time (since the estimator is unstable), a late solution can prove to be positively wrong; even an inexact, approximate solution obtained quickly is preferable to an exact solution which is late. The functioning of a complex system represents from this point of view a contest in speed with the environment. Here we have in mind such situations where the acceptable speed of the search of a solution can be reached only through use of an organization which to one extent or another have problems that are solved by practical activity.

Returning to the problem of the search for the minimum of a nonstationary function of many variables, we wish to clarify in this instance the role of hypotheses and tactics. It is convenient for us to divide these tactics into three groups.

The so-called *methods of blind search* belong to the first group. For these methods it is characteristic that the space points of the working parameters may be looked over either at random or in well-defined order.

Upon attainment of sufficiently small values of Φ, the search ceases until these values go beyond permissible limits. The tactics of blind search do not use any significant prior assumptions regarding the structure of the function Φ (with the exception of the assumption that the speed of movement in space of the working parameters and the measurement of the value of Φ are sufficiently great in comparison with the speed of the change in time of the same function), so that the organization of the estimator is used to the least degree. The results of an individual experiment in the subsequent search are not taken into consideration; the values of Φ do not decrease from experiment to experiment.

The second group consists of multiple *methods of local search* such as the gradient, the relaxational, the method of steepest descent, and the different varieties of all of these methods. A common feature of all the methods of this group is their localism: for preparation of the following experiment, the knowledge was used of the values of Φ in a small neighborhood of the previous experiment, and the working point continuously moves in the space of the parameters. The hypotheses on the continuity of the Φ function are based on the tactics of local search. Local tactics make it possible to carry out systematic reduction of the Φ values in the process of the search which is the essential advantage of them in comparison with the methods of blind search. The localism of the tactics, however, limits their effectiveness, making a constant danger of the "cyclization" of the search in some kind of secondary shallow "trough." With the small values of the gradient the search turns into a random wandering, and its effectiveness sharply drops.

We put the methods of *nonlocal search* in the third group. With the use of these methods the movement of the working point in the space of the parameters is not continuous; moreover, unlike the methods of blind search, the utilization of the information about the values of the function for the planning of a subsequent experiment is specified. We will describe here one of those such methods (the "ravines method") which we wrote up in greater detail in [14, 15]. The method of ravines makes it possible to use more profound characteristics of the estimator than its local behavior.

The following hypothesis is based on this method. We assume that the working parameters of the system $X_1, \ldots, X_n$ can be divided into two groups. The first group includes almost all these parameters and consists of those independent variables the change of which leads to a marked change in the values of the Φ function; consequently the selection of the values of these parameters is made relatively simply and quickly. We will call the parameters of the second group unessential.

The second group of parameters includes a small (for example one, two, or three) number of independent variables. These variables are either parameters of the number $X_1, \ldots, X_n$ or their functions. The change of the variables of the second group leads to a relatively small change of the values of the estimator. We will call these variables essential, and the number of them, the dimension of the ravine.

Of course such a subdivision of the parameters is by no means possible for every function. However, for functions encountered in practical problems such a subdivision apparently occurs very often. We will call *well-organized* those functions permitting the just described subdivision of independent variables. The hypothesis about the good organization of a function is also based on the method of ravines. This hypothesis indicates only the feasibility of the separation of the essential variables. The separation itself of these variables must be made, of course, automatically in the course of the search itself; separation of the parameters into essential and nonessential depends on the time and space point $X = (X_1, \ldots, X_n)$ of the working parameters.

We will turn now to a description of the search method itself. First the arbitrary point X_0 is selected and from it is carried out the descent by the gradient or by any other local method. The descent continues until the continuation of it ceases to reduce essentially the values of the estimator, until the quantity $\Delta\Phi/\Phi$ (the gradient sample) becomes less than some previously fixed number Δ. The size of Δ selected should not be too small. The fact is that in that domain where $\Delta\Phi/\Phi$ is small, the essential and non-essential variables become equivalent, i.e., the function ceases to be well organized. Consequently with subsequent use of the local method of descent not making any noticeable headway with respect to the essential variables, we will wander in a random manner changing nonessential variables.

Suppose now that the local descent leads us to point A_0. After this point,

X_1 is chosen in the vicinity of X_0 at a distance essentially exceeding the step of the gradient descent (for example, in a hyperplane, perpendicular to the gradient). From X_1 descent is made to point A_1. After points A_0 and A_1 are obtained, point X_2 is selected with the help of the so-called "ravine step." The points A_0 and A_1 are joined by a straight line on which is also selected point X_2 at a distance L from A_1. The quantity L is called the length of the ravine step. This quantity significantly exceeds the size of the gradient step and is selected experimentally, predetermining to a significant degree the success of the search. With the fixed quantity L we "envelop the high mountains" and "step over the small water divides."

After point X_2 is chosen, the gradient descent is carried out to point A_2; point X_3 is selected with respect to A_1 and A_2 exactly in the same way that point X_2 was selected with respect to points A_0 and A_1, after which the process is repeated.

Thus, points X_i are selected in places where small values of Φ are expected, after which occurs the local adaptation of the result; consequently the whole search is mainly in the domain of small values of the estimator Φ.

Points A_i lie, so to say, "at the bottom of the ravine" moving "along the ravine" in a straight line passing through points A_{i-1} and A_i, the system, generally speaking, can climb the "ravine slope" and therefore from point X_{i+1} it is essential to again make a descent.

We should note an important circumstance here. The fact is that according to the degree of advancement "along the ravine," its direction (i.e., separation of essential variables) is defined more accurately and the paths of the local descents become considerably smaller than the path of the ravine step, "adaptation," so to say, occurs. It stands to reason that "on the turns of the ravine" the paths of the local descents increase, and "relearning," so to say, takes place.

The computing method described here proved to be very effective in the solution of problems of the phase analysis of the dispersion of one proton after another [5, 6], and also in problems concerning the determination of crystal structures of organic compounds (amino acid) according to the data of x-ray structure analysis [11, 16].

We tried to trace in the example of the study of the physiological mechanisms of the control of movements [7–9, 15] many features of the tactics of the "ravines"—the alternation of nonlocal, rapid displacements in the space of the working parameters with relatively slow, local adaptation of the values of the estimator; the presence in connection with these different levels of control of at least two typical time characteristics; and the type of information which it is essential to store with the use of tactics of this kind.

2. Before proceeding to the next group of mathematical models which emerged in our work in the physiology of the central nervous system, it seems useful to us to emphasize certain distinctive features of structure of the complex control systems, the ideas about which serve as the basis for modeling: (a) the complexity of the system: the presence of a large number of relatively autonomous subsystems, and the difficulties of experimental study and description of the interaction of such subsystems; (b) the dependability of function, guaranteeing expedient behavior of the entire system·even in the case of a disablement of one or another of the subsystems; and (c) the diversity of problems solved by the complex control systems, and the impossibility of separating the specialized systems for the solution of each problem.

In the modeling of behavior of complex control systems there naturally arises a need of the separation of the simplest forms of such behavior, a search for a construction having expedient behavior in the simplest cases, and a structure of language suitable for describing the interaction of the simplest constructions, the collective behavior of which would make it possible to transmit the essential features of the behavior of the complex control systems. We would like to talk about one attempt at description of complex control systems by means of the separation of the elementary problems and the constructions.

In the choice of constructions of the simplest systems having expedient behavior, we used finite automata. With a finite automaton we term the object able in each movement of time $t = 1, 2, \ldots$ to receive a finite number of signals $S_1, \ldots, S_n$ and, depending on them, to change its own internal state. The automaton can carry out a finite number of operations $f_1, \ldots, f_k$. The choice of operation is determined by the internal state of the automaton; we assume that the automaton has m internal states $\phi_1, \ldots, \phi_m$; the number m we will call the memory capacity of the automaton.

If the automaton is placed in a certain environment, then its operations f cause responsive reactions of the environment S which emerges, in their turn, as input signals for the automaton. The automaton, so to say, uses them for acceptance of a solution concerning subsequent operations.

In the simplest case we will assume that all the possible reactions of the environment are received by the automaton as relating to one of two classes—favorable or unfavorable; these classes we will call payoffs and losses. In each of these classes the reactions of the environment are indistinguishable from each other. The expediency of the behavior is the increase of the number of favorable reactions and the decrease of the number of unfavorable reactions.

The role of the environment is the establishment of a connection between the automaton's operations and the signals feeding into its input. This connection, generally speaking, can be very complex, especially when a given automaton interacts with other automata. On the other hand, information obtained by the automaton is limited to data about the payoff or loss its least operation involved, so that the nature of the environment beforehand is not known to the automaton. It is therefore natural to choose a construction of the automaton so that its behavior has maximum expediency in the simplest cases, and then to study the behavior of automata singly and collectively in more complex environments.

Behavior of the automaton in a stationary random environment is the simplest of the problems emerging here. In such environments for each of the operations f_i ($i = 1, \ldots, k$), of the automaton there are determined mathematical expectations a_i of its payoff so that the set of quantities $a_1, \ldots, a_k$ give the stationary random environment. In this probability the payoffs and losses of the automaton are defined by the formulas

$$q_i = \frac{1 + a_i}{2}, \qquad p_i = \frac{1 - a_i}{2}.$$

The functioning of the finite automaton A in a stationary random environment is described by the finite Markov chain, and for the ergodic chains we can talk about the final probabilities of the states and the final (independent of the initial state) quantity $W(A)$ of the average payoff obtained by the automaton in a stationary random environment.

It is natural to compare the average payoff of such an automaton with the average payoff which a person, who (unlike an automaton) would be told beforehand the parameters $a_1, \ldots, a_k$ of the environment, could guarantee for himself. Such a person would, obviously, carry out only that operation which guarantees maximum payoff, and the average payoff for such a person would be equal to the maximum of the numbers $a_1, \ldots, a_k$.

It turns out that the payoff of any finite automaton is less than max $(a_1, \ldots, a_k)$, but we can construct the sort of sequences of finite automata $A_1, \ldots, A_n \ldots$ that

$$\lim_{n \to \infty} W(A_n) = \max (a_1, \ldots, a_k).$$

Such sequences are called *asymptotic optimums*. In the papers [19, 20, 22, 25, 26] devoted to modeling of the simplest forms of behavior, a series of such constructions are described.

Moreover, we can interpret number n of the automaton in asymptotic optimal sequence as its memory capacity. We will not dwell here on questions connected with the behavior of automata in environments the properties of which are not stationary [see 26], for we are primarily interested in problems connected with the collective behavior of automata.

The collective behavior of automata is the result of their interaction. We agreed to examine only the simplest automata, in the constructions of which there are not contained a priori data about the environment or about the other automata; information obtained by the simplest automaton is confined only to data about the payoff or loss which follows one or another operation. Consequently we have to consider only the kind of forms of interaction which can be realized in collective behavior of such elementary automata. The language of the theory of games is a convenient method of representation of such forms of interaction.

However, models of collective behavior of automata essentially differ from the point of view assumed in the theory of games. For example, in the theory of games it is customary to assume that the system of payoff functions assigning the game is reported to the players beforehand. Using this a priori information and applying arbitrary means of computing, the player carries out a choice of strategy. Strategies chosen in this way in the process of the game no longer change, so that the game is like a chess party beginning and ending with an analysis by the participant.

The models of collective behavior of the automata (the game of automata) [26] do not provide for the presence of any a priori information so that the strategies are chosen in the course of the game itself.

It is assumed that the game of automata consists of a sequence of parties. In this case the set $f(t) = (f'(t), \ldots, f^N(t))$ of operations (strategies) chosen at this moment by the automata $A^1, \ldots, A^N$ participating in the game is called the party $f(t)$ of the game Γ played at the moment of time t. The set $S(t + 1) = (S^1(t + 1), \ldots, S^N(t + 1))$ where $S^j(t + 1) = 0$ if the automaton A^j in this party won and $S^j(t + 1) = 1$ if this automaton in the party $f(t)$ lost is called the outcome $S(t + 1)$ of the party $f(t)$.

The task of the construction of the playing automata and of the probabilities $P(f,s)$ of the outcomes of the parties determines the game of the automata. By these probabilities we can determine the mathematical expectation $W^j(f)$ of the payoff of the automaton A^j in the party f. The system of payoff functions thus constructed determines a certain game in the sense of the theory of games equivalent to the game of the automata.

Thus, data about the payoff or loss as a result of the operations of a given party assigns the values of the input variables for the playing automata, determining the choice of strategy in the subsequent parties of the game. Moreover, the automata receive no data either about the operations of their own partners, or about the strategies which the partners have available, or even about the number of partners. For a given automaton the role of the remaining players of the game leads to the formation of a more or less complex random environment in which the automaton must possess expedient behavior. Consequently when choosing the constructions of the automata participating in the game, it is natural to require that these constructions guarantee expedient behavior in the simplest game—the game with a single player (the "game with nature"), i.e., in a stationary random environment.

It turns out that in a whole series of games of automata the simplest constructions of that sort guarantee expedient behavior. We will examine, as an example, the game of two persons with a zero sum. Let the matrix of the game be passed on beforehand to one of the partners, and let him choose the optimum strategy in the sense of the theory of games; let the automaton belonging to an asymptotic optimal sequence be the other partner. Then the automaton with sufficient storage capacity attains a payoff equal to the value of the game according to von Neumann.

If both partners in such a game are automata, then their payoff is also close in a certain sense to the price of the game.

Games involving many automata are, of course, the most instructive. We are interested here in the simplest games, namely, the type in which all the players of the game are equivalent. The so-called Gur game is the simplest example of such a game. In this game, N persons participate, each of whom has at his disposal only two operations. Moreover, the probability of payoff for each of the players is determined only by the share of participants using the first strategy. It is clear how people knowing the conditions of the game beforehand would behave in this game—they would arrange matters so that the first strategy would be used by that number of participants which guarantees maximum payoff.

The simplest automata participating in the game do not know beforehand the payoff functions of the game, the operations, or the payoffs of the partners in its individual parties; each automaton receives data about its own payoff or loss in a given party. Nevertheless, with sufficient memory capacity they also maximize their own payoff. The expediency of behavior of each automaton in the simplest problem guarantees the expediency of their collective behavior, replacing such hard-to-be-formulated phenomena as the "understanding about joint operations."

It is interesting to note that if with a fixed memory capacity the number of automata participating in the Gur game grows, the expediency of their collective behavior falls, the limit not differing from a random one. On the other hand, with any fixed number of participants of a game, the growth of the memory capacity of each automaton leads to a rise of the expediency; the mean payoff, moreover, approaches the maximum possible.

We will cite one more example of a game of automata which we shall call a game in distribution. The situation modeled by this game is typical, for example, for a problem about the choice by predatory animals of areas for rut. In this case the number of game falling to the share of a beast of prey is determined by the supply of game in the area chosen by them, and by the number of beasts of prey hunting in it at the same time. The use in a game of one or another strategy corresponds to the choice of an area for hunting, and to the amount of game—one or another value of the payoff function. The game in distribution is assigned k nonnegative numbers $a_1 \geqslant \cdots \geqslant a_k \geqslant 0$, called the *power of the strategies.*

In a game the automata $A^1, \ldots, A^N$ ($N < k$), participate, each of which has k strategies $f_1, \ldots, f_k$. The mathematical expectation of the payoff for the automaton, choosing in some party of this game the strategy f_j, equals a_j/m_j, where m_j is the number of automata using in this party strategy f_j.

We studied the game in distribution of the simplest automata by means of modeling on a computer. In this case the behavior of the automata with sufficient quantities of storage capacity did not differ from the behavior of people knowing beforehand the conditions of the problem and arranging their own actions—the automata (with probability close to 1) chose their strategies to the best advantage. Thus, for example, with a game in distribution of five automata and with $a_1 = 0.9$, $a_2 = a_3 = \cdots = a_7 = 0.33$, it turned out that in 99 percent of the parties the first strategy was chosen by two automata, and the rest by one automaton. Moreover, the average payoff of each automaton is 0.38. It is not difficult to prove that in this case it is not advantageous for a single automaton to change its strategy. Such behavior of the automata coincides with the behavior of people knowing the power of the strategies beforehand.

Nevertheless, when entering into an agreement about sharing payoffs, people would be able to arrange for an even larger payoff. Actually, if the total payoff is divided equally among all the participants of the game, then it would be advantageous to use one by one the first five strategies: then the payoff of each player would reach the size 0.44.

If we combine the payoffs of the automata in each party of the game in distribution and distribute them equally (game with a common cashbox) then the behavior of the automata also changes: in each party the first strategies are chosen, where each strategy is chosen by one automaton. In this case the average payoff changes, and with the growth of the memory capacity of each of the automata participating in the game aims for the maximum value (in our example, 0.44).

It is interesting to note that the growth of the average payoff with the introduction of a common cashbox is achieved only in the case of a sufficient memory capacity. If the memory capacity is small, introduction of a common cashbox lowers the average payoff. In other words, if the individual expediency of behavior of each player is not too large, an equalizing distribution of payoffs is directly disadvantageous.

The behavior of automata in a game in distribution has the characteristic features of dependability. Indeed, let us assume that automata participating in a game with a common cashbox can get out of operation. However, the remaining automata will as before play the most advantageous parties: regardless of which automaton got out of operation, the strategies with the greatest power will be used as before. The growth of the average payoff of each of the automata continuing the game will partially compensate for the drop of the total payoff. With the inclusion into the game of new automata they also are arranged in the most advantageous way. The behavior of such a collection of automata is similar to an ideally dependable machine in which the deterioration of the most responsible components are automatically compensated for at the expense of details less important.

For subsequent development, it is important to note that both the Gur game and the game in distribution are examples of the sort of collective automata which it is "easy" to control in the sense that for control it is enough to give only the payoff functions; for attainment of the optimal working behavior there is no necessity to keep track of the behavior of each individual automaton. In these examples the optimal working behavior is selected by the automata not having information about the operations of the other automata and not able directly to change these operations, so that the interaction of the automata is restricted to participation in the overall game. We note again that a somewhat complicated game in distribution can serve as a model for naturally arising problems about the most advantageous assignment of computing facilities with the need of simultaneous solution of a series of problems.

Of interest are the types of game in which the payoff functions of each automaton depend only on the strategy chosen by it and on the strategies of a limited number of other participants, its game "neighbors." The so-called *game on the periphery* in which the payoff of player A^j depends only on its strategy and the strategies of players A^{j-1} and A^{j+1}, its neighbors, can serve as the simplest example of such a game.

A comparatively quick selection of parties guaranteeing maximal payoff is characteristic for games with a limited number of neighbors. Moreover, the behavior of the automata possesses those features of dependability about which we made mention in the description of a game in distribution.

3. The first two sections of the present article were devoted to modeling certain features of the behavior of complex systems. In the construction of these models we tried to take into consideration concrete physiological structures guaranteeing behavior.

In this section we will, by necessity briefly, describe some attempts of modeling the simplest physiological structures the properties of which are similar to the simplest properties of excitable tissues and which make it possible perhaps to explain certain features of their functioning. Moreover, in contrast to the traditional models of the nervous system of McCulloch and Pitts, we will examine not a system of a large number of individual elements with a complex system of connections among them, but a continuous medium, considering as points adjacent to the given system those points which lie in its immediate geometric vicinity.

We observe that in a physiological experiment the isolation of an individual element is sometimes a difficult and not always sufficiently meaningful task. However, we will examine the simplest example of a continuous medium [13], calling an *active tissue* a medium possessing the following properties. (a) Each point of the medium is capable of instantaneous excitation. During time R after the moment of excitation, the point cannot be excited. The size of the interval R is called the *time of refractoriness*. The time since the moment of the last excitation is called the *phase* $\tau(x,t)$ of point x at moment t. If $\tau(x,t)$ is less than R, then we will say that the point is *refractory*. (b) Excitation can spread in the medium. The speed $s(x,t)$ of spreading of excitation in point x at moment t depends on the phase of this point $c(x,t) = \phi[\tau(x,t)]$. Excitation does not spread over a refractory area. (c) The point possesses spontaneous activity, meaning that within time T (the period of spontaneous activity) after the last excitation the point again becomes excited spontaneously. (We observe that property c is not obligatory, and we will also examine mediums without spontaneous activity.)

The spreading of excitation in active tissues possesses a series of interesting properties. For example, the process of the spreading of impulses of excitation along a homogeneous ring of active tissue is autosynchronized: regardless of the initial stages and initial arrangement of impulses in the ring, behavior is established so that the impulses are set in the ring equidistantly, and they spread at a constant rate. With periodic excitation of the end of a segment of active tissue the spreading of impulses within the limits also occurs at a constant speed regardless of the initial distribution of the phases [1, 2, 17, 24].

We will cite now the example of the work of a flat excitable tissue. We will assume for simplicity that the speed c of excitation is constant, and that the initial phase is the same for all the points and equals 0.

It is clear that all the points of such a tissue will get excited simultaneously with period T. We will suppose now that at moment $t_0 \geqslant R$, point x_0 will be excited from without. Then, obviously, excitation will spread from this point at speed c. At moment $t < T$ the excited point set forms a ring of a radius $c(t - t_0)$ with the center at x_0, and at moment T all the points will be excited that are outside the ring of the radius $c(T - t_0)$. At moment $t_0 + T$, point x again becomes excited spontaneously, and the process will be repeated periodically. Moreover, each point of the medium will get excited with period T; we can say that in this case interaction is absent in the system. It is not difficult to be convinced that with unrestricted initial distribution of the phases and the external excitations in the system, that sort of behavior is established during which each point will get excited spontaneously earlier than it receives excitation from its neighbors, i.e., behavior without interaction.

For an active tissue the points of which have different periods of spontaneous activity, in established behavior the period of excitation of any point of the medium will be equal to the minimum period of spontaneous activity—the medium will be synchronized with the most active point. The same mechanism of synchronization is realized in certain physiological objects. In the work of I. M. Gelfand, S. A. Kovalev, and L. M. Chaylakhyan [10] it was shown, for example, that the pacemaker of the heart is synchronized automatically with the active cell working with the greatest frequency.

The concept of active tissue was used in the construction of a series of models of physiological mechanisms. Thus, for example, in the work of I. P. Lykashevich [21] there is described the modeling on computers of a process of excitation spreading in the heart. This model makes it possible to learn the peculiarities of a series of pathologic working conditions of the heart. I. A. Keder-Stepanova and N. N. Rikko [18] (see Chapter 7 of the present volume) used heterogeneous active tissues for construction of models of the onset of volley activity of the respiratory center.

Recently, Yu. I. Arshavskiy, M. B. Berkinblit, S. A. Kovalev, V. V. Smolyaninov, and L. M. Chaylakhyan [3, 4, 23][1] studied the cable networks, i.e., the homogeneous networks, the edges of which have the properties of active tissues and the peculiarity of excitation conducted in structures of the type of nerve cells.

4. We described here different mathematical models somehow or other connected with our pursuits in physiology. We well understand all the heterogeneity of these models, which are associated only by the common character of their origin. All the same it seems to us that in these models there is some inner common character, perhaps inherent to those physiological mechanisms, which led to these models. We have in mind the principle of the least interaction. We are still very far from a definitive formulation of this principle; in fact, we link with these words only our hope for the construction of the sort of mathematical theory of complex control systems in which this principle would play the same role as the variational principles in analytical mechanics.

We would like to think that in this future theory there will be the same description and those preliminary mathematical models about which we are talking in the present paper. If nevertheless we try to talk here about some considerations connected with the principle of the least interaction, then this is only because such considerations were useful to us in our work.

We will describe a system as working expediently in some external environment if the system tends to minimize interaction with the environment. Moreover, as a rule, the definition of the function of interaction in

---

[1] See Chapters 2, 3, and 4.

a natural way rises from the properties and purpose of the system itself. For example, the deviation of parameters of an organism's inner medium from the optimal values can serve as a measure of the interaction of the organism with the medium.

For the models which we described earlier, these functions are different. For a system searching for the minimum of the function of many variables (section 1 of this chapter), the average value (for a certain interval of time) of the minimized function can serve as a measure of interaction. For automata having expedient behavior (section 2), interaction is measured by the average size of the loss. In models of active tissue (section 3), the monotonic function of the deviation of the average interval between two excitations from a period of spontaneous activity can serve as a measure of interaction for an element of the tissue. Thus all our models are examples of expedient systems. For such systems it is typical that the state with minimum interaction proves to be the most stable. In this sense expedient systems are, so to say, inertial—they tend to pass into a state with small interaction in order no longer to change states. We should note that similar reasons lie at the base of Ashby's principle of homeostasis.

For complex controlling systems, the typical structure permits the separation of individual, relatively automatic subsystems. The expedient behavior of systems of this kind is produced by the expedient behavior of the subsystems forming them. For each subsystem of that type all the remaining subsystems belong to the outside environment and the expediency of the subsystems appears in the minimization of interaction among them so that in stable conditions these subsystems function as if independently—autonomously. Finally, the work of each subsystem is determined by the outside environment and by the work of the whole complex system a part of which it is; but at each moment the subsystem solves its "individual," "particular" problem—it minimizes its interaction with the environment. Consequently the complexity of the subsystem does not depend on the complexity of the whole system. Moreover, the expediency of the whole system appears as the minimization of the total interaction of the system with the environment. With a change of the outside environment the previous working behavior of the control system no longer guarantees a minimality of interaction; there increases the interaction of both the whole system with the environment and also the individual systems of it among themselves. Moreover, the expediency of the system leads to a new stable behavior which guarantees a small bit of interaction in the changed environment.

We will illustrate what has been said with examples. We examine at first the most simple—homogeneous systems all the subsystems of which are equivalent. Such a system is the model of active tissue. We can examine the points of the medium forming the tissue as elementary subsystems of it. We have seen now that in such a tissue, independently from the initial distribution of the phases and loci of initial outside excitations, behavior is established during which all the points become excited spontaneously, i.e., they do not interact among themselves. The spreading of impulses in the ring of active tissue occurs at a steady speed so that interaction of all the points is alike and, with intelligent selection of the interaction function, also minimal.

In similar terms we can also describe such phenomena, important in physiology, as the synchronization of the working of individual elements and explain the necessity of appearance of special desynchronizing mechanisms (for example of the type of the Renshaw cells of the spinal cord).

Another example of homogeneous expedient systems are the homogeneous games of automata—for example the Gur game or the game in distribution described in section 2. Here we also see that the expedient behavior of individual automata guarantee the expediency of the behavior of the whole system of automata. In this instance when the outside environment (in the model, the values of the functions assigning the game) change, the behavior of the automata changes, reducing interaction. In the example of the game in distribution we have already spoken about the characteristics of dependability of the collective of automata playing. This characteristic is generally representative for complex systems in which expedient behavior is shaped out of the interaction of the subsystems also having expediency. For these systems, apparently, one more important peculiarity is typical: expedient joint behavior of the individual systems can be guaranteed even without the presence, so to say, of direct connections among them; in a series of problems for the attainment of such expediency there is enough of that elementary interaction which arises in games of automata. The fact that such direct connections are not obligatory also favors the rise of dependability of the control systems. The fact that such direct connections are not obligatory is important as well because it makes it possible to build complex systems of simple subsystems. Otherwise each subsystem would have a system of connections growing with the complication of the whole system; and its structure, which guarantees expedient behavior, would also become complex.

We note again that the principle of the least interaction is important also because it makes it possible to examine any system (or part of it) if it has expedient behavior as a single whole, which simplifies the study significantly.

The system for the search of the extremum of the function of many variables by the method of ravines can serve as an example of a heterogeneous control system. Actually, this system consists, roughly speaking, of two subsystems—the level forming the local search and the level producing the "step along the ravine." The first of these levels minimizes the value of the function, using, for example, the gradient method. The second minimizes the time required for the local search so that the first is the environment for the second level. In those domains where the characteristics of the function (i.e., separation of the variables into essential and nonessential) are stationary, the interaction of both subsystems is minimized, i.e., both the values of the function (on the average) and the length of the gradient descents become small. Where the function changes (for example, at the turn of the ravine), interaction increases. The expediency of the work of our subsystems makes the work of the whole system expedient, for it lowers the average values of the function in the process of the search.

Perhaps the concepts set forth here make it possible to clarify some general features of interaction of the nerve centers.

In the central nervous system there is a large quantity of individual nerve centers ensuring control of the effectors, so that each act of behavior (for example, motor activity) is the result of their joint activity. Moreover, the change of the functioning of one of the centers must inevitably involve a change of the activity of the remaining ones. If we assume that each nerve center represents an expediently working system, then our mathematical models make it possible to a certain extent to imagine the interaction of the nerve centers without resorting to a complex system of connections and coordination of their activity. Moreover, we can assume that for nerve centers of a certain level, afferentation "coming from below" replaces the effects of the environment, and the effects of the higher centers determine their individual tasks. The organization of this afferentation is, so to speak, defined in a system of payoff functions.

Here, of course, is the very important question of which "particular," "individual" problem each nerve center solves, which is the degree of interaction for it. We assume that the incoming afferentation performs this role for each nerve center, and the expediency of its work is the reduction of the stream of incoming afferentation into the nerve center. We venture to suggest that this role of afferentation is universal, that is, somehow or other it occurs for all nerve centers, and the nervous system as a whole is organized along the principle of the least interaction. For each external situation the problem of the nervous system is an output into the sort of working behavior in which afferentation is minimal. An important feature of the model described here is the relatively simple control: after the payoff functions are assigned, the necessity of detail control by the centers of the given level is eliminated: on the strength of the expediency of their behavior they themselves choose the best distribution of duties (compare with the game of distribution described in section 2). On the other hand, such a way of control is distinguished by great flexibility; corrections in the structure of the payoff functions can be introduced by a whole series of nerve centers having some relation to the problem and not coordinated among themselves.

REFERENCES

1.
Yu. I. Arshavskiy, M. B. Berkinblit, and S. A. Kovalev, *Biofizika* 1962:7, 449.
2.
———, M. B. Berkinblit, and S. A. Kovalev, *Biofizika* 1962:7, 619.

3.
———, M. B. Berkinblit, S. A. Kovalev, V. V. Smolyaninov, and L. M. Chaylakhyan, Chapter 2.

4.
M. B. Berkinblit, S. A. Kovalev, V. V. Smolyaninov, and L. M. Chaylakhyan, Chapter 3.

5.
I. M. Gelfand, A. F. Grashin, and L. N. Ivanova, *Zh. Experim. i Teoret. fiziki* 1961:40, no. 5.

6.
———, A. F. Grashin, and I. M. Pomeranchuk, *ZhETF* 1961:40.

7.
———, V. S. Gurfinkel, S. A. Fomin, M. L. Tsetlin, and M. L. Shik, *Materialy 2-go biofiz. Kongr.* (Abstracts of the second Biophysics Congress) in press.

8.
———, V. S. Gurfinkel, and M. L. Tsetlin, *Dokl. AN, SSSR* 1961:139, no. 5.

9.
———, V. S. Gurfinkel, and M. L. Tsetlin, *Biologicheskie aspekty kibernetiki* (in collection *Biological Aspects of Cybernetics*), Moscow: Izd-vo AN, SSSR, 1962.

10.
———, S. A. Kovalev, and L. M. Chaylakhyan, *Dokl. AN, SSSR* 1963:148, 973.

11.
———, I. I. Pyatetskiy-Shapiro, and Yu. G. Fedorov, *Dokl. AN, SSSR* 1963:152, no. 5.

12.
———, I. I. Pyatetskiy-Shapiro, and M. L. Tsetlin, *Dokl. AN, SSSR* 1963:152, no. 4.

13.
———, and M. L. Tsetlin, *Dokl. AN, SSSR* 1960:131, 1242.

14.
———, and M. L. Tsetlin, *Dokl. AN, SSSR* 1961:137, no. 2.

15.
———, and M. L. Tsetlin, *Uspekhi Matem. Nauk* 1962:17, 103.

16.
———, B. K. Vaynshteyn, R. A. Kayushina, and Yu. G. Fedorov, *Dokl. AN, SSSR* 1963:153, no. 1.

17.
————, and D. A. Kazhdan, *Dokl. AN. SSSR* 1961:141, 527.
18.
I. A. Keder-Stepanova, and N. N. Rikko, Chapter 7.
19.
V. I. Krinskiy, *Biofizika* 1964:9, no. 2.
20.
V. Yu. Krylov, *Avtomatika i telemekhanika* 1963:24, no. 9.
21.
I. P. Lykashevich, *Biofizika* 1963:8, no. 6.
22.
V. A. Ponomarev, *Biofizika* 1964:9, no. 1.
23.
V. V. Smolyaninov, Chapter 4.
24.
V. R. Telesnin, *Radiofizika* 1963:6, 624.
25.
M. L. Tsetlin, *Dokl. AN, SSSR* 1961:139, no. 4.
26.
————, *Uspekhi Matem. Nauk* 1963:18, 112.

I
The Spreading of
Excitation in Continuous
Media and the Electrical
Properties of
Branching Structures

2

An Analysis of the Functional
Properties of Dendrites in
Relation to Their Structure

Yu. I. Arshavskiy
M. B. Berkinblit
S. A. Kovalev
V. V. Smolyaninov
L. M. Chaylakhyan

According to the most prevalent theory about the physiology of the nerve cell [26, 29], excitation of the neuron develops in a limited region with a heightened excitability (the trigger zone), for instance, the axon hillock of the motor neuron. The neuron, from this point of view, integrates incoming excitatory and inhibitory synaptic effects. If, because of these effects, the lowering of the membrane potential reaches a critical level, defined as the threshold of the trigger zone, the neuron fires. In this sense the neuron can be considered a summation device, such that synapses located on the body of the neuron, that is, in the trigger zone region, are the most important. Away from the trigger zone, the value of the synapses located on the dendrites lowers due to electrotonic attenuation. As a result, the individual synapses located on the terminal branching of the dendrite should exert a virtually insignificant effect on the trigger zone.

Such a theory assigns to the dendrites a very modest role in the working of the neuron. Eccles [26] introduced the extreme view that the terminal branchings of the dendrites only perform the function of an ion reservoir, weakening the changes of the ionic concentration gradients during activity. We will note, however, that the structure of the dendrites is, apparently, energetically unsuitable for this purpose since, in proportion to their branching, the ratio of surface area to volume increases.

The majority of the other authors [18, 20, 43, 55, 67, 68], also regarding the neuron as a summator, think that synapses located on the dendrite cannot cause the triggering of the cell; rather, they modulate the excitability of the trigger zone. A quantitative evaluation [72] of the total effect on the trigger zone of the dendritic synapses at a distance from the body of the cell showed that their effect can be rather significant because of the large area of dendrites and, consequently, the large number of synapses located on them.

The shortcoming of such a theory of the neuron as a summator is that it essentially ignores the local structural features of the dendrites. In this article we will endeavor to examine the question of the functional significance of dendrites, based on an analysis of the formation and spread of excitation in branched structures. Here we started with two basic assumptions: (1) the dendritic membrane by its properties is not essentially different from the membrane of the body of the neuron, that is, it has electrical excitability and is capable of conducting excitation in an "all or nothing" pattern; (2) the functioning of at least a large portion of the synapses located on the dendrites is, essentially, the same as that of one of the synapses located on the body of the neuron. It is impossible to consider both these assumptions as strictly valid. Completely contradictory opinions regarding the nature of the dendritic membrane are well known. However, the results of a series of studies [15, 25, 30, 33, 37, 47, 60, 83] provide a basis for considering these assumptions quite probable.

The presupposed assumptions and the application of the classical cable theories during study of the functional features of dendrites have allowed two basic considerations, associated with the principal pattern of dendritic configuration, to be distinguished. (1) The tapering of the dendrites in proportion to the distance from the body of the cell leads to an increase of the postsynaptic potential (PSP) under one and the same presynaptic effect and thereby increases the probability of the formation of an impulse in the dendrite. Figuratively speaking, with identical matches it is easier to light a shaving than a log. (2) The spreading of an impulse originating in a dendrite encounters significant difficulty in the nodes of the branching; under certain conditions, it can be blocked. These and other considerations have already been discussed [2, 27, 29, 51, 52].

The systematic development of these considerations leads to some new inferences about the functional features of the neuron which do not allow us to consider this structure only as a simple summator. Moreover, apparently the cell uses a complicated dendritic apparatus thanks to which a restricted group of synapses, or even a single synapse, can excite the neuron. Here, the result of the synaptic effect depends on the combination of the triggered synapses, for example, whether they are located on the same or nearby dendritic branches or whether they are far from one another. Such a system is similar to a voting system with a large number of participators who have an unequal number of ballots. There is, also, a correction factor which takes into consideration their group behavior. The final result depends, of course, upon the total number of votes cast "for" or "against"; but also, in no lesser degree, it depends on precisely who votes and how the voters team up. Besides that, this system is multistaged, since the voting is carried out not only by the synapses but also by the dendritic branches which join at a single node of the branching.

There are tempting speculations about how, and in which functional tasks, these features of the neuron are used. However, the first question is that of the validity of the presupposed assumptions and some quantitative studies of the cited considerations.

1. DEPENDENCE OF THE SIZE OF THE POSTSYNAPTIC POTENTIAL ON THE LOCATION OF THE SYNAPSE

In this section an attempt will be made to evaluate the effectiveness of the synapses [1] located on different parts of the dendrites. We know [26, 52, 53] that the size of the PSP depends on the input resistance (R_{in}) of the postsynaptic structure, that is, on the resistance between the protoplasm and the outside environment measured at the site of the synapse. We will examine this question in greater detail.

1.1 DEPENDENCE OF THE SIZE OF THE POSTSYNAPTIC POTENTIAL ON THE INPUT RESISTANCE OF THE OBJECT

In Fig. 1, (A) represents an equivalent electrical diagram of a neuron. According to contemporary theories, the mechanism of action of an excitatory synapse is that at the end of the presynaptic fiber a mediator is given off which increases the permeability of the postsynaptic membrane to all ions.

In the equivalent diagram the result of the excitatory synaptic action can be expressed by including the postsynaptic resistance R_s (Fig. 1B). As a result, a current arises which causes a lowering of voltage on the input resistance of the cell, or in other words, the membrane potential "recharges." This lowering of voltage, or depolarization of the membrane, corresponds to the excitatory postsynaptic potential (EPSP). Proceeding

---

[1] Now and later, when speaking of the effectiveness of the synapse, we will not mean the effectiveness of its effect on the trigger zone, but the size of the PSP in the immediate area of the synapse location.

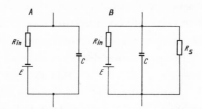

Fig. 1. Equivalent diagram of a neuron at rest (A) and during synaptic action (B). Symbols are defined in the text.

from the equivalent diagram, it is possible, according to Ohm's law, to indicate the size of the EPSP by the formula

$$V_s = I_s R_{in} E \frac{R_{in}}{R_{in} + R_s}. \tag{1}$$

Following are the symbols and definitions used in this article:

E the membrane potential
V_a the amplitude of the action potential
V_t the threshold potential
V_s the postsynaptic potential
I_s the postsynaptic current
C_m the membrane capacitance per unit area (in $\mu F/cm^2$)
R_m the membrane resistance per unit area (in $ohm \cdot cm^2$)
R_i the specific resistance of the protoplasm (in $ohm \cdot cm$)

$r_m = \dfrac{R_m}{\pi d}$ the membrane resistance per unit length (in $ohm \cdot cm$)

$r_i = \dfrac{4R_i}{\pi d^2}$ the protoplasmic resistance per unit length (in $ohm \cdot cm$)

R_{in} the input resistance
R_s the resistance of the activated postsynaptic membrane
$\rho = \sqrt{r_m r_i}$ the characteristic resistance (the entry resistance of a semi-infinite cable)
$\lambda = \sqrt{r_m/r_i}$ the length constant
$\tau = R_m \cdot C_m$ the time constant

l the length of the cable element (the distance between nodes of the branchings in the dendrites)

S the surface area of the body of the neuron

During activation of a large number of synapses, R_s is much greater than R_{in}. In this instance, the size of V_s for one and the same synaptic effect is directly proportional to R_{in}.

It is clear from what has been said that the amplitude of the EPSP must depend on the configuration of the object since under otherwise equal conditions the smaller the size of the cells, the larger the value of R_{in}.

The capacitance of the membrane was not considered in formula (1) so it is applicable only for steady-state conditions arising during a prolonged flow of constant current. Actually, the postsynaptic current is momentary [26], and therefore the EPSP does not reach the maximum amplitude for a given current. Consequently, the value of R_{in} being the same, the size of the EPSP should depend upon the speed of the buildup of the membrane potential. However, if the sizes of the cells change, the rate of the potential rise should remain constant. Indeed, for spherical cells, the rise of the membrane potential is defined by the formula

$$V_T = V_\infty(l - e^{-t/R_m C_m}) \tag{2}$$

where V_T is the membrane potential at a given moment of time t; V_∞ is the membrane potential under steady-state conditions.

For cylindrical cells, the rise of the potential is expressed as

$$V_T = V_\infty \, \text{erf} \, \sqrt{\left(\frac{t}{R_m \cdot C_m}\right)} \tag{3}$$

where the function erf (x) is determined through the integral of probability [50].

Since the temporal parameters connected with the sizes of the cell enter the above formulas only as the product $R_m C_m$, which does not depend on the sizes, the amplitude of the EPSP of different sized cells, during one and the same synaptic effect, is proportional only to R_{in}.

Experimental proof of the proportional dependence of the EPSP's

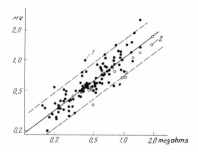

Fig. 2. Dependence of the amplitude of the miniature end-plate potentials on the input resistance of the muscle fibers [53]. (1) m. sartorius fiber; (2) m. extensor longus digitorum fiber. Both scales are logarithmic.

amplitude on R_{in} is obtained in the example of the miniature potentials of the end plate of the muscle fiber [53] (Fig. 2).

Thus, in order to determine the effectiveness of the synapses in relation to their location, it is essential to estimate R_{in} in the different parts of the neuron. Since the diameter of their branches decreases in proportion to the branching of the dendrites, the input resistance of the postsynaptic structure should naturally increase with the distance from the body of the neuron. The quantitative character of these changes will be given below.

1.2 METHOD FOR DETERMINING R_{in} IN DIFFERENT PARTS OF THE NEURON

It is natural to consider the dichotomous branching of the dendrites as a continuous structure composed of cylindrical cable elements (we will call the section of the dendrite between branching nodes a *cable element*). The input resistance at a certain point of the dendrite is equal to the inclusive parallel R_{in} of the segments of the cable elements branching out from this point. To figure them we use the well-known cable equations. The input resistance measurable at one end of the segment of a cable whose other end is loaded with a certain resistance r_n is equal to

$$R_{in} = \rho \frac{r_n + \rho \tanh l/\lambda}{\rho + r_n \tanh l/\lambda}. \tag{4}$$

Knowing the diameter of the cable element and its passive electrical properties R_m and R_i, it is easy to figure out the values of ρ and λ. There-

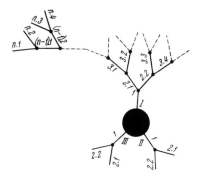

Fig. 3. Diagram of a dendritic tree and of the numbering of its branches used in the description of the calculations.

fore, to find R_{in} in the different parts of the neuron, the basic problem is to find r_n at the ends of each cable element. For this we use the method of "successive steps" similar to that which Rall used for a somewhat different purpose [71]. The resistances of the loads of the distal ends of the cable elements are found successively beginning with the end branches of the dendrites; the resistances of the loads of the proximal ends are determined beginning with the main trunk of the dendrite going directly out from the body of the neuron.

Since for simplicity's sake, we assume that dendrites are made out of cylindrical cable elements, r_n of the distal end of the terminal branch of nth sequence (as illustrated by the outlined diagram of the neuron in Fig. 3) is equal to the resistance of the membrane at the butt end of the cylinder. Hence, if the passive electrical parameters R_m and R_i of the cable elements are known, by equation (4) it is possible to figure out the R_{in} measurable at the proximal end of the terminal branch, as if it were detached from the rest of the dendrite. The combined parallel R_{in} of the two branches of sequence n are, in their turn, the resistance of the load for the distal end of the corresponding branch of sequence $(n - 1)$ etc. Thus, we can successively figure out r_n for the distal end of any cable element and, consequently, the R_{in} for all the dendrite as if it were detached from the neuron's body. We can go on from here to figure out the R_{in} of the neuron, measurable in the region of the body, as the resistance of the somatic membrane, parallel to which are the inclusive R_{in} of all the dendrites. Knowing this quantity is essential for the solution of the second half of the problem: the determination of the r_n of the proximal ends of the cable elements. The load of the proximal end of the branch of the first sequence is the R_{in} measurable in the neuron's body with no regard for the given dendrite. Now, we successively compute the R_n of the proximal ends of all the cable elements.

For real dendrites we are limited to computing the R_{in} in the nodes of branching and in the centers of the internodal segments. As the previous work clearly shows, the R_{in} in the node of branching is equal to the enclosed parallel input resistances of the three branches leaving it. The R_{in} in the center of an internodal segment equals the enclosed parallel input resistances of the distal and proximal halves of the given segment.

1.3. RESULTS

On microscopic sections of a cat's brain 100 μ thick processed by the Golgi method, there were located neurons the dendrites of which branch primarily on the surface of the slices. With the help of a drawing apparatus, the neurons were copied. At the same time the lengths and diameters of all the dendritic branches were carefully measured and the surface area of the body of the neuron was figured out. Such findings were obtained for the motor neuron of the spinal cord, the pyramidal cell of the motor cortex,[2] the Golgi cell and the stellate cell of the cerebellar cortex, and the neuron of the interstitial nucleus of the cerebellum. Knowing the diameters of the dendritic branches, and taking the specific values of R_m and R_i, we found the sizes of ρ and λ and, next, by the method described above, the R_{in} in the centers of the dendritic branches and in the nodes of the branching.[3]

In Fig. 4, a sketch of a cat's motor neuron is shown. The numbers indicate the size in megohms of the input resistances in the body of the neuron, in the middles and at the ends of the dendritic branches. The results obtained are more fully given in the Appendix, where there are sketches of the neurons, diagrammatic representations of them (Figs. A1–A5), and

---

[2] The motor neuron and pyramidal cell are sketched from preparations obligingly donated by G. P. Zhukov.

[3] The computation of R_{in} was done on the computer by L. E. Sotnikov, whom the authors wish to thank.

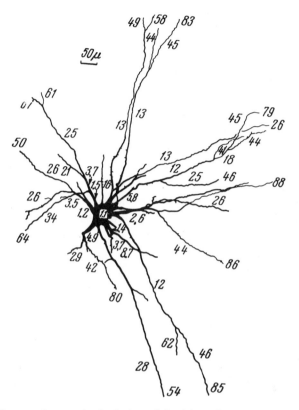

Fig. 4. Input resistances in the body and dendrites of a neuron. The numbers indicate the input resistance in megohms in the body of the neuron, and in the middle and on the ends of the corresponding dendritic branches with membrane resistance of 500 ohm · cm². Motor neuron of a cat.

also, Tables A1–A5 with the sizes of R_{in} in the different parts of the neuron.

As seen in Fig. 4 and Tables A1–A5, the R_{in} vigorously increases toward the periphery of the dendrites. This is especially pronounced in large neurons with long, vigorously branching shoots (the motor neuron and the pyramidal cell). In them, the input resistances of the thin dendritic branches exceed by some 10 times the R_{in} of the body.

The sizes of R_{in} cited in Fig. 4 and in Tables A1–A5 were obtained with the values $R_i = 100$ ohm·cm and $R_m = 500$ ohm·cm^2 [26]. This size of R_m is, apparently, sufficiently workable for the motor neurons, inasmuch as the input resistance of the soma that we figured out is near to the experimentally measured size. However, other neurons can have a different size R_m. Moreover, as regards the motor neuron, some researchers cite large sizes $R_m = 3500$–4000 ohm·cm^2 [20, 76].[4] Hence, for the motor neuron depicted in Fig. 4, the input resistances were figured with different values of R_m in a sufficiently wide range (100–5000 ohm·cm^2). These calculations show that the absolute values of the input resistances in different parts of the neuron naturally increase with the enlargement of R_m. It is important, however, to emphasize that the differences, which interest us, of the R_{in} in the body and in the dendrites depend substantially less on R_m. Although these differences decrease with an enlargement of R_m, the order of the differences remain the same.

It is necessary to note that some errors in the measuring of the lengths and diameters of the dendritic branches could be tolerated. To begin with, the precision of the measurement of especially thin dendritic branches was rather small. As a result of the histological analysis, the diameter measurable on the specimen could differ from the actual one. To simplify the calculations, the diameter of the branches between the nodes are assumed to be the same for the full length of it. Errors in calculating the area of the somatic membrane could be tolerated. Finally, although all the dendrites of the neurons chosen for measurements were located on the surface of the microscopic section, some dendritic branches nevertheless could prove to have been cut off. All this decidedly reduces the preciseness of the results obtained. However, it is doubtful if it can affect the overall qualitative character of the experiment.

The resulting data show that the local effectiveness of dendritic synapses is much higher than that of somatic synapses because the same presynaptic action should cause on the terminal branchings of the dendrites a PSP 10 times greater than that on the soma. However, it is not clear how the in-

---

[4] The question of reasons for such a divergence presents an independent interest and we will not consider it in this article, since only the differences in effectiveness of the synapses located on the body and dendrites of the neuron concern us here.

fluence of the synapses on the trigger zone will change with distance from the body of the neuron. To answer this question, we must compare the degree of increase of the R_{in} with the size of the electronic attenuation potential.

Calculations of the electronic attenuation of potential for the length of the dendritic branch toward the body, and from the body of the neuron, is expressed by the formula

$$V = V_0\left(\cosh\frac{l}{\lambda} + \frac{\rho}{r_n}\sinh\frac{l}{\lambda}\right)^{-1}. \tag{5}$$

We assumed that a single potential (V_0) was applied to one end of the branch, and the size of the potential v at the other end was calculated. From Tables A1–A5, it is apparent that for the length of a given dendritic branch, the degree of increase of R_{in}, measurable as the ratio of the input resistances in its proximal and distal nodes, is always less than the degree of the electrotonic attenuation of potential toward the cell body for the length of that branch. Consequently, the local increase of effectiveness of the dendritic synapses should be accompanied by a decrease of their influence on the trigger zone.

For greater clarity, in Fig. 5 there are depicted two dendrites of the same motor neuron as in Fig. 4. The numbers located in the middle of the dendritic branches represent the size of the electrotonic attenuation of potential toward the body of the neuron for the length of the given dendritic branch. The underlined numbers are the sizes of the input resistances in the body of the neuron, in the node of branching, and on the terminals of the dendritic branches. We can assume that the sizes of the input resistances are equal to the amplitudes of the PSP caused by the same presynaptic action (in certain arbitrary units), inasmuch as the PSP is proportional to R_{in}. Then, knowing the electrotonic attenuation, we can calculate the sizes of the potentials aroused in the body of the neuron by identical synapses located on different parts of the dendrites. The figures thus obtained are given in parentheses in Fig. 5. Obviously, these figures are always less than the size of the R_{in} in the soma, which is equivalent to the amplitude of the PSP caused by a synapse located directly on the body of the neuron. Moreover, the further on the dendrite the synapse is located, the less potential it evokes in the body.

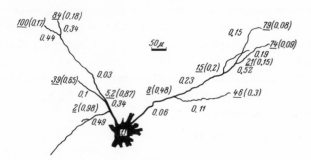

Fig. 5. Input resistances and electrotonic attenuation of the potential in different sections of the neuron. The same motor neuron as in Fig. 4; for the sake of clarity only two dendrites are represented. The number at the middle of the dendritic branches indicate electrotonic attenuation of the potential toward the body of the neuron for the extent of the corresponding branch; the underlined numbers at the nodes of branching and at the ends of the dendritic branches indicate the input resistances (in megohms) that are equivalent to the amplitudes of the postsynaptic potentials (in arbitrary units) caused by the same presynaptic effect; the numbers in parentheses indicate the changes of the potential in the body of the neuron (in the same arbitrary units) caused by the same synapses located in corresponding places of the neuron.

A comparison of the data in the last two columns of Tables A1–A5 makes apparent also the fact that the electrotonic attenuation of potential toward the body of the cell is substantially stronger than that from the body of the cell. This difference ties in with the fact that the r_n in the proximal ends of the dendritic branches is lower than that in their distal ends. As we shall see shortly, such behavior of the electrotonic attenuation of potential must be taken into consideration during an analysis of the rate of the spread of excitation along the dendrites in the orthodromic and anti-dromic directions; and also during evaluations both of the effect of the potential arising in the region of the dendrites on the body of the neuron, and, on the other hand, of the potential originating in the body of the neuron on the dendrites.

Thus, the effectiveness of the synapses located on sufficiently thin dendritic branches can be 10 times higher than the effectiveness of the synapses located on the body of the neuron. Hence, with our presupposed assumptions, there is sufficient activation of a small number of synapses, or even of one synapse, for the excitation of a thin dendritic branch. The question arises as to whether the excitation of an individual dendritic branch can lead to a triggering of the whole neuron. To answer it, we must analyze the conditions for the conduction of excitation along the dendrites.

2. PECULIARITIES OF THE SPREADING OF EXCITATION ALONG THE DENDRITES
Inasmuch as we assume that the membrane of the dendrites does not differ by its properties from the membrane of the remaining parts of the neuron, the peculiarities of the spreading of excitation along the dendrites can be explained only by their configuration: (1) the small thickness of the terminal dendritic branches, and (2) the branching of the dendrites [27, 29]. To answer the question posed in the preceding section, it is, of course, necessary to evaluate how these factors affect the dependability of the conduction of excitation. In addition, we are interested first of all in conduction in the orthodromic direction from the terminal dendritic branches toward the body.

We know that the dependability of the conduction of excitation is determined by the safety factor K, that is, the ratio of the amplitude of the action potential V_a to the threshold potential V_t:

$$K = \frac{V_a}{V_t}. \tag{6}$$

According to the theory of local currents, the amplitude of the spreading action potential, in its turn, is defined by the ratio

$$V_a = E \frac{R_n}{R_{n'} + R_n} \tag{7}$$

where E is the membrane potential during excitation, R_n is the input resistance of the nonexcited region (the resistance of the load), $R_{n'}$ is the input resistance of the excited region (the internal resistance of the generator).[5]

[5] For simplification, it is usually understood that E and the membrane resistance R_m are identical for the whole length of the excited region. However, it can be shown rigorously that equation (7) is correct and, for the case of the heterogeneous distribution of E and R_m along the excited region, how this occurs under real conditions.

Moreover, V_t depends on the rate of the prethreshold growth of the potential (phenomenon of accommodation): $V_t = f(dV/dt)$.

Thus, to evaluate the significance of the configuration factors, it is essential to analyze both the effect of the thickness and the branching of the fibers on the relationship of the resistances of excited and nonexcited regions, and the nature of the prethreshold growth of the potential.

We cannot give an exact mathematical theory for the spreading of excitation which would help to describe conduction in objects with a varied configuration, although there is hope in forming such a theory to use an analogy of the process of the spread of excitation with combustion, and to apply the existing mathematical apparatus [39]. Subsequently, there will even be adduced some qualitative considerations about the peculiarities of the conduction of excitation in fibers of a different diameter and in the node of branching.

2.1. CONDUCTION OF EXCITATION IN FIBERS OF DIFFERENT DIAMETERS

In the literature the assumptions were made that a critical thickness of the nerve fibers exists below which conduction is impossible [6, 26]. However, there is basis to doubt the accuracy of such an assumption.

We know that the rate of excitation is proportional to the length constant λ and inversely proportional to the time constant τ [13, 22, 48, 73, 76]. As we have already said, in fibers with identical membrane properties, τ does not depend on the diameter. Hence, it is easy to show that the safety factor also does not depend on the diameter. Actually, in a length of fiber with a substantially larger λ, the ratio of the resistances of the nonexcited and excited regions, that is, the ratio $R_n/(R_n + R_{n'})$ depends on resistances of the membranes of the nonexcited and excited regions and on the ratio l/λ for the excited region.

Naturally, the resistances of the membranes of the excited and non-excited regions do not depend on the diameter. The length of the excited region with the very same duration of excitation at each point of the membrane is proportional to the rate of conduction, which is proportional to λ. Therefore, the ratio l/λ for the excited region also does not depend on the diameter. Thus, even the ratio $R_n/(R_n + R_{n'})$ does not depend on the diameter.

It is necessary to show that a change of the diameter does not influence even the rate of the prethreshold growth of the potential during the spreading of excitation. The rate of the prethreshold growth of potential on each point of the fiber is defined as the product of the gradient of the potential along the length of the fiber times the rate of the conduction of the excitation:

$$\frac{dV}{dt} = \frac{dV}{dx} \cdot \frac{dx}{dt}. \tag{8}$$

Since the potential gradient along the length of the fiber is inversely proportional to λ, and the rate of conduction is directly proportional to λ, their product should not depend on the diameter.

Thus, although the rate of conduction lowers with a reduction of the diameter, the dependability of conduction, determined by the size of the safety factor, does not change at the same time.[6]

The conduction of excitation in fibers of the same diameter as that of the terminal dendritic branches (up to 0.4 μ) was experimentally shown in a study of the peripheral nerve fibers of group C [38].

It is necessary to note that the conclusion about the absence of a critical diameter was made on the analysis of only the cable properties of the excited fibers. Moreover, there was not taken into account the circumstance that with the lessening of the diameter, the ratio of the surface area to the volume of the fiber increases. Because of this fact, extremely disadvantageous conditions arise for the maintenance of normal ionic gradients. This may worsen the conditions for the conduction of impulses in thin fibers, especially with a deterioration of the overall state of the nerve system, or under a rhythmic emergence of impulses. But it is extremely difficult to evaluate the real significance of this factor since the active and passive ionic flows into a unit of the area of the dendritic membrane and their changes in different experimental conditions are unknown.

---

[6] In an article devoted to an analysis of the mechanism of transformation of rhythm in a nerve fiber, we arrived at the conclusion that there must be a critical rate below which conduction of excitation becomes impossible [5]. From the above it is clear that by itself the rate of conduction is unimportant, but the reasons leading to its decrease are important.

2.2 CONDUCTION OF EXCITATION AT THE NODE OF BRANCHING

If, with the branching of the nerve fiber, the total area of the section of the fiber and the area of the membrane increases, the R_{in} in the region of the node of branching decreases, that is, the R_n for the spreading action potential decreases. As is apparent from equation (7), this will lead to a lessening of the amplitude V_a, and consequently, to a lowering of the safety factor in proportion to the approach of the impulses to the node of branching [9, 56]. Assuming, moreover, that the other parameters of equations (6, 7), and especially E, V_t, and R_n, are constant, we can quantitatively estimate the decrease of the safety factor in the node of branching, if the ratio of the resistances of the membranes of the excited and nonexcited regions is known. However, the size of the decrease of the safety factor obtained during such an evaluation may be appreciably underestimated. This is because during the spread of excitation along the fiber, the sizes of $R_{n'}$, E, and V_t in their turn are functions of V_a. Thus with the decrease of V_a the rate of excitation lowers, a fact which leads to the reduction of the dimensions of the excited region, that is, to the increase of its input resistance $R_{n'}$. On the other hand, with the reduction of V_a, E also will diminish, since the size of the membrane potential during excitation is determined by the changes of ionic permeability, which are functions of the potentials [48, 49]. Finally, in contrast to what should occur during a change of the diameter of the fiber, the lowering of the rate of conduction (dx/dt) in the region of the node of branching can be sharper than the increase of the potential gradient along the length of its fiber (dV/dx). This will lead to a lessening of the rate of the prethreshold growth of the potential which, in the presence of accommodation, should cause an increase of V_t. Thus, the reduction of V_a, due to the lowering of R_n in the region of the node of branching, entails changes in R_n, E, and V_t in the direction leading to the further lowering of V_a and the safety factor. The exact quantitative evaluation of the changes in the different parameters of excitation during transmission through the node of branching is involved in the solution of the system of differential equations of Hodgkin and Huxley for the case of heterogeneous cables.

A worsening of conditions of conduction to the point of complete blocking of the impulses in the region of the node of branching in muscle afferent fibers [51] and the axons of mollusks [82] was experimentally observed. It is significant that in the experiments on muscle afferents, a block was observed only during orthodromic spreading of impulses out of the thin terminal shoots to the afferent fiber, that is, toward the sharpest fall of the input resistance. During antidromic spreading of the impulse out of the thick afferent fiber to the end terminals, a block never set in.

From the estimates of the input resistances in different parts of the neuron cited in the preceding section, it is apparent that the R_{in} of the dendrites lessen proportionately as the body of the cell is approached. The sharpest changes of R_{in} occur at sites of dendritic branching. From all that has been said, it is clear that precisely in these parts there is a high probability of a blockage of the spreading impulse. Depending upon the specific conditions of branching, an excitation which emerged in an individual dendritic branch may lead to a triggering of the whole neuron, or it can attenuate in one of the nodes of branching located in its path.

It is obvious from these estimates that the conditions of conduction along the dendrites in the orthodromic and antidromic directions are different. The data presented in the last two columns of Tables A1–A5 show that the electrotonic attenuation of potential in the proximal direction is substantially stronger than in the distal direction. This means that the rate of conduction of excitation in the orthodromic direction is lower than the rate of conduction in the antidromic. The decline and blocking of conduction in the nodes of branching can be explained by a series of peculiarities of the propagation of excitation along the dendrites already described in other writings (the different capacity for conduction in the orthodromic and antidromic directions, "decreased" conduction, the extremely low rate of propagation of excitation, etc.

3. PROPERTIES OF THE DENDRITIC MEMBRANE AND SYNAPSES ACCORDING TO DATA FOUND IN THE LITERATURE

The interpretations of the results presented in sections 1 and 2 depend substantially on the properties of the dendritic membrane and the synapses located on the dendrites. Thus, if the dendritic membrane is unexcited, the calculations of the input resistances and the electrotonic attenuation in the dendrites must apparently be taken into consideration only during an evaluation of the effect of the different synapses on the trigger zone. However, the most interesting conclusions about the functional peculiarities of the neuron follow from results obtained through the accuracy of the assumptions that we have adopted, that the properties of the dendritic membrane and of the synapses located on them do not essentially differ from the properties of the membrane and the synapses of the body of the cell.

At the present time, it is virtually impossible to subject either of these assumptions to direct experimental testing, especially if the question is about the terminal branchings of the dendrites. Existing circumstantial data lead some researchers to exactly opposite conclusions. Nevertheless, on the basis of the data described in the literature, the assumptions that we have adopted appear to us to be the most probable.

3.1. PROPERTIES OF THE DENDRITIC MEMBRANE

In a study of the electrical responses of the cortex to the stimulation of different peripheral or central structures, some authors concluded that these responses are dependent either wholly or partly on the activity of the apical dendrites of the pyramidal cells [8, 10–12, 17–21, 23, 24, 43–46, 62, 63, 67, 68, 69, 75]. This especially has to do with the response of the cortex to direct electrical stimulation which is frequently described in articles as the "dendritic potential." On the basis of an analysis of these responses, the conclusion is repeatedly made that the dendrites are characterized by specific properties different from the properties of the body and axon of the neurons.

It is necessary, however, to note that the methods used in the study of the "dendritic potential" are inadequate for forming an opinion on the properties of separate neurons, let alone their parts. In conditions where a complex structure containing a network of presynaptic fibers, neural bodies, and dendrites is stimulated, and where a sufficiently large electrode is applied for the recording of a response, all the interpretations of the results obtained cannot be reliable. Even if we agree with the opinion that the response to direct stimulation of the cortex is solely the dendritic potential, its characteristics do not provide a basis for conclusions about the specific properties of the membrane of the dendrites. They can be explained by the conditions of the conduction of excitation in a complex branching structure (see section 2).

The inadequacies of the method are indicated by the contradictoriness of the conclusions regarding the properties of the dendrites, to which different researchers came, on the basis of studying the electrical responses of the cortex in similar, methodical conditions. According to Chang [17–21], the stimulation of the cortical surface causes excitation of the dendrites which spreads at the rate of 1 to 2 m/sec. Bishop and his colleagues [10–12, 23, 24] came to the conclusion that during a single stimulation the spreading of excitation along the dendrites occurs only in an antidromic direction; during orthodromic stimulation there arise only local responses or an excitation spreading with a decrement. The work of Grundfest, Purpura, and Okudzhava [43–46, 62, 63, 67, 68, 69] expresses an extreme point of view. Chiefly on the basis of the pharmacological analysis of cortical responses, the conclusion was drawn that dendritic potentials are solely postsynaptic potentials; the dendritic membrane itself is electrically inexcitable and incapable of active conduction of excitation.

It is necessary, however, to note that the results of the pharmacological analysis of the "dendritic potentials" are full of contradictions. The conclusion about the electrical inexcitability of the dendritic membrane was made, first of all, on the basis of the fact that tubocurarine suppresses the responses of the cortex down to complete disappearance [62, 63, 67–69]. On the other hand, in those very papers, the conclusion was made that the "dendritic potential" is the sum of the excitatory and inhibitory postsynaptic potentials sensitive to different synaptic toxins.

It is incomprehensible why tubocurarine, which as we know blocks only cholinergic synapses, completely eliminates a response of the cortex. Possibly this contradiction is explained by the fact that the disappearance of cortical responses under the effect of tubocurarine is determined not by its specific effect on the postsynaptic membrane, but is the result of an overall depressant effect of curare. To support this hypothesis one can cite the findings of Ochs [61] that the disappearance of cortical responses during injection of tubocurarine is observed only during a fall of the blood pressure below a certain level. A series of additional critical observations in connection with the concept developed by Grundfest, Purpura, and Okudzhava are cited in the article by Eccles [27] and in the discussion on the report by Okudzhava [62].

Thus, the reasons cited to support the viewpoint that the membrane of the dendrites differs basically from the rest of the membrane of the neuron, do not seem convincing to us. On the other hand, there is a series of studies which show the feasibility of the spreading of excitation along the dendrites, as well as the ability of the dendrites to respond to electrical stimulation. We will dwell on some of the studies that are the most convincing for our viewpoint.

Fatt [33] and Nelson et al. [60], analyzing the spread of electrical fields originating during antidromic stimulation of a single motor neuron of a cat, came to the conclusion that excitation spreads along the dendrites for a distance of at least 300 μ. Although the recording of the potential was obtained extracellularly, in so far as excitation of a single motor neuron took place, those objections made previously concerning the studies of electrical responses of the cortex are irrelevant to these studies.

Terzuolo and Araki [83], recording the antidromic excitation of a cat's motor neuron with two intracellular microelectrodes, sometimes observed action potentials rising with different latent periods. By all indications, the microelectrode recording the action potential with a short latent period was in the body of the cell, and the electrode recording the action potential with a larger latent period was in the dendrite. Thus, the fact shows that excitation spreads from the body of the cell to the dendrites.

A series of authors came to a conclusion regarding the feasibility of the spreading of excitation along the apical dendrites of the pyramidal cells of the hippocampus [1, 2, 25, 37]. During orthodromic excitation of the pyramidal cells, a slow potential is usually recorded against a background of which a spikelike response arises. The recording of the potentials at different depths showed that this spikelike response spreads along the dendrites in both directions from the point of its origin.

The most demonstrative data attesting to the electrical excitability of the dendrites and the spread of excitation in an "all or none" pattern, were presented in the work of Hild and Tasaki [47]. In experiments on a tissue culture of the cerebellum, they had the opportunity to stimulate and record the activity of the dendritic branches of the Purkinje cells and other large neurons at a distance up to 100 μ from the body of the cell. It was shown that excitation can spread along the dendrites in both directions (Fig. 6).

Although the work cited attests to the feasibility of the spread of excitation only along sufficiently thick dendritic branches (to an extent of several hundred microns) there is no basis to think that the membrane of the terminal branches should, in this respect, differ from the remaining dendritic membrane.

The question arises whether or not there are experimental results indicating the possibility of the onset of excitation in the dendrites during natural synaptic activation of the neuron. It is difficult to answer this question. In the conditions of recording of the potentials from the cell body, without special analysis, it can be difficult to discern whether triggering of the neuron takes place because of activation of somatic or dendritic synapses. If in the soma region a zone exists with increased excitability, the excitation arising in the dendrites, as it approaches the body of the cell, will first of all cause a triggering of this region. At the same time there should be recorded in the body of the neuron a slow growth of the potential replaced by a sharp rise on reaching the threshold of the trigger zone. The slow growth of the somatic potential, determined by the spread of excitation along the dendrites, is apparently difficult to distinguish from the excitatory postsynaptic potential. This distinction will be especially difficult if, at the same time in the conditions of the experiment, the synapses located on both the body and on the dendrites are activated.

The most precise data attesting to the possibility of excitation arising in the region of the dendrites were obtained by Eccles et al. [30] during a study of a cat's motor neurons chromatolyzed as a result of a preliminary cut of the anterior roots. Two circumstances, apparently, contributed to the exposure of the site of onset of excitation in the chromatolyzed neurons—an overall attenuation of the synaptic effect as a result of which the difference in the effectiveness of the somatic and dendritic synapses is more sharply revealed, and the lessening of the difference in the excitability of the membrane of the initial segment and the somatodendritic

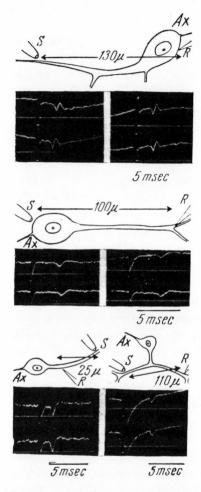

Fig. 6. Conduction of excitation along dendrites. In each example at the top is a diagram of the distribution of the stimulating (S) and recording (R) electrodes. Ax is the axon; below are the extracellularly recorded action potentials. Preceding each action potential is seen an artifact of stimulation. Neurons from a 10–19 day culture of a rat's cerebellum [47].

membrane. During reflex stimulation of gradually increasing strength, an abrupt increase of the prethreshold depolarization is recorded in the body of the chromatolyses motor neurons, while each new increase of potential

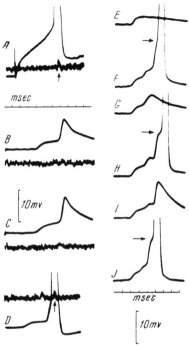

Fig. 7. Occurrence of "partial spikes" during orthodromic excitation of a chromatolyzed motor neuron. (A–D) at top, intracellular recordings of a motor neuron m. flexor digitorum longus, the axon of which had been cut for 16 days; below, recording from a filament containing the axon of a motor neuron. (A) response to a depolarizing shock of current; (B–D) responses to maximum stimulation of the afferents of group I n. flexor digitorum longus; (E–J) responses to an afferent discharge, showing a wide range of variability of the "partial spikes" which are superimposed on the EPSP (E) and which cause a complete response when depolarization reaches a critical level (about 13 mv) shown by arrows on records (F, H, and J) [30].

("partial spikes" to use the authors' terminology) is realized in an "all or none" pattern (Fig. 7). Hyperpolarization of the body of the motor neuron does not affect the "partial spikes," which indicates their origination in the region of the dendrites. One can surmise that the "partial spikes" reflect the excitation of the individual dendritic branches, which, being blocked, do not reach the body of the neuron. During the stimulation of the afferent fibers of different muscle nerves, the action potential of chromatolyzed motor neurons could arise under a different level of pre-threshold depolarization (Fig. 8); sometimes the prespike synaptic potential was practically absent altogether (Fig. 8C). This fact also speaks to the possibility of the onset of excitation in the region of the dendrites. Actually, during onset of an impulse in the immediate region of the soma, the amplitude of the EPSP is equal to the level of the threshold potential. The further from the body of the cell the spreading excitation arises, the less prethreshold depolarization will be recorded. At the time of the emergence of a spreading impulse far enough from the body of the neuron, the excitatory postsynaptic potential preceding it can generally not be seen. In other words, in conditions where the difference in the excitability of the axon hillock and the somatodendritic membrane is reduced, the picture may be close to that which emerges during the recording of a spreading impulse, for instance, in the axon. During the hyperpolariza-tion of the body of a motor neuron, which impedes the emergence of the action potential in it, the stimulation of the sural nerve arouses a "partial spike" (Fig. 8D) which, as already pointed out, apparently is an electro-

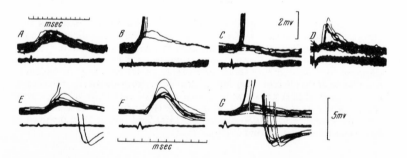

Fig. 8. Synaptic excitation of a motor neuron during stimulation of different afferent nerves. Recording from two motor neurons (A–D and E–G) for the m. plantaris, the axons of which were cut for 20 and 44 days, respectively; (A) and (E), (B) and (F), (C) and (G) responses to stimulation of the nerves of the flexor digitorum longus, plantaris, and gastrocnemius-soleus, respectively. (D) response to the same stimulation as in (C), but against a background of the transmission of a hyperpolarizing current with an intensity $7 \cdot 10^{-9a}$ across the motor neuron [30].

tonic expression of the impulse evoked in the dendrites and blocked on its way to the body of the cell.

One might think that the formation, observed by Eccles et al., of impulses in the region of the dendrites takes place in normal neurons. Chromatolysis only assists the exposure of the emergence of the site of origin of the excitation.

There still exists a series of facts, although much less convincing which possibly are also accounted for by the onset of excitation in the region of the dendrites. Spencer and Kandel [81], during intracellular recording of the activity of pyramidal neurons of the hippocampus originating in response to the afferent stimulation, occasionally observed "fast potentials" arising in an "all or none" pattern. According to the authors' opinion, the "fast potentials" reflect the onset of excitation in the dendrites. However, such an assumption is rather hypothetical and needs further verification.

The spasmodic growth of prespike depolarization during gradual increase of strength of afferent stimulation [50, 66] is described in a series of articles. Though it is not clear from the diagrams cited in these articles, the spasmodic growth of the potential is realized in an "all or none" pattern. Therefore we can think that the spasmodic growth of the potential observed in these studies can arise as a result of the stimulation of some afferent paths activating neurons with a different latent period.[7]

3.2. STRUCTURAL AND FUNCTIONAL CHARACTERISTICS OF THE DENDRITIC SYNAPSES

As we know, some histologists have concluded that the density of the synaptic contacts is greatest on the cell of the body and decreases progressively on the dendrites in proportion to the distance from the soma, whereupon the terminal branching of the dendrites are generally deprived of synaptic contacts [7, 14]. Eccles considered this morphological data as circumstantial confirmation of the viewpoint that the generation of the spreading impulses is realized only in the trigger zone, located on the region of the cell body. Actually, the effect of dendritic synapses on the cell body should be weakened due to the electrotonic attenuation of the potential. For the terminal dendritic twigs, this attenuation will be very

[7] More recently two more articles [70, 84] appeared, indicating the possibility of the origin of excitation in dendrites.

sharp, a fact which makes us conclude as absurd the existence of synapses on them.

On the other hand, there is a large number of articles in which it is shown that the synapses can be located all the way along the dendrites down to the terminal branching [16, 31, 77–79]. Moreover, some authors specially note that the density of the synaptic contacts on the dendrites is as great as that on the cell body [4, 54, 74, 85, 86]. In this respect the work of Kositsyn [54] is especially significant. He discovered that in the reticular neurons of the medulla oblongata the density of the synaptic contacts does not change for a distance of at least 400 μ from the soma (Fig. 9). It is possible that the conclusion of some authors about the reduction of density of the synaptic contacts on the dendrites is a result of defects in their methods [3, 54, 74, 85, 86].

Electron microscopic research of neurons showed that somatic and dendritic synapses have an identical ultramicroscopic structure: the presence of vesicles in the presynaptic endings, a thickening of the presynaptic and postsynaptic membranes, and a series of other characteristic features [29, 42, 64, 65].

Some authors emphasize the morphological difference between the axosomatic and axodendritic synapses. If there is an overwhelming majority of synapses on the bodies of neurons it is synapses of the end feet type; on the dendrites, moreover, there are spiny synapses. It was once thought that only spiny synapses were characteristic for the dendrites of highly specialized cells (the pyramidal cells of the cortex of the large hemispheres and the Purkinje cells of the cerebellar cortex). However, it has recently been shown that in these cells, side by side with spiny synapses are also synapses of the end feet variety [4, 79, 86]. Spiny synapses are generally not characteristic of the dendrites of other cells, for example the neurons of the spinal cord and medulla [87]. The average area of the synaptic contacts on the body and on the dendrites is approximately the same size [55, 85, 86]. Thus, at least a part of the dendritic synapses have the same morphology as do the synapses on the cell body.

Of course it is impossible on the basis of morphological data alone to conclude whether the function of somatic and dendritic synapses is the same. On the other hand, as already stated, because of the obvious

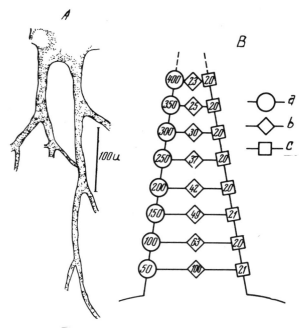

Fig. 9. (A) dendrites of a giant cell of the reticular formation with synaptic plaques; (B) diagram illustrating the density of the distribution of the synaptic endings on the dendrites of cells of the reticular formation; (a) distance from the cell body (in μ); (b) percent expression of the diameter of the dendrite at a specific level (for 100%, the average size of the diameters of the dendrites at a level 50 μ from the body of the cell is assumed); (c) the number of synaptic endings on 100 μ^2 surface of the dendrite [54].

methodical difficulties, there are no real physiological studies of the mechanism of the synaptic action on dendrites. The most convincing record of an excitatory postsynaptic potential arising in the region of the dendrites of a frog's motor neuron is found in the work of Brookhart and Fadiga [15]. Anatomical [59] and physiological data [15, 32, 57] show that in the frog the descending fibers of the lateral columns end monosynaptically on the soma, and the posterior root afferents end on the dendrites of the motor neurons. Under treatment of Nembutal in a concentration which suppresses the spike activity of the neurons, one could record exclusively excitatory postsynaptic potentials arising during stimulation of both paths. It was shown in these conditions that the postsynaptic potentials arising in the region of the soma and in the region of the dendrites have similar characteristics.

It is necessary to note that when we speak about the fundamental similarity of the function of somatic and dendritic synapses, we do not have in mind, for instance, a similarity of the chemical nature of the mediators. Confirmation rests on the fact that synaptic conduction in both cases is realized by a chemical path during which the function of the mediator consists of the nonspecific rise of the penetrability of the membrane, which is shown on the equivalent diagram as the inclusion of the R_s (see Fig. 1B).

The facts cited in this section do not, of course, fully prove the accuracy of our two assumptions. However, they do speak to their sufficiently large probability. Granting our presupposed assumptions, we will see below what basic conclusions regarding the functional possibilities of the neuron follow from the analysis described of the local features of the onset and spread of excitation.

4. FUNCTIONAL FEATURES OF THE NEURON (SOME CONCLUSIONS FROM THE PRECEDING ANALYSIS)

1. In view of the features of the dendritic synapses described, it is possible to think that there are two types of excitation of the neuron. The generally acknowledged type of excitation is realized through the trigger zone. At the same time the neuron works as a summation device. For stimulation of the neuron in this way, the relative arrangement of the excitatory synapses is unimportant, only the activation of a sufficiently large number of them is essential. Moreover, the synapses located in the region of the trigger zone have the greatest value; the value of the synapses located on the dendrites gradually diminishes due to electrotonic attenuation. Since for such a triggering of the neuron only the lowering of the trigger zone potential to a definite size is essential, though the effect of exactly which synapses being unimportant, it is naturally called nonspecific.

The other type of neuronal excitation is realized through the dendrites.

For it, not so much the number but the relative location of the activated synapses is important. The activation of a few synapses, or even one synapse, located locally on a fine enough dendritic branch can be more effective than the activation of a much larger number of synapses located diffusely on different parts of the neuron. Therefore, with this type of stimulation, a sharp nonequivalence of the interneuronal connections should be apparent. Since the second type of excitation depends essentially upon exactly which synapses of the neuron are activated, we can call it specific. Obviously, the specific type of excitation is much richer in its functional possibilities than the nonspecific, since with it one group of neurons may affect the other neurons in a far larger number of ways.

In accordance with the concept proposed concerning the two types of neuronal excitation, the importance of the synapses located on the different parts of the dendrites should have a dual appraisal: according to the size of the depolarization which they cause in the trigger zone, and according to the size of the depolarization in the immediate region where the synapse is located. As already shown in Fig. 5, as the distance from the body of the neuron increases, the local effectiveness of the synapses begins to exceed their effect on the trigger zone. This conclusion is correct even taking into account the fact that the excitability of the trigger zone is 3 to 5 times higher than the excitability of the rest of the membrane.

In connection with the described functional differences of the somatic and dendritic synapses, it is pertinent to pay attention to the following circumstance. During the subliminal action of single dendritic synapses, the changes of excitability that arise should bear a more or less local character because of the sharp electrotonic attenuation of potential toward the cell body. On the other hand, the attenuation of potential away from the cell body is substantially weaker (see Tables A1–A5 in the Appendix). Therefore, a subliminal change of the soma potential will show an amply effective influence on the whole surface of the neuron, including the distant dendritic branches. Consequently, the regulating influence on the excitability of the neuron, consisting of a more or less even change of the membrane potential of the entire cell, naturally takes place with the help of the synapses located in the region of the body, and not in the region of the terminal branchings of the dendrites.

2. The features of the dendritic synapses that have been described let us make a series of hypotheses about the mechanism of the "spontaneous" rhythmic activity observed in many neurons.

In the anatomical literature, cases are described where the axon of a given neuron produces a recurrent collateral going to the terminal dendritic parts of the same or neighboring neurons. Such recurrent collaterals are described in the greatest detail for the pyramidal cells of the cerebral cortex [80]. If it is assumed that the triggering of a cell may occur only in the region of the body, the significance of such a recurrent collateral would be vague, since it terminates in the region of the terminal dendritic branching and the effect of its synapses on the body should be insignificant. According to the viewpoint developed in this article, the recurrent collateral terminates where the effectiveness of the synapses is greatest, and therefore it can play an important role in the maintenance of the rhythmic activity of the cell.

Another hypothesis concerning the "spontaneous" activity mechanism of the neurons was expounded by Katz and Miledi [52]. Taking into account that the amplitudes of PSP depend on the R_{in}, they concluded that in sufficiently thin dendritic branches, even miniature potentials arising during the secretion of a single "quantum" of the mediator, can be supraliminal. In this case, they would serve as a source of the "spontaneous" activity of the neurons. Such a hypothesis is not confirmed for the motor neurons on which Katz and Miledi conducted their experiments. Possibly this is explained by the low excitability of the somatodendritic membrane of motor neurons [26]. However, the mechanisms assumed by Katz and Miledi may occur in other neurons.

3. The special features of the spreading of impulses in the nodes of branching acquire considerable significance during a specific type of activation of the neuron through the dendrites.

As described in detail above, conduction of the excitation arising in an individual dendritic branchlet is impeded in the node of branching where the safety factor decreases. This makes the nodes of branching a convenient place for regulating the conduction of excitation along the dendrites with the aid of synaptic influences. Possibly increase of the density of synaptic contacts on the nodes of branching of the dendrites [54] is involved with exactly this process.

It is especially necessary to dwell on the possibility of the inhibition of neurons as a result of the corresponding effect on the nodes of branching. A rather large amount of recently compiled experimental data, obtained during the intracellular recording of activity of different neurons, shows that inhibition can take place without any changes of the membrane potential or excitability of the body of the neuron. Usually such examples of inhibition are interpreted as presynaptic inhibition [28], which in several cases actually was proved.

However, another explanation of this phenomenon is also theoretically possible [34]. If the inhibitory synapses are localized in the region of the dendrites at a great enough distance from the soma, they may block the spreading of excitation at the nodes of branching of the dendrites without causing at the same time any change in the body of the cell. For example, the stimulation of certain afferent pathways to the hippocampus causes hyperpolarization in the region of the apical dendrites of the pyramidal neurons, which is accompanied by inhibition of the activity of the cells [36]. Taking into account the length (700 μ) and the thickness (3 μ) of the apical dendrites of the pyramidal cells, it is conceivable that the inhibitory potentials effectively suppressing the spreading of impulse along the dendrites should have practically no effect on the soma.

Such inhibition in the nodes of branching of the dendrites ought, through its own physiological significance, to differ both from the post-synaptic inhibition localized in the cell body and from presynaptic inhibition. In contrast to postsynaptic inhibition localized on the body, during postsynaptic inhibition localized on the dendrites, the whole neuron should not be inhibited, but only part of it. This inhibition will not affect excitation arising in the other parts of the neuron. In contrast to presynaptic inhibition, which is considered an agent for selectively disconnecting specific afferent paths drawing near to a given neuron, the inhibition localized on the dendrites will suppress all the paths approaching the given part of the dendrite.

4. If the excitation approaching a node of branching along one of the

branches passes through it, it will not only spread further toward the body of the cell but also in an antidromic direction. Naturally, if impulses emerge at this time in other dendritic branches, they will be blocked by the excitation coming to meet them. In this way, if excitation arising in one of the dendritic branches can spread to the whole neuron, it will block the excitations arising in all the other parts of the neuron. By means of this mechanism the leading afferent path is, in a sense, singled out. The neuron will always trigger from the branch excited with the most frequency at a given moment, which is similar to the sinus node of the heart: the rhythm is determined by the cell most frequently functioning [40, 41].

5. With a specific type of dendritic branching or state of the membrane, excitation arising in an individual branch may be blocked in the node of branching. However, if excitation approaches a node along two branches simultaneously, conduction is distinctly facilitated. Therefore, if the excitation arising in one dendritic branch cannot spread through the node of branching, it can pass through it during the simultaneous excitation of the two neighboring branches. Obviously, thanks to such characteristics, the node of branching may be considered the simplest logical element.

It is necessary to note that for transmission of excitation through the node of branching, there is no need for the strict synchronization of the arrival of the impulses along the two branches. In the first place, during the blocking of an impulse an electrotonic potential will develop behind the region of the block that lasts, as is known, longer than the impulse. This potential will facilitate the conduction of both the next impulse coming along the same branch (temporal summation) and the impulse arriving along the neighboring branch (spatial summation).

The duration of the facilitation will be determined by the time of abatement of the electrotonic potential. In the second place, during the approach of excitation at different times along two adjacent branches to the node of branching, a process will occur that will lead to the decrease of the interval between them. This occurs as a result of a peculiar "tightening up." The impulse, as the first approaching the node of branching, will cause in front of itself a subliminal electrotonic depolarization which must spread even to the adjacent branch. This should lead to a certain rise of the rate of the spreading of the excitation coming along the second branch.

Thus both these mechanisms increase the probability of a concurrence of impulses arriving at the node along two branches.

The functional features that we have assumed for the neuron permit us to think that the branching dendritic structures has a wide range of possibilities for the realization of different logical functions. Thanks to this assumption, in a single neuron having such a structure, it is possible to carry out logical functions for the accomplishment of which at least several neuron summators would have been required.

APPENDIX

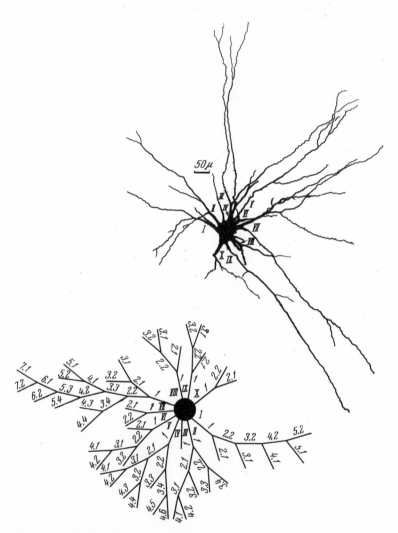

Fig. A1. Motor neuron of the cervical section of the spinal cord of a cat (Golgi method) and its diagrammatic representation.

Table A1. Input Resistance and Electronic Attenuation of Potential in Different Sections of the Motor Neuron of a Cat (Fig. A1). Surface area of neuron body, 10,000 μ^2. Membrane[1] resistance of neuron body, 5 megohms.

Number of branch or node of branching[2]	$l.\ \mu$	$d.\ \mu$	R_{in} at center of branch or body of neuron	R_{in} at node of branching or at end of branch	Electrotonic attenuation of potential toward cell body	Electrotonic attenuation of potential away from cell body
CELL BODY			1.1			
I 1	40	9	1.2	1.5	0.67	0.97
2.1	70	4	4.2	7*	0.21	0.95
2.2	60	4	3.5	5.1	0.25	0.82
3.1	250	2	26	50*	0.04	0.40
3.2	80	2.3	12	17	0.22	0.73
4.1	25	1.5	25	32*	0.53	0.96
4.2	85	2	26	36	0.39	0.81
5.1	20	1	47	60*	0.59	0.95
5.2	30	1	52	71*	0.48	0.93
II 1	50	7	1.5	2	0.48	0.93
2.1	40	3.5	3.7	5.2	0.34	0.88
2.2	75	2	11	18	0.07	0.64
3.1	280	2	25	40	0.04	0.25
3.2	130	2	21	39*	0.10	0.74
3.3	30	1.5	26	36*	0.47	0.98
3.4	180	1.7	34	64*	0.07	0.25
4.1	40	1	61	84*	0.44	0.92
4.2	60	1	67	100*	0.34	0.87
III 1	160	2.3	16	30*	0.02	0.68
IV 1	20	11	1.1	1.2	0.84	0.97
2.1	30	6.8	1.5	1.8	0.64	0.95
2.2	25	4.5	1.8	2.4	0.45	0.89
3.1	75	4	4.5	7.1	0.22	0.89
3.2	420	3	13	17	0.03	0.15
3.3	420	3	13	26*	0.04	0.29
3.4	540	3	13	21	0.04	0.10
4.1	30	2	12	17*	0.41	0.96
4.2	60	2	16	26*	0.24	0.81
4.3	210	1.5	45	83*	0.08	0.41
4.4	150	1.5	44	76*	0.14	0.61
4.5	70	1.5	38	58*	0.32	0.88
4.6	50	1.5	35	49*	0.40	0.94
V 1	190	4	5.8	8	0.06	0.44
2.1	170	2	25	46*	0.10	0.62

Table A1 *(continued)*

Number of branch or node of branching[2]	$l.\ \mu$	$d.\ \mu$	R_{in} at center of branch or body of neuron	R_{in} at node of branching or at end of branch	Electrotonic attenuation of potential toward cell body	Electrotonic attenuation of potential away from cell body
2.2	190	3	12	15	0.23	0.43
3.1	50	2	18	21	0.52	0.74
3.2	140	1.5	41	74*	0.13	0.64
4.1	130	1.5	44	74*	0.19	0.68
4.2	160	1.5	45	79*	0.17	0.57
VI 1	20	5	1.5	1.9	0.53	0.96
2.1	120	4	6.1	11*	0.16	0.87
2.2	40	4	3.5	5*	0.37	0.96
VII 1	20	11.5	1.1	1.2	0.87	0.98
2.1	400	3	12	19	0.02	0.17
2.2	95	5.5	2.6	3.8	0.24	0.77
3.1	100	1	62	111*	0.11	0.70
3.2	250	1.5	46	85*	0.09	0.32
3.3	10	4	4.1	4.4	0.86	0.97
3.4	125	2	17	23	0.08	0.47
4.1	30	2.3	6.8	8.8	0.42	0.82
4.2	20	2	7.2	9.9	0.41	0.93
4.3	80	1.5	41	63*	0.31	0.85
4.4	410	1.5	45	88*	0.03	0.10
5.1	280	1.5	44	86*	0.02	0.26
5.2	330	2	26	53*	0.04	0.25
5.3	40	2	15	20	0.44	0.90
5.4	30	1.5	19	28*	0.32	0.94
6.1	35	1.5	29	38	0.52	0.93
6.2	15	1	30	39*	0.52	0.98
7.1	20	1	50	62*	0.62	0.96
7.2	15	1	47	57*	0.70	0.98
VIII 1	15	12	1.1	1.2	0.88	0.99
2.1	50	4.5	2.6	4.1*	0.25	0.98
2.2	30	7	1.4	1.7	0.62	0.96
3.1	70	4.5	3.7	5.8*	0.28	0.96
3.2	180	3.5	8.7	16*	0.07	0.72
IX 1	120	4.5	3.7	6	0.12	0.70
2.1	30	2.3	9.5	13*	0.48	0.98
2.2	120	3	10	13	0.28	0.60
3.1	490	2	28	54*	0.02	0.09
3.2	65	2	22	32*	0.38	0.92

Table A1 (*continued*)

Number of branch or node of branching[2]	$l.\ \mu$	$d.\ \mu$	R_{in} at center of branch or body of neuron	R_{in} at node of branching or at end of branch	Electrotonic attenuation of potential toward cell body	Electrotonic attenuation of potential away from cell body
X 1	70	3.2	4.9	8.2	0.11	0.82
2.1	80	1.5	29	52*	0.14	0.85
2.2	200	1.5	42	80*	0.05	0.44

[1] All calculations were carried out with $R_m = 500$ ohm·cm² and $R_i = 100$ ohm·cm.
[2] In this and all the subsequent tables, a node of branching is marked with the same number as the branch branching out from it in the direction of the body. For terminal branches the sizes of the R_{in} measurable on the distal ends are given. These numbers are marked with asterisks.

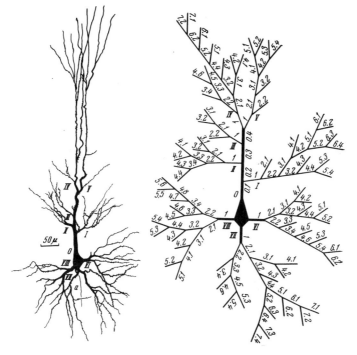

Fig. A2. Pyramidal cell of the motor region of the cerebral cortex of a cat (*a*-axon).

Table A2. Input Resistances and Electrotonic Attenuation of Potential in Different Sections of the Pyramidal Cell of the Motor Region of the Cerebral Cortex of a Cat (Fig. A2). Surface area of neuron body, 200 μ^2. Membrane resistance of neuron body, 25 megohms.

Number of branch or node of branching	$l. \mu$	$d. \mu$	R_{in} at center of branch or body of neuron	R_{in} at node of branching or at end of branch	Electrotonic attenuation of potential toward cell body	Electrotonic attenuation of potential away from cell body
CELL BODY			41			
0.1	51	6.6	3.6	3.7	0.80	0.85
0.2	12	6	3.7	3.8	0.83	0.87
0.3	24	5.5	4	4.2	0.83	0.93
0.4	99	4	5.8	6.5	0.38	0.59
I 1	9	4	4	4.3	0.85	0.98
2.1	60	0.7	64	123*	0.03	0.83
2.2	18	2.7	5.4	6.6	0.60	0.93
3.1	6	1.3	8.6	11	0.56	0.93
3.2	22	1.3	14	21	0.27	0.85
4.1	78	0.7	80	152*	0.05	0.75
4.2	24	1.3	18	24	0.35	0.83
4.3	91	0.7	92	170*	0.08	0.68
4.4	15	0.7	33	43	0.41	0.84
5.1	7	0.7	31	38	0.59	0.96
5.2	19	1.3	29	34	0.63	0.90
5.3	66	0.7	96	157	0.22	0.81
5.4	66	0.7	96	157*	0.22	0.81
6.1	15	0.7	54	71*	0.54	0.96
6.2	27	0.7	60	95*	0.38	0.96
6.3	91	0.7	100	176*	0.12	0.68
6.4	85	0.7	98	171*	0.14	0.71
II 1	18	2.7	5	6.3	0.56	0.95
2.1	42	1.3	22	35	0.14	0.83
2.2	18	2	9	12	0.52	0.95
3.1	12	0.7	48	62*	0.57	0.98
3.2	99	0.7	103	183*	0.12	0.64
3.3	12	0.7	24	36	0.31	0.95
3.4	31	0.7	36	54	0.15	0.70
4.1	24	0.7	61	87*	0.40	0.97
4.2	12	0.7	49	63*	0.58	0.98
4.3	37	0.7	86	132*	0.44	0.76
4.4	30	0.7	46	77*	0.15	0.71

Table A2 (*continued*)

Number of branch or node of branching	$l.\ \mu$	$d.\ \mu$	R_{In} at center of branch or body of neuron	R_{In} at node of branching or at end of branch	Electrotonic attenuation of potential toward cell body	Electrotonic attenuation of potential away from cell body
III 1	30	1.5	12	20	0.18	0.84
2.1	27	0.7	42	60	0.26	0.78
2.2	72	0.7	83	150*	0.10	0.77
3.1	24	0.7	84	109*	0.53	0.97
3.2	72	0.7	110	174*	0.27	0.78
IV 1	30	2.7	7.6	8.5	0.63	0.82
2.1	55	2.7	10	11	0.55	0.72
2.2	18	0.9	19	29	0.25	0.86
3.1	182	2.7	16	21	0.52	0.57
3.2	145	1.3	39	42	0.07	0.27
3.3	9	0.7	37	45	0.61	0.95
3.4	66	0.7	86	148*	0.15	0.81
4.1	158	0.9	73	142*	0.06	0.43
4.2	121	0.9	72	132*	0.09	0.58
4.3	252	0.7	111	217*	0.03	0.15
4.4	24	1.3	44	46	0.75	0.85
4.5	36	0.7	79	117*	0.36	0.93
4.6	22	0.7	68	92*	0.47	0.97
5.1	164	0.7	64	82*	0.56	0.99
5.2	22	0.7	51	53	0.58	0.63
6.1	115	0.7	113	197*	0.14	0.56
6.2	3	0.7	51	52	0.92	0.95
7.1	121	0.7	114	200*	0.14	0.53
7.2	103	0.7	112	192*	0.17	0.62
V 1	15	2.7	7.2	7.9	0.77	0.93
2.1	91	2.7	12	15	0.37	0.70
2.2	91	0.7	85	164*	0.03	0.68
3.1	158	2	22	25	0.22	0.36
3.2	30	0.7	47	79*	0.18	0.95
4.1	22	2	27	28	0.82	0.94
4.2	72	1.3	38	47	0.31	0.59
5.1	194	0.7	110	211*	0.03	0.26
5.2	182	0.7	110	210*	0.04	0.30
5.3	152	0.7	113	208*	0.09	0.40
5.4	133	0.7	113	203*	0.11	0.48
VI 1	15	5.4	3.6	3.8	0.86	0.97
2.1	18	2.7	5.1	6.3	0.58	0.95

Table A2 (*continued*)

Number of branch or node of branching	$l. \mu$	$d. \mu$	R_{in} at center of branch or body of neuron	R_{in} at node of branching or at end of branch	Electrotonic attenuation of potential toward cell body	Electrotonic attenuation of potential away from cell body
2.2	12	2.7	4.7	5.5	0.67	0.97
3.1	43	1.3	21	33	0.15	0.77
3.2	12	1.3	11	15	0.39	0.91
3.3	35	0.7	80	157*	0.02	0.71
3.4	30	2	9.6	13	0.36	0.89
4.1	73	0.7	92	159*	0.16	0.77
4.2	91	0.7	99	176*	0.13	0.68
4.3	103	0.7	93	178*	0.05	0.62
4.4	36	0.7	42	57	0.16	0.63
4.5	85	0.7	85	161*	0.06	0.71
4.6	7	1.8	14	16	0.82	0.97
5.1	55	0.7	102	155*	0.32	0.87
5.2	145	0.7	116	208*	0.12	0.43
5.3	85	0.7	86	162*	0.69	0.71
5.4	66	0.7	67	104	0.09	0.61
6.1	36	0.7	133	169*	0.58	0.93
6.2	18	0.7	121	140*	0.72	0.96
VII 1	12	4	3.7	4	0.82	0.96
2.1	15	2.7	5	5.8	0.64	0.94
2.2	43	2.7	7	9.7	0.37	0.90
3.1	82	1.3	31	48	0.07	0.61
3.2	43	2.7	8.5	11	0.47	0.90
3.3	12	1.3	14	17	0.52	0.92
3.4	72	0.7	75	144*	0.05	0.78
4.1	85	0.7	107	178*	0.19	0.71
4.2	55	0.7	95	148*	0.28	0.86
4.3	99	0.7	90	173	0.04	0.64
4.4	12	0.7	20	26	0.32	0.76
4.5	24	0.7	36	49	0.26	0.74
4.6	99	0.7	93	176*	0.06	0.64
5.1	12	0.7	35	41	0.55	0.85
5.2	30	0.7	45	56	0.31	0.66
5.3	52	0.7	94	144*	0.30	0.87
5.4	127	0.7	113	202*	0.12	0.50
6.1	30	0.7	61	76	0.41	0.76
6.2	64	0.7	94	154*	0.22	0.82
6.3	99	0.7	113	191*	0.19	0.64
6.4	50	0.7	85	110	0.36	0.70

Table A2 (*continued*)

Number of branch or node of branching	$l.$ μ	$d.$ μ	R_{in} at center of branch or body of neuron	R_{in} at node of branching or at end of branch	Electrotonic attenuation of potential toward cell body	Electrotonic attenuation of potential away from cell body
7.1	55	0.7	117	168*	0.39	0.86
7.2	33	0.7	106	140	0.52	0.94
7.3	18	0.7	127	148*	0.74	0.96
7.4	36	0.7	138	174*	0.59	0.93
VIII 1	18	5.4	3.6	3.9	0.84	0.97
2.1	18	3.3	4.7	5.6	0.69	0.96
2.2	24	2	6.1	8.1	0.45	0.93
3.1	22	2	8.6	11	0.44	0.92
3.2	24	1.3	14	22	0.22	0.88
3.3	55	1.3	26	40	0.15	0.74
3.4	18	1.3	15	21	0.35	0.94
4.1	12	1.3	15	19	0.55	0.92
4.2	109	0.7	94	181*	0.04	0.52
4.3	85	0.7	91	165*	0.09	0.71
4.4	15	0.7	37	50	0.42	0.92
4.5	55	0.7	88	141*	0.24	0.86
4.6	85	0.7	101	174*	0.16	0.71
4.7	30	0.7	53	85*	0.25	0.95
4.8	18	0.7	38	54	0.36	0.90
5.1	128	0.7	101	194*	0.05	0.50
5.2	121	0.7	100	191*	0.05	0.53
5.3	15	0.7	66	82*	0.60	0.97
5.4	36	0.7	84	121*	0.39	0.93
5.5	12	0.7	67	81*	0.68	0.98
5.6	36	0.7	88	125*	0.41	0.93

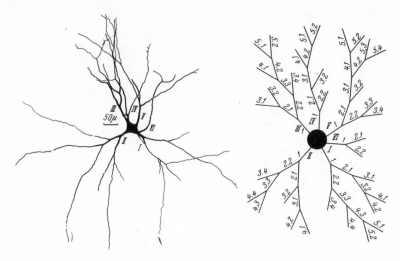

Fig. A3. Neuron of the nucleus interpositus of the cerebellum of a cat.

Table A3. Input Resistances and Electrotonic Attenuation of the Potential in Different Sections of a Neuron of the Nucleus Interpositus of the Cerebellum of a Cat (Fig. A3). Surface area of neuron body, 3350 μ^2. Membrane resistance of neuron body, 15 megohms.

Number of branch or node of branching	$l. \mu$	$d. \mu$	R_{in} at center of branch or body of neuron	R_{in} at node of branching or at end of branch	Electrotonic attenuation of potential toward cell body	Electrotonic attenuation of potential away from cell body
CELL BODY			3.1			
I 1	30	5.5	3.5	3.8	0.75	0.94
2.1	78	2.7	8.2	12	0.25	0.73
2.2	78	2.7	8.1	11	0.25	0.69
3.1	85	1.8	27	42*	0.24	0.86
3.2	260	1.8	32	52	0.09	0.32
3.3	12	1.8	13	14	0.76	0.91
3.4	290	1.2	60	117*	0.02	0.18
4.1	18	0.9	64	76*	0.68	0.96
4.2	36	0.9	73	96*	0.51	0.94
4.3	24	1.8	17	19	0.68	0.88
4.4	50	1.2	36	59*	0.22	0.92
5.1	222	1.2	60	115*	0.06	0.31
5.2	85	1.8	32	47*	0.34	0.86

Table A3 (*continued*)

Number of branch or node of branching	$l. \mu$	$d. \mu$	R_{in} at center of branch or body of neuron	R_{in} at node of branching or at end of branch	Electrotonic attenuation of potential toward cell body	Electrotonic attenuation of potential away from cell body
II 1	302	4.3	6.8	8.1	0.03	0.29
2.1	60	2.4	12	15	0.42	0.77
2.2	36	3.3	9	10	0.72	0.89
3.1	27	1.5	20	25	0.50	0.84
3.2	302	1.2	60	118*	0.02	0.16
3.3	27	2.4	11	13	0.69	0.89
3.4	340	0.9	77	152*	0.005	0.04
4.1	66	0.9	61	102*	0.20	0.83
4.2	133	1.2	60	104*	0.14	0.60
4.3	350	1.5	45	87*	0.02	0.16
4.4	328	1.5	45	87*	0.03	0.18
III 1	60	4.6	4.4	5.6	0.49	0.88
2.1	85	2.2	14	21	0.24	0.76
2.2	50	1.8	13	19	0.21	0.75
3.1	66	1.2	47	75*	0.25	0.87
3.2	60	1.2	45	72*	0.25	0.89
3.3	133	1.2	46	54	0.12	0.33
3.4	260	0.9	77	152*	0.013	0.17
4.1	85	0.9	86	131*	0.32	0.75
4.2	55	0.9	73	94	0.42	0.75
5.1	18	0.7	111	131*	0.76	0.96
5.2	30	0.7	120	151*	0.61	0.95
IV 1	18	3.9	3.7	4.4	0.69	0.96
2.1	9	3.9	4.7	5	0.89	0.99
2.2	218	1.2	57	113*	0.008	0.32
3.1	121	1.8	21	30	0.08	0.50
3.2	12	1.8	7.6	10*	0.49	0.99
4.1	72	0.7	90	156*	0.15	0.78
4.2	103	1.2	46	51	0.24	0.40
5.1	118	0.9	86	141*	0.22	0.59
5.2	85	0.9	84	130*	0.30	0.75
V 1	66	3.9	5	6.6	0.38	0.81
2.1	24	2.4	8.4	9.9	0.59	0.89
2.2	206	1.5	40	70	0.04	0.37
3.1	121	1.8	22	27	0.16	0.49
3.2	72	1.8	23	31*	0.24	0.89

Table A3 (*continued*)

Number of branch or node of branching	*l.* μ	*d.* μ	R_{in} at center of branch or body of neuron	R_{in} at node of branching or at end of branch	Electrotonic attenuation of potential toward cell body	Electrotonic attenuation of potential away from cell body
3.3	18	0.9	82	94*	0.74	0.96
3.4	12	0.9	78	87*	0.18	0.99
4.1	30	1.2	34	38	0.55	0.73
4.2	60	1.2	44	58	0.34	0.72
5.1	145	0.9	81	143*	0.13	0.48
5.2	85	0.9	76	123*	0.20	0.75
5.3	18	0.7	77	97*	0.59	0.98
5.4	99	0.7	115	192*	0.19	0.64
VI 1	187	1.8	26	43	0.13	0.41
2.1	55	0.9	72	106*	0.36	0.88
2.2	36	0.9	65	89*	0.47	0.94

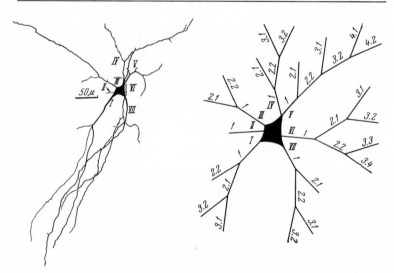

Fig. A4. Golgi cell of the cortex of the cerebellum of a cat.

Table A4. Input Resistances and Electrotonic Attenuation of the Potential in Different Sections of a Golgi Cell of the Cortex of the Cerebellum of a Cat (Fig. A4). Surface area of neuron body, 1160 μ^2. Membrane resistance of neuron body, 43 megohms.

Number of branch or node of branching	$l.\,\mu$	$d.\,\mu$	R_{in} at center of branch or body of neuron	R_{in} at node of branching or at end of branch	Electrotonic attenuation of potential toward cell body	Electrotonic attenuation of potential away from cell body
CELL BODY			5.1			
I 1	85	3.7	6.9	8.3	0.44	0.70
2.1	6	3.7	8.5	8.4	0.94	0.98
2.2	40	1.2	25	43*	0.18	0.95
3.1	318	1.7	37	75*	0.04	0.22
3.2	305	2.5	18	34*	0.08	0.36
II 1	20	1.2	14	23*	0.22	0.96
III 1	50	2.2	10	15	0.28	0.83
2.1	43	1.2	33	51*	0.28	0.94
2.2	297	1.3	56	108*	0.03	0.19
IV 1	50	2.2	10	14	0.28	0.80
2.1	115	1.3	48	86*	0.13	0.70
2.2	160	1.3	46	67	0.08	0.37
3.1	20	1.2	75	84*	0.79	0.97
3.2	56	1	89	119*	0.50	0.89
V 1	20	3.7	5.9	6.6	0.75	0.97
2.1	15	1.7	10	14*	0.49	0.99
2.2	15	2.5	7.8	9	0.71	0.97
3.1	10	1.2	13.5	18*	0.5	0.996
3.2	20	2.5	11	12	0.71	0.96
4.1	55	1.3	33	55*	0.22	0.91
4.2	50	1.2	34	57*	0.19	0.92
VI 1	15	2.5	6	6.6	0.69	0.89
2.1	30	2.5	9.5	11	0.56	0.95
2.2	55	2.5	9	12	0.43	0.71
3.1	25	1.5	18	26*	0.44	0.96
3.2	80	1.8	24	38*	0.25	0.87
3.3	90	1.8	25	41*	0.23	0.85
3.4	290	2.1	25	45*	0.09	0.34
VII 1	60	3.1	7.5	9.6	0.43	0.81
2.1	425	1.6	38	76*	0.01	0.1
2.2	20	2.5	11	12.5	0.73	0.91
3.1	297	1.2	57	112*	0.02	0.12
3.2	10	1.2	17	22*	0.58	0.996

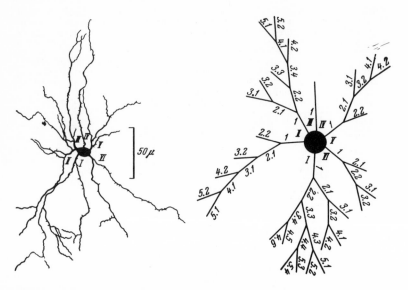

Fig. A5. A stellate cell of the cortex of the cerebellum of a cat.

Table A5. Input Resistances and Electrotonic Attenuation of the Potential in Different Sections of a Stellate Cell of the Cortex of the Cerebellum of a Cat (Fig. A5). Surface area of neuron body, 450 μ^2. Membrane resistance of neuron body, 110 megohms.

Number of branch or node of branching	$l. \mu$	$d. \mu$	R_{in} at center of branch or body of neuron	R_{in} at node of branching or at end of branch	Electronic attenuation of potential toward cell body	Electronic attenuation of potential away from cell body
CELL BODY			17			
I 1	10	1	22	25	0.61	0.89
2.1	95	0.75	89	143	0.10	0.55
2.2	62	0.75	64	80	0.16	0.50
3.1	12	0.5	170	200*	0.71	0.97
3.2	5	0.5	152	161	0.86	0.98
3.3	25	0.75	91	101	0.65	0.82
3.4	3	0.5	86	91	0.86	0.99
4.1	16	0.5	196	235*	0.67	0.98
4.2	10	0.5	184	209*	0.77	0.99
4.3	8	0.75	107	114	0.87	0.99
4.4	3	0.75	103	105	0.95	0.98
4.5	20	0.5	137	185*	0.48	0.97

Table A5 (*continued*)

Number of branch or node of branching	$l.\mu$	$d.\mu$	R_{in} at center of branch or body of neuron	R_{in} at node of branching or at end of branch	Electronic attenuation of potential toward cell body	Electronic attenuation of potential away from cell body
	4.6	32	0.5	158	232*	0.36	0.92
	5.1	25	0.5	167	226*	0.48	0.94
	5.2	12	0.5	142	171*	0.66	0.99
	5.3	40	0.5	181	269*	0.35	0.88
	5.4	16	0.5	142	181*	0.57	0.98
II	1	5	0.75	21	25	0.63	0.92
	2.1	6	0.75	29	33	0.71	0.93
	2.2	75	0.5	162	307*	0.05	0.67
	3.1	7	0.75	38	43	0.72	0.92
	3.2	40	0.5	122	213*	0.16	0.88
	4.1	120	0.75	100	149	0.12	0.40
	4.2	40	0.5	129	220*	0.17	0.88
	5.1	40	0.5	218	303*	0.43	0.88
	5.2	15	0.5	183	219*	0.67	0.97
III	1	10	1	22	26	0.61	0.91
	2.1	12	0.75	37	47	0.51	0.94
	2.2	7	1	29	33	0.77	0.96
	3.1	70	0.5	170	307*	0.11	0.70
	3.2	20	0.5	95	143*	0.32	0.97
	3.3	8	0.75	42	50*	0.67	0.99
	3.4	88	0.75	88	132	0.13	0.54
	4.1	15	0.5	158	185	0.64	0.94
	4.2	20	0.5	175	223*	0.57	0.97
	5.1	15	0.5	217	253*	0.72	0.96
	5.2	10	0.5	208	232*	0.80	0.99
IV	1	155	1	69	130*	0.08	0.47
V	1	7	1.3	20	22	0.77	0.95
	2.1	6	1	24	27	0.77	0.96
	2.2	30	1	39	58*	0.36	0.97
	3.1	72	0.75	88	154*	0.13	0.78
	3.2	34	0.75	55	79	0.26	0.79
	4.1	42	0.75	113	155*	0.48	0.91
	4.2	25	0.5	134	194*	0.39	0.95
VI	1	6	1.3	19	27	0.76	0.95
	2.1	132	1	68	125*	0.09	0.56
	2.2	75	1	54	86	0.17	0.72
	3.1	12	0.5	115	144*	0.59	0.99
	3.2	33	0.5	154	230*	0.34	0.92

REFERENCES

1.
P. Andersen, *Acta Physiol. Scand.* 1959:47, 63.
2.
————, *Acta Physiol. Scand.* 1960:48, 178.
3.
J. Armstrong, K. C. Richardson, and J. Z. Young, *Stain Technology* 1956:31, 263.
4.
————, and J. Z. Young, *J. Physiol.* 1957:137, 10 pp.
5.
Yu. I. Arshavskiy, M. B. Berkinblit, S. A. Kovalev, and L. M. Chaylakhyan, *Biofizika* 1964:9, 365.
6.
T. Barakan, G. Downman, and J. Eccles, *J. Neurophysiol.* 1949:12, 393.
7.
M. L. Barr, *J. Anat.* (London) 1939:74, 1.
8.
I. S. Beritashvili and A. I. Roythak, *Zh. Vissh. Nervn. Deyat* 1955:5, 17.
9.
M. B. Berkinblit, S. A. Kovalev, V. V. Smolyaninov, and L. M. Chaylakhyan, Chapter 3.
10.
G. H. Bishop, *Physiol. Rev.* 1956:36, 376.
11.
————, *Elektroenceph. Clin. Neurophysiol.* suppl. 1958:10, 12.
12.
———— and M. H. Clare, *J. Neurophysiol.* 1953:16, 1.
13.
H. A. Blair, *J. Gen. Physiol.* 1934:18, 125.
14.
D. Bodian, *Cold Spring Harbor Symp. Quant. Biol.* 1952:17, 1.
15.
J. M. Brookhart and E. Fadiga, *J. Physiol.* 1960:150, 633.
16.
S. P. Cajal, *Histologie du systeme nerveux de l'homme et des vertèbres*, vols. 1–2, Paris, 1909–1911.
17.
H. T. Chang, *J. Neurophysiol.* 1951:14, 1.
18.
————, *Cold Spring Harbor Symp. Quant. Biol.* 1952:17, 189.

19.
———, *J. Neurophysiol.* 1955:18, 332.
20.
———, *Problemy Sovremennoy fiziologii nervnoy i myshechnoy sistem,* sb. posv I. S. Beritashvili (*Problems of Contemporary Physiology of the Nerve and Muscle Systems,* collection dedicated to I. S. Beritashvili), Tbilisi, 1956.
21.
———, *Handbook Physiol. Neurophysiol.* 1959:1, 299.
22.
L. M. Chaylakhyan, *Biofizika* 1962:7, 639.
23.
M. H. Clare and G. H. Bishop, *Elektroenceph. Clin. Neurophysiol.* 1955a:7, 85.
24.
———, *Amer. J. Psychiat.* 1955b:111, 811.
25.
B. G. Cragg and L. H. Hamlyn, *J. Physiol.* 1955:129, 608.
26.
J. C. Eccles, *The Physiology of Nerve Cells,* Baltimore: Johns Hopkins Press, 1957.
27.
———, in Second International Meeting of Neurobiologists, Amsterdam, *Structure and Function of the Cerebral Cortex,* ed. D. B. Tower and J. P. Schade, New York: Elsevier, 1960, p. 192.
28.
———, *Seitai no kagaku,* 1963:14, 62.
29.
———, *The Physiology of Synapses,* Berlin, 1964.
30.
———, B. Libet, and R. R. Young, *J. Physiol.* 1958:143, 11.
31.
C. Estable, in *Brain Mechanism and Learning: A Symposium,* ed. J. F. Delafresnaye, Oxford: Blackwell, 1961.
32.
E. Fadiga and J. M. Brookhart, *Amer. J. Physiol.* 1960:198, 693.
33.
P. Fatt, *J. Neurophysiol.* 1957:20, 27.
34.
K. Frank, in collection *Elektronika i kibernetika v biologii i meditsine,* Moscow: I.L. 1963, p. 157.

35.
V. Freygang, in collection *Elektronika i kibernetika v biologii i meditsine*, Moscow: I.L., 1963, p. 140.
36.
Y. Fujita and Y. Nakamura, *Jap. J. Physiol.* 1961:11, 357.
37.
————and H. Sakata, *J. Neurophysiol.* 1962:25, 209.
38.
H. Gasser, *J. Gen. Physiol.* 1950:33, 651.
39.
I. M. Gelfand, *Uspekhi Matem. Nauk* 1959:14, 87.
40.
————, S. A. Kovalev, and L. M. Chaylakhyan, *Dokl. AN, SSSR* 1963: 148, 973.
41.
———— and M. L. Tsetlin, *Dokl. AN, SSSR* 1960:131, 1242.
42.
E. Gray, *J. Anat.* 1959:93, 420.
43.
H. Grundfest, *Physiol. Rev.* 1957:37, 337.
44.
————, *Elektroenceph. Clin. Neurophysiol.* Suppl. 1958:10, 22.
45.
————, *Handbook Physiol. Neurophysiol.* 1959a:1, 147.
46.
————, *Symposium on Evolution of Nervous Control from Primitive Organisms to Man*, Washington, D.C., 1959b, p. 43.
47.
W. Hild and J. Tasaki, *J. Neurophysiol.* 1962:25, 277.
48.
A. Hodgkin and A. Huxley, *J. Physiol.* 1952:116, 449.
49.
————, A. Huxley, and B. Katz, *J. Physiol.* 1952:116, 424.
50.
———— and W. Rushton, *Proc. Roy. Soc.* [*Biol.*] 1946:133, 444.
51.
B. Katz, *J. Physiol.* 1950:111, 248.
52.
———— and R. Miledi, *J. Physiol.* 1963:168, 389.
53.
———— and S. Thesleff, *J. Physiol,* 1957:137, 267.

54.
N. S. Kositsyn, *Dokl. AN, SSSR* 1962:147, 477.
55.
P. G. Kostyuk and A. I. Shapovalov, in *Sovremennye problemy elektro-fiziologicheskikh issledovaniy nervnoy sistemy (Current Problems of Electrophysical Research of the Nervous System)*, Izd-vo Meditsina, 1964, p. 31.
56.
S. A. Kovalev and L. M. Chaylakhyan, in collection *Elektrofiziologiya nervnoy sistemy*, Materialy IV vses. elektrofiziol. Konf. Tezisy dokladov PhD (*The Electrophysiology of the Nervous System*, material of the IV All-Union Electrophysiological Conference, Ph.D. thesis reports) 1963, p. 163.
57.
K. Kubota and J. Brookhart, *J. Neurophysiol.* 1963:26, 877.
58.
C. Li, *J. Cell. Comp. Physiol.* 1963:61, 165.
59.
C. N. Lin and W. Chambers, *Anat. Rec.* 1957:127, 326.
60.
P. Nelson, K. Frank, and W. Rall, *Fed. Proc.* 1960:19, 303.
61.
S. Ochs, *Amer. J. Physiol.* 1959:197, 1136.
62.
V. M. Okudzhava, in *Strukturnye i funktsional'nye osobennosti korkovykh nevronov*, Gagrskie besedy (*Structural and Functional Characteristics of Cortical Neurons*, Gagra symposia), vol. IV, Tiblis, 1963a, p. 323.
63.
————, *Aktivnost' verkhushechnykh dendritov v kore bol'shikh polushariy (Activity of the Apical Dendrites in the Cerebral Cortex)*, Tbilisi, Izd-vo AN Gruz. SSR 1963b.
64.
S. Palay, *J. Biophys. Biochem. Cytol.* 1956: suppl. 2, 193.
65.
————, *Exp. Cell Res.* 1958; suppl. 5, 275.
66.
C. Phillips, *Quart. J. Exp. Physiol.* 1959:44, 1.
67.
D. Purpura, in A. Brodal, *Reticular Formation of the Brain Stem*, Springfield, Ill.: Thomas, 1957.

68.
———, in *Mekhanizmy tselogo mozga* (*Mechanisms of the Brain*), Moscow, I.L. 1963, p. 9.
69.
——— and H. Grundfest, *J. Neurophysiol.* 1956:19, 573.
70.
D. P. Purpura and R. J. Shofer, *Nature* 1965:206, 833.
71.
W. Rall, *Exp. Neurol.* 1959:1, 491.
72.
———, *Biophys. J.* 1962:2, 145.
73.
N. Rashevsky, *Cold Spring Harbor Symp. Quant. Biol.* 1936:4, 90.
74.
J. Rasmussen, in *New Research Techniques of Neuroanatomy*, ed. W. F. Windle, Springfield, Ill.: Thomas, 1957, p. 27.
75.
A. I. Roytbak, *Bioelektricheskie yavleniya v kore bol'shikh polushariy* (*Bioelectrical Phenomena in the Cerebral Cortex*), Tbilisi, 1955.
76.
W. Rushton, *Proc. Roy. Soc.* [*Biol.*] 1937:124, 210.
77.
S. A. Sarkisov and G. I. Polyakov, in *Tsitvarkhitektonika kory bol'shogo mozga cheloveka* (*The Cytoarchitectonis of the Cerebral Cortex*) Moscow: Medgiz, 1949, Chapter 4, p. 102.
78.
J. Szentágothai, in *Struktura i funktsiya nervnoy sistemy* (*The Structure and Function of the Nervous System*), Trudy Konf., Moscow: Medgiz, 1962, p. 6.
79.
Ye. G. Shkolnik-Yarros, in *Strukturnye i funktsional'nye osobennosti korkovykh neyronov* Gagrskie besedy (*Structural and Functional Characteristics of Cortical Neurons*, Gagra symposia), vol. IV, Tbilisi, 1963, p. 35.
80.
D. Sholl, *The Organization of the Cerebral Cortex*, London, 1956.
81.
W. Spencer and E. Kandel, *J. Neurophysiol.* 1961:24, 272.
82.
L. Tauc and G. Hughes, *J. Gen. Physiol.* 1963:46, 533.
83.
K. Terzuolo and T. Araki, in *Sovremennye problemy elektrobiologii*

(*Contemporary Problems of Electrobiology*) Moscow: Izd-vo Mir, 1964, p. 225.
84.
P. D. Wall, *J. Physiol.* 1965:180, 116.
85.
R. Wyckoff and J. Young, *Proc. Roy. Soc.* [*Biol.*] 1956:144, 440.
86.
J. Young, *Elektroenceph. Clin. Neurophysiol.* Suppl. 1958:10, 9.
87.
G. P. Zhukova, *Arkhiv Anat. Gistol. i Embriol.* 1961:41, 58.

3

The Electrical Behavior of the
Myocardium as a System and the
Characteristics of the Cellular
Membrane of the Heart

M. B. Berkinblit
S. A. Kovalev
V. V. Smolyaninov
L. M. Chaylakhyan

Important progress in the solution of basic problems of electrophysiol-ogy—the generation of biopotentials and the spread of excitation—was achieved after the classic works of Hodgkin, Rashton, Katz, Cole, Curtiss, and others who identified the cable structure of the nerve and muscle fibers and who showed that excitation spreads by means of local currents. In these works there was shown the quantitative dependence of the sizes, actually measured in an electrophysiological experiment and characteriz-ing the entire fiber, on the properties of the membrane and the geometric parameters of the fiber. This made it possible to study the properties of the membrane at rest and during excitation (examined first hand in only a limited number of preparations under the conditions of very specific methods, for example, the clamp method) by means of the measurement of the system parameters of the fiber. On the other hand, it was proved that for the consideration of some questions, in particular that of the conduction of excitation, those very system parameters such as the length constant are extremely significant.

An adequately strict quantitative study of generation and spread of excitation in the myocardium is also impossible without information on its passive electrical properties. However, the theory of such properties for the complex branched structures which form the myocardium requires special elaboration. Just as for the well-examined linear fibers, the prob-lem essentially is to determine the dependence of the system parameters on the electrical characteristics of the membrane and the myoplasm of the cells of the myocardium as well as on its geometric structure. The con-sideration of this question seems extremely important to us, and in this paper primary attention will be paid to it. Taking into account informa-tion about the passive electrical properties, we will also examine certain aspects of the problems of conduction of excitation and generation of biopotentials by the myocardial cells.

1. ELECTRICAL PROPERTIES OF CLOSED SYSTEMS

1.1. THE MYOCARDIUM—A FUNCTIONAL SYNCYTIUM

In an investigation of the electrical properties of the myocardium it is essential to take into consideration the basic patterns of its structure. Electron microscopic work [37, 72, 73, 74, 83, and many others] showed the existence of double transverse membranes in the region of the intercalated disk of the myocardial fiber. The disk membranes pass without interruption into the surface membranes from which they do not essentially differ in regard to structure. In this way, the myocardial fiber does not appear morphologically continuous; it is a combination of elongated cells closely adjoining one another of a length of approximately 100 μ. Neighboring cells are separated by a gap of the order of 100–200 A, while the disk membranes of these cells form fingered protuberances going behind one another.

Thus, from the point of view of histologists the widespread concept of the myocardium as a syncytium is inapplicable. This circumstance greatly hampers analysis of the electrical properties of the myocardium and resolution of the question about the mechanism of conduction of excitation from one myocardial cell to another. Can we consider the myocardial fiber electrically continuous and apply to the whole fiber a theory of local currents, or are the myocardial cells functionally divided and the impulse transmitted synaptically?

The hypothesis about the synaptic interaction of the myocardial cells, developed by Sperelakis [54, 85–87] and others, seems highly unlikely. These authors see one of the basic arguments for the absence of electrical continuity in the nature of the changes of resistance to the current passed between two microelectrodes in conditions when either one or both microelectrodes are led into the myocardial cells of a frog's ventricle [85]. In the case of the intracellular position of both electrodes, the resistance between them on an average twice exceeded the resistance measured when only one microelectrode was inside the cell. That sort of relationship was maintained in electrode spacings from 0.5 to 10 mm. This means that the current that is passed between the electrodes located inside the cell spreads not by the myoplasm but by the surrounding external fluid, a circumstance which according to the opinion of the authors clearly testifies to the high resistance of the disk membranes and the divisibility of the muscle of the heart into electrically individual, minute cellular elements. We should point out, however, that such a conclusion does not follow from the results presented, which do not in the least contradict the concept about the electrical continuity of the myocardial fiber if by chance its length constant is substantially less than 0.5 mm. In the section on the analysis of the passive electrical properties of the myocardium (section 1.3.) it will be shown that in the experimental conditions described (point application of current) the length constant of the myocardium is of the order of 100–200 μ. This means that with the quoted microelectrode spacings there should in any case be measured a doubled input resistance. We noted earlier the possible sources of methodological errors in the described experiments [64].

Other arguments from supporters of the synaptic transmission of excitation between the myocardial cells are based on blocking of conduction and on the prepotentials preceding the emergence of action potentials, which are often observed during perfusion of the heart by hypertonic solutions [86]. The same prepotentials were recorded also on the embryonic cells of the myocardium growing in the tissue culture [87]. The authors are inclined to consider these fluctuations of the membrane potential as postsynaptic potentials caused by the secretion of mediator from neighboring cells. However, exact proof of such a viewpoint is lacking. It is easy to explain the described prepotentials as a difficulty of the electrotonic interaction of neighboring cells which arises on account of the compression of the cells and the expansion of the intercellular gap during hypertonia of the medium [106].

In exactly the same manner the "prepotentials" observed in the myocardial tissue culture cannot on principle directly testify for whatever transmission mechanism of excitation there might be—synaptic or electric. From the viewpoint of electrical transmission, the concept about the electrotonic nature of these prepotentials is not contradictory; their connection with the region of intercellular contact is uniformly regular, since the absence of protoplasmic connection of the cells in this region necessarily hampers the flow of current from one cell to another, at least to some extent. Such a "weakness" of the contact region can be underscored by the most diverse conditions. For instance, a reduction of the excitability of the myocardial cells could lead to the appearance of the prepotentials described.

A priori, it is not excluded that in the embryonic cultivated tissue the difficulty of electrical transmission of impulse can be linked with certain peculiarities of the electrical properties of the disk membranes (higher resistance than in completely differentiated cells) or with the incomplete organization of the contact region of the cells which are growing together (an incomplete area of contact, or a greater width of the gap). A combination of both causes may occur. We will note that the properties themselves of the disk membranes can depend essentially on the conditions of the contact of the cells, for instance, on the degree of contact of the disk membranes with the external solution or on the interaction of the cellular surfaces adjoining one another. Thus, the phenomenon—very well known for the myocardium—of "healing," consisting of the rapid reestablishment of the resting potential after injury of the regions near the site of removal [see, for example, 100] possibly explains the abrupt increase of resistance of the disk membrane in event of the destruction of the neighboring cells.

Finally, the supporters of synaptic transmission consider as one more argument for their point of view observation of pairs of myocardial cells which have grown together in the tissue culture [87].

The injury of one of the cells of such a pair during introduction of a heavy microelectrode, which led to the practically complete fall of the resting potential of this cell and to a deadlock of contractions, did not stop the contractions of the adjoining cell. The resting potential of the latter also maintained their normal size. Unfortunately, the author did not subject this observation to the quantitative examination essential for conclusions about the electrical properties of the disk membranes. In the account of the experiments the data necessary for such calculations are also absent. However, even if the results of the quoted experiment clearly testified to a relatively high resistance of the disk membranes, it would be easy to explain them by the above-mentioned considerations concerning the incompleteness of the differentiation of the region of intercellular contact (especially in the early stage, when the two cells are only growing together).

We will note that the hypothesis about the high resistance of the intercalated disk is directly contradicted by the experiments of Crill, Rumery, and Woodbury [32], who recorded on a myocardial tissue culture almost the same changes of the membrane potential during intracellular application of current both when if by chance the current and recording electrodes were located in the same cell, and at that time when the microelectrodes were clearly divided by a visible intercellular boundary.

Thus, any direct proofs of the electrical isolation of the cells of the myocardium are lacking. On the other hand, there are clear experimental proofs of the electrical continuity of the myocardial fiber.

The condition of electrical continuity, as a matter of fact, leads to the requirement that the potential applied to one myocardial cell electrically spreads beyond cellular dimensions, that is, the structure and properties of the intercellular contact should guarantee the flow from one cell to another of an appreciable share of the current that is sufficient for the excitation of the cell. A reliable experimental way to clear up the question of electrical continuity is the direct measuring of the electrotonic distribution of the potential along the fiber.

Both the published data [92, 100, 106] obtained on different preparations—the Purkinje fibers of ungulates, the auricles of a frog and rat—and our experimental material (section 1.3)—obtained on the ventricle of a frog [12], show that the electrotonus spreads far beyond the cellular dimensions.

For example, according to Weidmann's data the length constant for the Purkinje fibers of ungulates is about 2 mm; moreover, the dimensions of the cells of these fibers measured on the specimens kindly given to us by E. S. Kirpichnikov proved to be about 130 μ, that is, the reduction of voltage on the length of fiber, corresponding to the cellular dimensions, is altogether 5 percent. It is clear that the reduction of voltage in the contact region, determined by the loss of current across the gap and the resistance of the disk membranes, will be still less. According to our data, the myocardial fibers of a frog's ventricle have similar characteristics of the cellular contacts. It is apparent that the facts enumerated clearly show electrical continuity of the heart fibers.

In connection with the cellular structure of the heart such a result may seem somewhat unexpected; however, it is possible to name several preparations where cells clearly divided morphologically form an electrically continuous totality. Such properties were discovered belonging to the smooth muscle [75] of syncytial formation, located in the region of the myoneural junctions of the ascarid [9], the electromotor neurons of some fish [10], and the cells of the salivary glands of the fruit fly [68]. Naturally, the electrical continuity of the myocardial fiber requires specific properties, the so-called disk membranes, which divide among themselves the neighboring cells.

In connection with this, of considerable interest are the experimental and theoretical analyses of the electrical properties of the disk membrane and the determination of conditions during which there occurs so negligible a drop of the potential in the region of cellular contacts as follows from the data on electrotonus. Individual attempts to measure experimentally the resistance of the disk membranes by the movement of labeled ions through the region of contact [103] are well known. According to this data, the specific resistance of the disk membranes is extremely low in comparison with the outer membrane of the order of 1 or 10 ohms/cm^2.

The approximate calculations of Woodbury and Crill [106] showed that in order for electrical transmission of excitation from one myocardial cell to another just to be able to be accomplished, the disk membrane should have a resistance of the order of 1–10 ohm·cm^2.

Of interest is the question of whether just the hypothesis about the low resistance of the disk membrane for the quantitative description of the experiments with the electrotonus is sufficient. In connection with it there arose the task of analyzing the potential attenuation in a cable with periodic inhomogeneities corresponding to the cellular contacts. A detailed treatment of this problem is presented in another place [13].

The most simple model of cellular contact was chosen: the cells bounded by flat disk membranes, the width of the intercellular gap approximately 100 A, the specific resistance of the gap fluid close to the specific resistance of Ringer's solution.

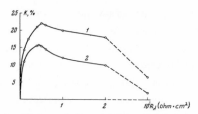

Fig. 1. Graph of the change of the coefficient K of the voltage transmission from cell to cell with different values of resistance of the disk membrane R_d. (1) resistance of external membrane in calculations, 1,000 ohm·cm²; and (2) the same, but for 100 ohm·cm².

Assigning reasonable values of resistance of the external membrane in a wide range, it was possible to calculate during which resistance of the disk membrane there was secured the maximum transmission of voltage, applied to one myocardial cell, into the neighboring cell.

The analysis led to certain interesting conclusions. It turned out that there exists for the transmission of voltage an optimum resistance of the disk membrane which lies among the reasonable values of resistance of the external membrane in the range 0.5 to 1 ohm·cm² (Fig. 1). However, and this is the most important, even the optimum values of resistance of the disk membrane in the simplest model of cellular contact taken by us could not guarantee a transmission of even 50 percent of the voltage from one cell to another. The results of the calculations were confirmed by measurements carried out on a physical model of the cellular contact zone with flat disk membranes.

It is easy to see that the results obtained sharply contradict the above-mentioned experiments with electrotonic spread of the potential along the myocardial fibers. This means that the simplest model with disk membranes, flat and similar in electrical properties, does not satisfy the conditions really existing in the region of contact of the myocardial cells. Two possible assumptions inevitably come to mind: (1) The disk membrane is nonuniform. There are sections in it with high resistance and "windows" with low resistance; in addition to which it is advantageous for such windows to be located in the central portions of the disk membrane. (2) The fingered protuberances of the adjoining disk membranes coming one behind the other play a significant role.

A combination of these two possibilities is also probable.

At present there is not any information about the active properties of the disk membranes connected with electrogenesis and the active transfer of ions. Possibly in their structure they are similar to the membrane of the endoplasmic reticulum or other intracellular membranes, for example, the nuclear membrane, the specific resistance of which is very low, of the order of 1 ohm·cm². As already mentioned above, certain circumstantial data relating to the rapid healing of a cut in the muscle fibers of the heart gives us reason to propose that the disk membrane possesses the remarkable feature to increase sharply its specific resistance at the time of destruction of the adjoining cells. The mechanism of this phenomenon, unfortunately, has not as yet been thoroughly studied. Woodbury [107] proposed a hypothesis explaining the low resistance of disk membranes by the accumulation of potassium ions in the intercellular gap, and the increase of resistance during "healing" as the outflow of potassium into the surrounding solution. One might also conjecture that the small resistance of the disk membranes is explained by a low concentration of ions of calcium in the intercellular gap (especially in the center of it), which is determined by the fact that the calcium, as it penetrates the cell, combines, as we know, but the rate of diffusion in the narrow space between the cells is low. Such an assumption, in just the same way as Woodbury's hypothesis, easily explains the discontinuity which we assumed of the disk membranes and the change of their electrical properties during "healing." Putting aside for the time being these assumptions, which require careful verification, we can suppose that disregarding a concrete mechanism such properties of the myocardial disk membranes have, apparently, a great biological significance. In the descriptive words of one of the prominent physiologists, this allows the cells of the heart "to work together but to die singly."

1.2. THE SELECTION OF A MODEL

The material presented in the preceding paragraph shows that theoretically in place of a column of myocardial cells we may examine one equivalent cable. However, this circumstance by itself still does not allow us to apply the usual cable equations to the myocardium. These equations would be applicable if it was possible to ignore the branching of the myocardial fibers, that is, if the distance between the nodes of branching was substantially to exceed the length constant of the myocardial fiber.

However, such a condition scarcely exists in the functioning myocardium. We will show below that the length constant of the myocardial fibers of a frog's ventricle is not less than 560 μ, and that the distance between the nodes of branching should be measured in millimeters, which seems scarcely probable. Moreover, if we were to use the values which we obtained of the input resistance (2 megohms) and the length constant (560 μ) (see section 1.3) and calculate by the usual cable equations the specific resistance of the myoplasm, then we will obtain a value 55 ohm · cm which, apparently, is unacceptable. It not only is lower than the corresponding value for the skeletal muscle of the frog (250–260 ohm · cm [36–38, 42–51, 53–62]) and the Purkinje fibers of warm-blooded animals (105–153 ohm · cm [28, 100]), but is less even than the specific resistance of the extracellular fluid (the resistance of Ringer's solution is not less than 80 ohm · cm), while for all the thoroughly studied preparations the specific resistance of the protoplasm exceeds by 1.5–3 times the resistance of the external media.

Finally, the assumption about the large distance, in comparison with the length constant, between the nodes of branching contradicts the experiments of Woodbury and Crill [106], who investigated the distribution of the potential in the auricle of a rat by means of intracellular polarizing and recording microelectrodes. They showed that the electrotonic spreading of the potential takes place both along the fibers and— which is especially important—across them, while the potentials measurable in different directions coincide in order of sizes at similar distances from the point of polarization. Obviously, this situation is possible only when the length constant is greater than the distance between the nodes of branching.

All the facts and considerations mentioned show that it is impossible to examine the myocardium as a totality of relatively independent elements to each of which according to individuality are applicable cable equations. Thus it is essential to consider each point of the myocardial fiber as sufficiently near the node of branching, and it is impossible not to take in consideration in any model the presence of branching.

Woodbury and Crill [106] proposed a two-dimensional generalization of the cable structure. The tissue of the auricle is examined by them as one flat cell, bounded by parallel membranes. In this way the authors to some degree take into account the nearness of the nodes of branching. However, in such a model there is lost one of the characteristic features introduced by the branchings into the structure of the myocardium, and that feature is the presence of inhomogeneities which, as will be shown in section 2, can play a significant role during the spread of excitation along the myocardium. As the authors themselves emphasized, the model proposed by them contradicts their own experimental material; that is, the theoretically predicted length constant exceeds by approximately an order of magnitude that measured in experiments on the auricle of a rat. It is difficult to consider it acceptable on the strength of the considerations presented, and because of the inadequacy of the two-dimensional model of the real structure of the myocardium.

A need arises in the investigation for models closer to the actual structure of the myocardium. Considering the electrical continuity noted above of the myocardial fiber, it is reasonable to take as such models networks built of the equivalent cable. Extremely diverse variations of such networks are possible (Fig. 2). Obviously, the closed syncytium is closer than

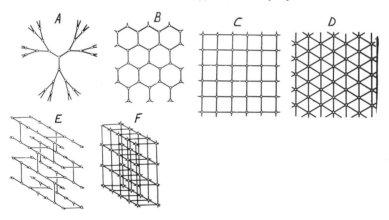

Fig. 2. Types of syncytia. (A) open, dichotomous; (B, C, D) plane closed (hexagonal, square, triangular); (E, F) three-dimensional closed (dichotomous and cubical).

the open to the structure of the myocardium. As shown in Fig. 2, among the closed syncytia differences exist, first of all according to the number of branches going off from each node. Although the histological data available to us do not give exact information about the real structure of the myocardium, dichotomous branching appears, nevertheless, the most probable since such a type of branching is seen most frequently in histological specimens. The hexagonal network is a rectilinear closed structure with dichotomous branching. Such a network, apparently, is the best two-dimensional model of the myocardial tissue. Naturally, however, the real volume of the myocardium adds significant additional complexities. We will return to this question below.

It is natural to examine for the networks the dependence of the system parameters, which characterize their passive electrical properties, on the electrical properties of the elements out of which the network is constructed. A brief summary of the results obtained during the solution of this problem will be presented below. We will examine the nature of the potential distribution in the networks and the dependence of the input resistance (R_{in}) of the system on the properties of the cable elements for different types of syncytia. We will compute the R_{in} as the resistance between the protoplasm and external medium, calling the *cable element* of the system that portion of fiber (with length l and diameter d) between two adjacent nodes of branching. The fundamental electrical constants of the cable elements are the resistance of the membrane per unit of area (R_m) and the specific resistance of the protoplasm (R_i). The capacitive properties of the myocardium will not be examined.

The problem of the input resistance of a closed hexagonal syncytium was already examined by George [43], who used a model consisting of two parts: the center composed of a section of the hexagonal network, and the peripheral composed of a two-dimensional generalization of the cable similar to that assumed by Woodbury and Crill. The same drawbacks are, to some degree, inherent to the proposed model as were inherent to the plane models of Woodbury and Crill. However, the chief drawback characteristic of any plane model replacing the whole hexagonal syncytium or only part of it consists of the fact that as long as the problem about the system parameters of such a syncytium is unsolved there is lacking a criterion for the determination of the suitable characteristics of a plane model; for example, Woodbury and Crill select for their model the same characteristics which they assumed belong to the myocardial fiber, and George finds them proceeding from certain geometric deliberations, but this choice cannot be argued. Finally, both models are wholly unsuitable for examination of three-dimensional structures.

1.3. AN ANALYSIS OF THE PASSIVE ELECTRICAL PROPERTIES OF THE MYOCARDIUM

During solution of the problem concerning the connection between the input resistance of closed networks and the parameters of cable elements, ideas were used about the uniformity and the endlessness of the syncytium and the applicability to each of its elements of equations valid for a segment of cable.

The problem was solved for a series of types of syncytia: among the plane types for the square, triangular, hexagonal; and among the three-dimensional for the cubical. The formula expressing the dependence of the input resistance R_{in}, for example on the properties of the cable elements on the node of the hexagonal syncytium, has the following form:

$$R_{in} = (\tfrac{3}{4})\rho \sinh(2l/\lambda) \int_0^1 [\tfrac{3}{4}(1 + 3 \cosh 2l/\lambda) - \cos \pi x]^2$$
$$- 2(1 + \cos \pi x^0)^{-1/2} \, dx$$

where $\rho = \sqrt{(r_m \cdot r_i)}$; $\lambda = (r_m/r_i)$; and r_m and r_i are the resistances of the membrane and the protoplasm per unit of fiber length, respectively. The method of calculation is described in sufficient detail in an article by V. V. Smolyaninov in Chapter 4 of this volume and in [14].

In order to make the results easily visible, we will present them in graphic form.[1] At the same time it is easy to select as units of measurement of the input resistance the characteristic resistance of the cable element (ρ). In Fig. 3, there is shown the dependence of the input resistance of different syncytia expressed in ρ on the ratio l/λ. It is apparent from the graphs that if the distance between the nodes of branching exceeds two or more times the length constant, then the input resistance of any type of syncytium

[1] For the plotting of the graphs in Figs. 3 and 9, the values of the absolute integrals were obtained on the electronic computer by L. E. Sotnikov. We express gratitude to her for this task.

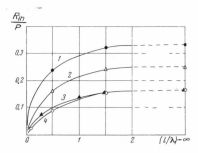

Fig. 3. Graph of the dependence of the input resistance of various syncytia in terms of ρ on the ratio l/λ (1) for hexagonal, (2) for square, (3) for triangular, (4) for cubical. The input resistance of the nonbranching cable is equal to 0.5 ρ.

strives for ρ/n where n is the number of branches going out from the node. The input resistance sharply falls with the decrease of the ration l/λ, since at the same time there increases the area of membrane and the total cross section of the fiber through which the current flows. We will examine somewhat later the question about the dependence of the input resistance of the syncytia on the resistance of the membrane.

In a general way the problem about the distribution of the potential was solved for the closed square syncytium (see Chapter 4). Naturally, the character of the distribution of potential is determined by the configuration of that contour (or region) of the syncytium along which are applied the source points of current.[2] For example, in the graph in Fig. 4, the quoted change of potential attenuation in a square syncytium (the source points of current were applied on the nodes of the syncytium on the sides of a square contour) depend on the dimensions of the side of the contour N (N is the number of internodes on the side of the contour).

It is apparent from the graph that the attenuation constant[3] can

[2] The calculations were performed on the electronic computer by A. V. Kholopov and L. E. Sotnikov for the syncytium with the ratio l/λ equal to 1/6; the distance at which the potential fell was reckoned by the perpendicular to the middle of the side of the square.

[3] Although the exponential fall of the potential on the membrane with removal from the source of current is characteristic for only a homogeneous cable, we will conditionally characterize the degree of attenuation in the syncytial structure as the *constant of attenuation*, that is, that distance at which the applied potential decreases e times.

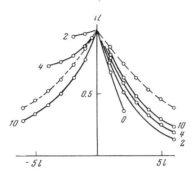

Fig. 4. Curves of the fall of the potential (for a square syncytium with the ratio $l/\lambda = 1/6$) from sources of current applied along the sides of a square contour, and for a nonbranching fiber (the dotted line). The numbers next to the curves correspond to the different sizes of the contour (i.e., to the number N). On the contour the potential is assumed per unit. Along the axis of the abscissa is plotted the distance from the side of a square in units of l (the curves on the right are the distribution of the potential out from the contour; the ones on the left, into the contour).

significantly increase with expansion of the front of the sources of current. It becomes clear that the attenuation constant has the smallest value for a single source point of current, and according to the degree of growth of the front of the sources applied along a straight line, the attenuation constant also grows (Fig. 4) tending toward a certain limit. These two values of the attenuation constant are of the greatest interest since they can be measured experimentally; we will introduce for these values the symbols λ_T (for current) and λ_L (for straight line). The final parameter is especially important for the determination of the electrical properties of the syncytial structure, since for a square and hexagonal syncytium it differs little from the length constant of the cable elements, which we will label λ_C.

Actually during such an application of current it is possible to examine the syncytium as a system of parallel cables, the equipotential points of which are joined by crosspieces (Fig. 5A). Then one may cut it, as shown by the dotted lines, into separate cables with offshoots loaded at the ends with infinite resistances (Fig. 5B). In addition to this even the maximum differences (when $l \ll \lambda_C$) of the syncytium and the nonbranching fiber

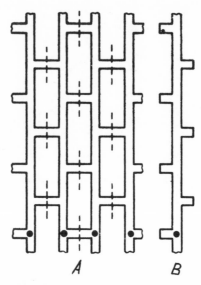

Fig. 5. Diagram illustrating that the potential, applied to the syncytium along a line, falls approximately the same way as in a cable. The dots show the distribution of the active sources.

with the same properties that belong to the cable elements are small; for a hexagonal network $\lambda_L = 0.82\lambda_C$, and for a square over $\lambda_C = 0.7\lambda_C$. The potential, applied along a plane, is distributed in a similar manner in a three-dimensional syncytium.

We will proceed to a brief description of the results of experiments in a study of the electrical structure of the myocardium of a frog's ventricle. The experiments are presented in greater detail elsewhere [15]. As we pointed out above, the structure of the myocardium is closest to the hexagonal syncytium. It is pertinent to note here that although the heart tissue appears as a three-dimensional structure, we can evidently, with justification examine a plane model. This is explained by the fact that in the identification of the system properties of the syncytium, the quantity of branches going out from one node is the decisive factor; thus, in Fig. 3 it is obvious that the plane, triangular, and three-dimensional cube syncytia behave very similarly. The assumption of a rectilinear structure of the syncytium is an obvious idealization. The need for such an assumption is connected with the inadequacies of the histological data. However, even if there were a complete three-dimensional reconstruction of the network of the myocardial fibers, even then an estimate of all the details would be in any case impossible.

Thus, we choose a hexagonal network. The diameter of the ventrical fibers of a frog, according to histological data [52, 102, 108] is in the neighborhood of 10 μ. The specific resistance of the protoplasm fluctuates rather weakly for the preparations studied, usually exceeding 1.5 to 3.5 times the resistance of the intercellular fluid. On the basis of the data of other authors [28, 36, 62] who measured R_i on similar preparations we chose for the R_i the value 200 ohm·cm. Thus, for the determination of the structure of the myocardial fiber it is necessary to find R_m and l.

These sizes can be found with sufficient approximation by the values λ_L and R_{in}. Actually, having determined λ_L, we will find the value close to λ_C and by the equation $\lambda_C = \sqrt{(r_m/r_i)}$ we will find R_m. This makes it possible to find ρ and by the graph Fig. 3 to determine l, since R_{in} is known.

Direct measurement of λ_L with the help of intracellular electrodes for the sources disposed as a front, is not possible. However, certain indirect methods are possible. It was shown [see, for example, 66] for homogeneous linear fibers that in the event of the approach lengthwise along the fiber to the point of registration of a source of current with a steady rate V, it is possible to determine the length constant by the formula $\lambda = V \cdot \sqrt{(t_K^2 + t_K \tau_m)}$. Here t_K is the rate constant of the change of the potential at the point of registration, that is, the time during which the potential changes e times; τ_m is the time constant of the membrane.

Since the action potential can be examined as a sort of moving source of current, the length constant of the excited linear fibers can be determined, knowing the rate of the conduction of excitation and the rate of growth of the initial prethreshold phase of the base of the action potential, if τ_m is known. For a homogeneous fiber this method of measuring λ is convenient in those cases when it is difficult to move the intracellular electrodes (polarizing and recording) lengthwise along the fiber. The use of a similar method in the case of the syncytial structures is very tempting since here it has the still greater advantage that like a natural path the arrangement of the sources of current emerge in a line (or along the surface, if one has in mind a three-dimensional structure of the myocardium) which corresponds to the front of the spread of the wave of excitation. In this instance instead of λ the value λ_L will enter the formula presented. The measuring of the size of V and t_K on the ventricle of a frog does not present difficulty. However, significant difficulties arise in the evaluation of τ_m for syncytial structures.

Two ways are theoretically possible for the exact evaluation of the time constant of the membrane τ_m and λ_L of the myocardial cells. One of them is the simultaneous measuring of V and t_K at two different temperatures. In this case, two pairs of values exist for V and t_K, and accordingly it is possible to set up two equations of the sort indicated above with the two unknown λ_L and τ_m. The second way is the general solution of the problem concerning the character of the determination of potential from a point source of current in syncytial structures of different types. In this case, for the evaluation of τ_m there are important experiments with rectangular shocks of current during intracellular recording.

In reference [15], during the evaluation of λ_L, the effect of the capacitance of the membrane of the myocardial cells was not taken into consideration. In this case the formula presented takes the form: $\lambda_L = Vt_K$. It was also used during the calculations. In this instance values of λ_L too low were naturally obtained and it is possible to speak about the evaluation of even lower boundaries of λ_L. During the experimental measurement of λ_L by the characteristics of the action potential, it is essential to arrange the intracellular recording electrode a sufficient distance from the stimulating electrodes so that excitation would approach it already as a wide front with a steady-state rate. The average figures for V and t_K were, respectively, 5 cm/sec and 11.2 msec. From them there was figured out the lower boundary of the values of λ_L which was 560 μ. Knowing the lower boundary of λ_L, one may assert that λ_L is not less than 560 μ, and having used the equation $\lambda_C = \sqrt{(r_m/r_i)}$ to obtain for the resistance of the membrane the minimum value 2500 ohm·cm². One must point out that the figures obtained belong, strictly speaking, not to the myocardial cells, but to the equivalent cable. The actual resistance of the membrane of the myocardial cells is higher than 2500 ohm·cm² not only because we assumed for λ_C a clearly understated value, but also because a part of the current flows out through the gap into the regions of intercellular contact.

The input resistance, determined as $R_{in} = U_0/I_0$, was measured with the help of two cemented glass microelectrodes with a distance between the ends 2–5 μ by the method of rectangular shocks [for details see 15]. The high values of the input resistances, which because of several reasons were estimated as the most reliable, on the average constituted 2 megohms. The characteristic resistance ρ of the cable element with $R_m = 2500$ ohm·cm², $d = 10$ μ, and $R_i = 200$ ohm·cm will be 14 megohms. Then the input resistance amounts to 0.143 ρ and from the graph in Fig. 3 it is easy to find the ratio l/λ, equal to 0.17, and the distance between the nodes of branching l, equal to 100 μ[4].

Thus, the experiments carried out make it possible on the basis of the theoretical analysis of the properties of closed networks to determine to a certain approximation the electrical structure of the myocardium of a frog's ventricle. It seems natural later on to subject to a like analysis other preparations possessing a similar structure.

In the literature there are some data on the passive electrical properties of the myocardium. For the first time in 1951 a rather complete and accurate analysis was carried out for the Purkinje fibers of ungulates (100). It was found that the resistance of the surface membrane equals 1200–1900 ohm·cm² the length constant is 2 mm, and the capacity about 11–12 mf/cm². This data can be considered as reliable as that obtained on other preparations (nerve and muscle fibers) since it was

[4] As was already pointed out, the disregard of the volume of the membrane of the myocardium's cells in the definition of λ_L may lead to a marked change of the parameters determined. The preliminary results of experiments on the determination of the time constant of the membrane by means of the method described above of the experimental change of the rate of conduction (see p. 93) provides the basis to think that τ_m of the cells of a frog's ventricle lies in the range from 10 to 25 msec. This means that λ_L can fluctuate in the range 700–1100 μ, i.e., at the worst with the simplified method we lowered it by a factor or two. In this case the resistance of the membrane will be 10,000 ohm·cm². However, the calculated size of l at the same time will decrease negligibly and be about 90 μ. Also, the size of the input resistance R_m strongly affects l. If the input resistance were twice more than that measured by us (not 2, but 4 megohms), then l for the model selected would be about 1 mm. Thus, it turns out a rather exact determination of the input resistance is very essential.

possible to work on portions of the Purkinje fibers with almost no branching. After the first attempts (1956) of an evaluation of the electrical characteristics of the myocardial fibers of other types (the auricle and the ventricle) the impression developed that these characteristics differed greatly from the parameters of the Purkinje fibers. Trautwein, Kuffler, and Edwards [92], working with an extracellular electrode on a bundle of fibers of the auricle of a frog, obtained the following figures: the resistance of the membrane, 200–300 ohm·cm², the length constant, 0.13–0.2 mm. The reason for such low values is now rather clear; it lies in the fact that in this work the bundle of auricular fibers were examined as a bundle of uniform cables running in parallel; the syncytial structure of the myocardium was, to all intents and purposes, not considered. This procedure resulted in the inadequacy of the systematic procedure and methods of calculation. So, if we use our experimental data for the input resistance and calculate R_m and λ by the cable equations, then we will obtain accordingly 200 ohm·cm² and 0.1 mm, that is, sizes close to the figures of Trautwein and the others.

In reference [88] Sperelakis and others obtained for the input resistance of the auricle a size in the neighborhood of 12 megohms, which exceeds 6 times that obtained by us. We have already indicated earlier the possible reasons (inadequacies of the systematic procedure) leading to values so large.[5]

In the work of Johnson and Tille [59] the values of the input resistance for the ventricular fibers of a rabbit are cited. These values are in the neighborhood of 50 to 100 kilohms, which, on the other hand, is 20 to 40 times less than the values obtained by us. It is impossible to explain, apparently, such divergences by the difference of the objects. If we assume for the myocardial tissue of a rabbit's ventricle the same structure of the syncytium as for a frog's ventricle ($l = 100 \mu$, the diameter of the fibers is 15μ and $R_1 = 200$ ohm·cm), then, according to the data of Johnson and Tille we will obtain for the resistance of the membrane of the ventricular fibers extremely low sizes in the neighborhood of 25 to 100 ohm·cm²,

---

[5] The size 1–2 megohms is shown by these authors for the input resistance of the fibers of the sartorius muscle of a frog, which also exceeds 6–7 times the sizes of the input resistances obtained in a series of works on this preparation.

which are 25 to 100 times lower than the values obtained by us for a frog's ventricle and by a series of authors [42, 62] for other muscle fibers.

It seems to us that such low values of input resistance (50–100 kilohms), cited by Johnson and Tille are inaccurate. They are connected with the arbitrary assumption that R_{in} at the peak of the action potential is negligibly small. The authors were not able to measure the input resistance by a direct method because of the large (which they themselves even recorded) coupling resistance between the channels of the double-barrelled electrode which they were using (it dropped about 1 V at the time the currents were applied by them, if the electrodes were in the solution). Therefore, for the evaluation of the input resistance the authors used the method proposed by Frank and Fuortes [38], which is based on the use of the formula

$$\frac{dE_s}{dI} = R_r - R_s$$

where E_s is the amplitude of the spike; I is the current being conducted; R_r, the input resistance at rest; R_s, the input resistance at the peak of the spike. If we assume that $R_s = 0$, then it is possible to find R_r by the changes of the amplitude of the action potential under the effect of the current being conducted through one of the channels of the microelectrode. However, this assumption appears unacceptable for the heart. In the third section of the article we will show that in comparison with at rest the input resistance of the myocardium in the ascending phase of the action potential can decrease only 1.5–2 times. Incidentally, the observations of the same authors about the possible rapid reduction of membrane resistance toward the end of the ascending phase of the myocardial action potential also contradicts this assumption. Thus, the sizes of the input resistance cited by Johnson and Tille may be understated by many times.

In the work of Woodbury and Crill [106] on the auricle of the rat, there was conducted for the first time an experimental analysis of the distribution of the potential from the source point of current with the help of two intracellular microelectrodes. Knowing the strength of the current flowing through the polarizing electrode (I_o in microamperes), and the value of the displacement of the membrane potential at distance x

(V_z in millivolts), the authors could express the distribution of current in terms of the ratio V_x/I_o, that is, in kilohms. Naturally, when the x are close to zero the input resistance of the system is found in this way. With the smallest values of x, cited by the authors ($\approx 50\ \mu$), the ratio V_z/I_o exceeded 400 kilohms. From the character of the curve cited by the authors it can be anticipated that when $x = 3$ to $5\ \mu$ (the approximate distance between the electrodes in our experiments), the value of V_z/I_o will be close to the value obtained by us for the input resistances of the myocardium of the frog's ventricle. Thus, the data of Woodbury and Crill coincide in order of magnitude with ours.

There was shown in these experiments by Woodbury and Crill that the character of the distribution of the potential near the source point of current in myocardial tissue sharply differs from an exponential dependence. A substantially faster drop of the potential was even observed in its measurements along the trabecula, to say nothing about the distribution in the perpendicular direction. In this instance, as we have already pointed out above, there is, strictly speaking, no point in estimating the length constant, since the fall of the potential by a factor of e upon withdrawal of the source of current will occur at different distances. However, the authors conditionally estimate this parameter on a section $50\ \mu <$ $x < 200$ and cite the size as $130\ \mu$. It is interesting to note that the character of the potential distribution experimentally obtained by Woodbury and Crill quantitatively coincides with our calculations for the point source of current carried out for a hexagonal network, which gave for λ_T a value in the neighborhood of $100\ \mu$.

Thus, the two experimentally measured parameters of the myocardial tissue of a rat's auricle, namely, the input resistance and the length constant near the source point of current, coincide with our experimental data and theoretical calculations made for the myocardium of a frog's ventricle. If the structure of the auricle is similar to that which we suggest, then the basic electrical parameters of the auricular fibers of a rat are, apparently, of the same order that belong to the fibers of a frog's ventricle.

Woodbury and Crill, on the basis of their plane model of the myocardium, about which we briefly spoke earlier, note that it is possible to explain the experimental results obtained by them (principally the values of λ_T) if one assumes either a very low resistance of the membrane (40 ohm·cm²), or a very high resistance of the disk membranes, or, finally, an appreciable resistance of the extracellular fluid. However, each of these assumptions, as the authors themselves note, is improbable. Therefore, the authors see the cause for this discrepancy in the too-large conditionality of the model they assumed in comparison with the real structure of the myocardium. Everything said above leads us to the conclusion that the model of the heart syncytium chosen by us is, apparently, more adequate to the real structure of the myocardium, although it undoubtedly requires further refinement.

2. SOME CONSIDERATIONS ABOUT THE SPECIAL FEATURES OF THE SPREAD OF EXCITATION ALONG THE MYOCARDIUM

2.1. THE EFFECT OF THE DISTRIBUTION OF THE POTENTIAL AND THE INHOMOGENEITIES

A rigorous analysis of the features of the spread of excitation in the syncytial structure is a separate task that exceeds the limits of the present work. However, even now on the basis of qualitative considerations and approximate calculations, it is possible to obtain a series of conclusions the character of which, apparently, will not change with more precise analysis.

The special features of the spread of excitation along the myocardium are determined by two factors: (1) the nature of the distribution of the potential in the syncytium in relation to the geometry of the source of current, (2) the local heterogeneities of the myocardium as an excitable medium.

We will briefly examine each of these factors.

1. The rate of the spread of excitation in the syncytium depends on the dimensions and configuration of the originally stimulated region. This involves the character of the distribution of the potential in the syncytium described in section 1.3. We will recall that the conditional "length constant" is small when the region excited is small, and according to the degree of growth of the excited region, it approaches the length constant of the cable element.

Since the rate of the spread of excitation is proportional to the length constant [24], the rate of the spread of a front of excitation of sufficient length will always be higher than the initial rate of the spread of excitation which developed in a small region; at the onset of excitation in a small region its rate will increase in proportion to the increase of the dimensions of the front. For this same reason, the concave sections of the front should spread more quickly than those formed like a straight line, and the convex sections, more slowly. Thus, the front of excitation will tend to "straighten itself," which is a direct consequence of the theory of local currents which was applied to the problem of the spread of excitation in the syncytium.

It is evident that the rate of spread of such a "straightened" front is the steady rate of the spread of excitation along the myocardium. Approximate calculations show that the initial rate of the spread of excitation which emerged in a small region can be 2 to 3 times lower than that sort of steady rate which is reached only when the radius of the excited region is in the neighborhood of a millimeter, if the parameters of the medium are close to those obtained by us for the ventricle of a frog.

The possibility is tempting that there are certain facts of the physiology of the heart to explain the dependence of the rate of spreading on the dimensions of the originally excited region. Thus, the rate of spread in the conducting region of the heart, where excitation develops in a region of limited dimensions, is substantially less than in the other regions of the heart. However, this point is complicated by a whole series of factors: by the special features of the characteristics of the action potentials of the cells of this region, by the diameter of the fibers, by the special features of the structure, etc. Nevertheless, it is relatively simple to subject to experimental verification the hypothesis expressed here.

2. It is convenient to begin an analysis of the effect of the local heterogeneities of the myocardium with an examination of the differences of the conduction of excitation in a linear conductor from conduction in an idealized compound element of the syncytium—a separate node of branching. In Chapter 2, this question is examined in sufficient detail; we will recall here only the basic results of the examination. With the approach of the impulse of excitation to the node of branching

the rate of conduction decreases and, moreover, the safety factor also decreases.

It is clear that the marked heterogeneity leading to the deterioration of conditions of conduction in each node of branching should occur in a wide-mesh syncytium where the distances between the nodes are greater than the length constant of the fiber or comparable to it. In a close-mesh syncytium, where the distances between the nodes are less than the length constant, the differences of the input resistances in the sections of fibers between the nodes and on the nodes of branching are extremely small and therefore the heterogeneity in the region of the node will be unimportant. The conditions of conduction in such a syncytium is nearly the same at any point. The very significant heterogeneity and poor conditions of conduction should combine during the transition of excitation from a single fiber or from a wide-mesh syncytium into a fine-mesh one, since a low load of resistance is determined, in this case, not only by the nodes of branching nearest the boundary but even by the many adjacent ones. These conclusions seem significant to use because histological data positively indicate the presence in the heart of zones with a different syncytial formation [2, 33, 81].

Can the decrease of the safety factor determined by the branching significantly affect the conducting of excitation along the myocardium? Experiments with intracellular stimulation of different regions of the heart [39, 40] (see section 2.2) showed that excitation of an extremely limited region always causes extrasystole, and so the nodes of branching do not generally appear as an obstacle for the spread of excitation. However, under certain functional or structural conditions, the described characteristics of the nodes can be very important. There can appear with such functional conditions a general reduction of the excitability of the specimen, for example, because of refractoriness during high-frequency stimulation. Moreover, on the nodes of branching and on the boundaries of the different zones of the syncytium, temporary blocks of conduction can develop. These blocks, singly or in groups, can bear an alternating character, as was shown in experiments on separate nerve fibers with different artificially produced heterogeneities [3–6, 66]. Similar results were obtained on the striae of the myocardium [8, 79]. A comprehensive review of the periodic transformations of impulses is presented by M. B. Berkinblit in Chapter 5. It is possible that such blocks can cause certain forms of pathology.

It is known that the presence of a temporarily inexcitable section of myocardium can lead to the rise of circular paths of excitation which probably are the basis of flutter and fibrillation [67, 104]. The role of such regions could be to serve as boundaries between zones with different syncytial structures during conditions favorable for the blocking of an impulse. It is well known that experimentally the pathological disturbances of the type of flutter and fibrillation are evoked much more easily in the auricles than in the ventricle. In addition to that, the data of V. A. Shidlovskiy and N. S. Daue [80] provide the basis of assuming that precisely the circular tissue divides into layers of wide-mesh and close-mesh syncytium; we note that the tissue of the ventricle is apparently more homogeneously constructed. The decrease of the safety factor at the nodes of branching could be significant both in the presence of a particular structure of branching of the myocardial fibers, and especially during the convergence of many minute fibers toward the node of branching. Then there is a great probability that the impulse, approaching along only one of them or along a small number of all those converging at the node of the fibers, cannot excite the node.

A series of morphological papers [44, 71, 89] provide some foundations to assume the presence of such structures in the region of the transition of the tapering multiple auricular fibers into elements of the atrioventricular node. If we assume that several auricular fibers unite into one atrioventricular[6] one (forming a distinct fiber; Fig. 6), then without additional assumptions about the properties of the atrioventricular cells it is possible to explain at least some characteristics of the conduction of excitation in this region.

Such characteristics are

1. The difficulty of retrograde conduction [94]. The conduction of excitation in the normal direction is assured from the viewpoint of this hypothesis by the overall effect on the node of many auricular fibers,

[6] Proceeding from several different considerations, Hoffman and Cranefield [52] propose a similar diagram for the structure of the atrioventricular node.

Fig. 6. Diagram of a hypothetical element of the atrioventricular region.

whereas the retrograde impulse always approaches the node of branching along only one fiber, although it does have a larger diameter than the auricular ones. It is easy to select such a correlation of diameters and a number of fibers converging at the node so that the retrograde impulse would have a very low safety factor and would surmount the node of branching only during high excitability of the membrane, for example, during a widely spaced sinus rhythm, which is what was observed in the experiments cited above.

2. The increase of the atrioventricular lag after a cut of part of the auricular trabeculae and the complete cessation of conduction with an insufficient number of trabeculae [65, 82]. Other experiments exist that also show the presence of summation in the atrioventricular region. In microelectrode analyses of the fibers of the atrioventricular node by a series of authors [32, 69, 70] there was uncovered the stepped shape of the ascending phase of the action potentials of these fibers, which becomes more pronounced in conditions favoring a block of atrioventricular conduction.

3. It is possible that the proposed structure also explains at least part of the normal atrioventricular lag which is determined by the asynchronous arrival of excitation along the auricular fibers approaching the node.

The special features of atrioventricular conduction enumerated led M. G. Udelnov [93, 95] to the concept that the atrioventricular node is a functional synapse. One may assume that the properties of conduction in the specially arranged continuous syncytial structure are the concrete mechanism of such a synapse.

In conclusion we will note that the proposed method of analysis of passive electrical properties of branched cable structures and of the conditions of conduction of excitation in them is applicable, apparently, not just to the heart. Probably it is interesting to use it, for example, also in the study of diffuse nerve system such as belong to the Coelenterata. Apparently it is pertinent to use an essentially similar analysis for examination of the functional characteristics of the dendrites of the nerve cells (Chapter 2). Such an approach—one might call it structural—appears useful to us during examination of a series of questions since it can substantially limit the number of assumptions about the specific properties of the separate elements to which one has to resort in order to explain certain physiological facts.

2.2. ON THE SYNCHRONIZATION OF WORKING OF THE CELLS OF THE SINUS NODE

During excitation of a small area of excitable tissue, the further spread of excitation is hampered since the high inner resistance of the excited section leads to a decrease of the action potential and the rate of the spread of excitation. We know that even for a linear fiber it is essential to excite a certain minimal region in order to obtain a spreading impulse. For the myocardium with its branched structure, the presence of the nodes of branching where the safety factor is lowered, the dimensions of the critical region should be deliberately greater than for a nonbranching fiber. In connection with this, the question naturally arises whether it is sufficient to induce the excitation of one cell in order to obtain a spreading excitation.

On the other hand the same question independently arises in connection with the work of I. M. Gelfand and M. L. Tsetlin [41] in which was examined a continuous excitable medium the points of which possess spontaneous activity. In that work it was shown that in such a medium the rhythm of the entire medium is determined by the fast element after a certain transitional behavior; that the points, the period of which is near of the leading element, begin to work virtually independently of one another; that in such a medium storage can be realized; etc. Some conclusions of that paper could have been directly applied to the sinus node of the heart, except that for this application it was necessary to show that the sinus node possesses the same character as the medium examined theoretically, that is, that an excitation aroused in a particular cell is capable of exciting the neighboring cells.

For the answer to this question, experiments were conducted with intracellular stimulation of the heart in which were recorded both the potential of the stimulated cell and the electrocardiogram of the heart [39, 40]. These experiments showed that as soon as the shock of the depolarizing current appeared adequate for excitation of an individual cell, then the heart responded with an extrasystole. These experiments were carried out on different regions (sinus node, auricle, ventricle) of a frog's heart and in all cases they produced identical results.

Thus it is possible to take for granted that excitation of an individual cell of any section of the heart leads to the onset of spreading excitation, and consequently, the critical region for the heart syncytium is smaller than cellular dimensions. We will note, by the way, that the conclusion has certain significance even for clinical physicians: the dimensions of the ectopic center, which is the source of extrasystoles, can be microscopically small. The obtained results make it possible to apply to the sinus node the conclusions obtained by Gelfand and Tsetlin during examination of the model of an excitable medium. In the case of the sinus node the fastest cell, having a certain period T, appears as the leading one. It also determines the working rhythm of the heart. Other spontaneously active cells having a similar period become excited by the wave of excitation approaching them. Owing to the successive waves of excitation arriving with the period T, that is, with virtually the same period as that of these cells, they are triggered at the very moment when the next wave comes to them. Just this circumstance makes it possible to record a significant number of "actual pacemakers" in the heart. Those very cells which have the longest period of spontaneous triggering are excited by the incoming wave and are recorded as "latent pacemakers."

The described nature of the working of the system guarantees its high dependability. When the fastest cell gets out of order or when its working slows down, all the remaining cells having the shortest period, as usual, are excited after a time T, thus guaranteeing the systole. In this way, although the rhythm of working of the heart is set by the fastest cell, a disorder of this cell has almost no effect on the working of the heart (with the exception of the variant when this cell has an abnormally fast rhythm in comparison with all the rest).

3. THE ELECTRICAL PROPERTIES OF THE MEMBRANE OF HEART CELLS

The unique form of the action potential of the cells of the heart apparently, is connected with the functional peculiarities of the myocardium. In an analysis of the generating mechanism of these cells one may point out these special features or, on the other hand, pay special attention to the generality of properties of the membrane of heart cells and of the other excitable tissues for the study of the universal generating mechanisms of biopotentials.

3.1. THE PURKINJE FIBERS

In the study of the properties of heart membranes, the Purkinje fibers of ungulates occupy a place similar to that of the giant axon of the squid in the study of the membrane of nerve fibers. This similarity, in the first place, the relatively large diameter of these fibers (up to 100 μ) and in the second place their very slight branching. Therefore, the data on the electrical properties of the cells of the heart obtained in these fibers are quite reliable.

The generation of the resting potential of the Purkinje fibers is apparently provided by the same mechanisms as for other excitable fibers [35]. It was shown by many authors that the resting potential of the Purkinje fibers is determined by the concentration gradient of potassium. The sizes of the resting potentials are close to the potassium equilibrium potential but never reach it. Just as for other cells, this is explained apparently by the appreciable permeability of the membrane for other ions besides potassium.

The action potential of the Purkinje fibers is very distinctive (Fig. 7), with a prolonged depolarization and a complex form of the repolarization phase. The questions arise: Can contemporary membrane theory (and, in particular, the sodium hypothesis) explain the special features of the action potential of Purkinje fibers? and What additional assumptions are needed for this explanation?

Unfortunately, voltage-clamp experiments carried out on the Purkinje fibers [101] did not give as well-defined results as in the classical experiments of Hodgkin, Huxley, and Katz [47–51]. Therefore, we will examine at first two other types of experiments, which together give sufficiently complete information about the generating mechanisms of the action

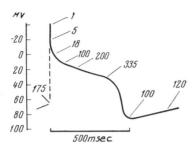

Fig. 7. Schematized representation of the action potential of the Purkinje fibers. The numbers are the sizes of the membrane resistance (in percentages of the diastolic) [102].

potential. These experiments are on the effect of changes of the ionic medium on the action potential and on the measuring of the action potential.

Draper and Weidmann [35] showed for the Purkinje fibers the dependence of the amplitude and the growth rate of the depolarization phase of the action potential on the concentration gradient of sodium. This dependence satisfies the requirements of the sodium hypothesis. As for other preparations, the membrane potential during excitation does not reach the sodium potential of equilibrium, which apparently is determined by a rather great permeability of the membrane to other ions.

A study of the impedance of the Purkinje fibers in different phases of the action potential was carried out by Weidmann [99]. He discovered a significant fall of the membrane resistance during the depolarization phase. At the peak of the action potential the resistance of the membrane was approximately 1/100 of its diastolic value. During the "plateau" the resistance substantially increased, and toward the end of this phase it exceeded the diastolic value by approximately 3 times. The phase of rapid repolarization is accompanied by a decrease of the membrane resistance (Fig. 7). The rise of the membrane resistance during diastole is connected with the automatic activity often observed in specimens of the Purkinje fibers.

A comparison of the just described two groups of experiments shows that the ascending phase of the action potential of the Purkinje fibers is linked with an abrupt selective increase of the permeability of the membrane for sodium ions. In addition to this, the experiments on the Purkinje fibers in conditions close to those of voltage-clamp experiments point to the great similarity of the properties of their "sodium channel" with the properties of the latter in nerve fibers [47–51]. Thus, for example, Weidmann [101] showed that the amplitude and growth rate of the action potentials of the Purkinje fibers decrease during the preliminary decrease of the resting potential, such that they are the stronger the lower the concentration of sodium outside the cell. It is easy to explain these observations by the concepts of the "inactivation" of sodium conductivity.

However, the observed changes of the membrane potential and the resistance during the process of repolarization cannot be explained by the system of change of potassium and sodium permeability of the membrane developed for the giant axon of the squid. It is easy to show that the changing temporal characteristics of the processes of the rise of sodium and potassium permeability can produce different forms of the action potential, among them even the characteristic one for the Purkinje fibers. However, there then remains unexplained the fact of the significant increase of the resistance of the membrane of the "plateau" level. If we then assume that at the start of the "plateau" the sodium permeability substantially lowers, then it is difficult to understand by means of what there is maintained during the "plateau" the prolonged depolarization of the membrane. Thus, additional assumptions are required for an explanation of the repolarization of the Purkinje fibers.

Any hypothesis about the mechanism of repolarization of the action potential of the Purkinje fibers and other cells of the heart must, in addition to the data presented, also explain a series of other facts obtained from study of the heart. We will briefly enumerate them: (1) the phenomenon of "all or nothing" repolarization [99]; (2) shortening of the action potential during the lessening of the interval between stimulations [21, 90]; (3) shortening of the action potential by the effect of acetylcholine [20, 53] during the rise of the extracellular concentration of potassium [17, 18, 23] and the decrease of the extracellular concentration of sodium [18, 35, 53]; (4) the phenomenon of overlap: the end of the repolarization phase takes a completely identical course in time for an action potential of a different length determined by the influence of the factors enumerated above [105]; (5) the special features of the volt-ampere characteristic of the membrane; and (6) the "pacemaker" potential.

Below we will examine the four most complete hypotheses of repolarization.

1. Brady and Woodbury [18] used the basic idea of Hodgkin and Huxley that the action potential is described by the successive changes of the membrane permeability for Na^+ and K^+, depending only on the potential and time. Brady and Woodbury suggested the presence in the Purkinje fibers of two groups of sodium conductivity—the fast and the slow, which are provided by two types of carriers. In order to explain the high resistance of the membrane during the "plateau," the authors assume a lowering of potassium conductivity during depolarization. From this point of view the action potential is described as follows. The ascending phase is determined by the rise of sodium conductivity; at the same time the fast group of sodium carriers are inactivated over a period of several milliseconds. However, because of the slow group, the sodium conductivity during the "plateau" somewhat exceeds the potassium, which at this time is assumed to be reduced; that is why depolarization is maintained. In connection with the continuing sodium inactivation the potential slowly falls, until it reaches a critical level during which the potassium permeability sharply rises. Even to the end, which developed owing to this phase of fast repolarization, the fast group of sodium carriers are reactivated, the slow group continues gradually to be activated into diastole in a time of the order of seconds.

2. Noble [76] also proceeds from the assumption that the permeability is a function of the membrane potential and time. He proposes that the increase and inactivation of the sodium conductivity of the Purkinje fibers can be described by the same equations (with different numerical coefficients) which Hodgkin and Huxley [50] used for the characteristics of those processes belonging to the axon of squid. The potassium conductivity consists of two channels. Conductivity of the first channel falls very quickly with depolarization, almost instantaneously.

Conductivity of the second potassium channel is theoretically similar to the potassium conductivity of the giant axon, that is, it increases with depolarization, but it is weaker and 100 times slower. From this point of view the action potential is described thus. During supraliminal depolarization the sodium conductivity increases sharply and the action potential reaches the peak at the same time the conductivity of the first potassium channel falls. The drop to the "plateau" is defined as the sodium inactivation. The plateau is maintained due to a certain excess of sodium conductivity. At this time the conductivity of the second potassium channel slowly increases. This determines slow repolarization on the plateau which leads to the increase of the conductivity of the first potassium channel. When the flux of potassium to the outside exceeds the stream of sodium to the inside, there begins fast repolarization which takes place as a regenerative process.

3. Hoffman and Cranefield [52] started from different initial premises. They assume that the permeability of K^+ is inversely proportional to the forces acting on the potassium, that is, to the electrochemical gradient. Since during diastole the electric field balances the concentration gradient, the sum of the forces acting on K^+ is close to zero and the permeability of K^+ is great. At the height of the action potential both forces work in one direction, tending to increase the flux of K^+ to the outside; therefore, the permeability for K^+ should sharply fall. Some accumulation in the outside fluid of K^+, coming out of the cell during the "plateau," raises the potassium conductivity which, in its turn, results in the gradual approach of the membrane potential to the resting potential. We will note that the increase of the outflow of K^+ from the cells of the heart during activity was shown by Wilde and O'Brien [96, 97]. However, the idea about the accumulation of potassium in the intercellular space is not a completely obligatory element of the hypothesis under consideration. Any factor causing the lowering of the membrane potential during the "plateau" will thereby increase the permeability of K^+. Finally, starting with some moment, the interaction process of the permeability and the potential is self-intensifying, just as this occurs during the ascending phase of the action potential. This moment is the start of the rapid repolarization phase.

4. Woodbury, and Johnson and Tille (for the working myocardium) [60], proposed a hypothesis which differs fundamentally from those stated above. According to this hypothesis, the permeability of the membrane for ions of potassium and sodium during the repolarization phase does not depend on the membrane potential; it depends only on time. The explanation of the fast phase of depolarization from this viewpoint does not differ from generally accepted ideas. The authors of the hypothesis think that then there occurs the fast (over a period of milliseconds) inactivation of the sodium channel, and subsequently the behavior of the conductancies, which determines the membrane potential, depends only on the size of depolarization which originally emerged, and on time. Thus it is assumed that any changes of the membrane potential during the repolarization phase do not affect the permeability of the membrane, just as the permeability of the postsynaptic membrane does not depend on the membrane potential of the postsynaptic structure but is determined only by the action of the mediator. Naturally, in this instance the character of the change of permeability is connected with the kinetics of the change of concentration of the mediator near the postsynaptic membrane. Woodbury notes that an assumption about the secretion of some chemical substance is not necessary for the hypothesis. A purely physical process can play a similar role, for example, the change of the pores of the membrane. This point of view explains the complex nature of repolarization of heart fibers by the corresponding kinetics of some of the hypothetical factors which determine the ion conductivity of the membrane.

We will try to evaluate critically the stated hypotheses. Their differences are associated with the interpretation of the repolarization phase whereas in the explanation of the fast depolarization phase they are all identical.

The fourth hypothesis is based on extremely powerful assumptions which, as it seems to us, clearly contradict experimental facts obtained on the Purkinje fibers. If the ion permeability in the repolarization phase does not depend on the membrane potential, then there should be observed a linear dependence of the current flowing across the membrane and on the shift of the potential on it. However, on the Purkinje fibers of a small goat's heart there was clearly shown by Weidmann [99] not only the presence of strong and nonlinearity, but also the possibility of obtaining in response to hyperpolarizing shocks of current of large enough size a regenerative repolarization, that is, an "all or nothing" repolarization response.

Woodbury [105] on the basis of his own hypothesis critically evaluates the results of Weidmann [99], noting the high values of hyperpolarizing current necessary for the rise of an "all or nothing" response. He came to the conclusion that even if there exists some fraction of membrane conductivity depending on the potential, then it all the same cannot influence the course of the repolarization phase under natural conditions, since for its development as an "all or nothing" response hyperpolarization exceeding the level of the resting potential is necessary.

However, this conclusion by Woodbury seems to us not well enough founded. Actually in the early stage of plateau the threshold of the "all or nothing" repolarization response is higher than the resting potential, yet the data of Weidmann clearly show that this threshold substantially lowers with the presentation of hyperpolarization shocks in the later stages of the plateau. So, in Fig. 26 from the book by Weidmann [102], when the hyperpolarizing shock is fed, as an example, at the halfway point of the plateau, it is very obvious that the threshold of the "all or nothing" response lies in the range of 20 to 25 mV for a spacing between the recording and polarizing electrodes of 1.2 mm. This spacing is equal to approximately half of the length constant of the Purkinje fibers at that moment of plateau, and consequently the real threshold of the hyperpolarizing electrode should accordingly be equal to 30–40 mV. At a later stage of the plateau the threshold of this reaction is apparently still lower. From this illustration (and a series of others) a clearly expressed nonlinearity is visible which begins to be apparent even with the hyperpolarizing shocks in the 10–15 mV range; this phenomenon most probably characterizes a subliminal reaction of the phenomenon being considered.

Thus there are no grounds to deny the possibility of the natural manifestation of this reaction at the end of plateau.

Also speaking in favor of this are the results of Noble and Hall [78] who, from the positions of their own hypothesis, had theoretically analyzed the conditions of the emergence of the "all or nothing" repolarization phenomenon during point (and consequently inhomogeneous) hyperpolarization of the fiber, as in the experiments of Weidmann, and during homogeneous hyperpolarization of a rather large portion of fiber, that is, under conditions close to those of the completion of natural repolarization. He showed quantitatively that in the first case it was significantly more difficult to cause the "all or nothing" repolarization phenomenon since at a certain distance from the polarizing electrode the displacements of the potential on the membrane are relatively small. On account of this the threshold for natural repolarization at the end of the plateau should be substantially still lower than for hyperpolarization artificially induced by means of an intracellular microelectrode.

All that has been said shows that the conductivity of the membrane during the repolarization phase of the Purkinje fibers is a function of the membrane potential, and the basic premise of Woodbury's hypothesis does not correspond to the facts.

Here we will not examine in detail the third hypothesis since a critical analysis of it already was carried out earlier [64]. We will note only that this hypothesis, worse than the first two, copes with the explanation of a whole series of facts accumulated recently, and is not developed quantitatively at all. Its basic assumption is not substantial enough.

Probably the initial premise of this hypothesis developed in connection with the facts obtained by Adrian [1] on the frog's sartorius, which led him to the conclusion that the permeability of the membrane for a flux of ions of potassium directed outward is inversely proportional to the forces causing this movement. However, this phenomenon, observable under very special conditions, permits an ambiguous interpretation, and it is impossible to examine one of the possible assumptions as a basic principle of ionic dynamics. In particular, as on a series of objects (among them on the sartorius) it was shown under normal conditions that the permeability for ions of potassium, at least in the first seconds, conform to the same rules which belong to the axon of the squid.

The initial premises of the first and second hypotheses do not differ fundamentally from the assumptions of the sodium theory. The solution of equations quantitatively describing these premises adequately reconstructs the action potential. Both hypotheses satisfactorily explain the behaviors of impedance, the "all or nothing" repolarization phenomenon, the shortening of the action potential during the lessening of the interval between stimulations, and a series of other phenomena analyzed in specific papers [2, 76]. In this respect the hypotheses are of equal value.

However, the initial premises of Noble's hypothesis seem more probable. Certain experiments on the Purkinje fibers provide a basis to assume the existence of two channels of potassium conductivity [55]. (In this work it is shown that during depolarization of the Purkinje fibers in sodium-free solutions, the conductivity of the membrane at first falls and then slightly increases.) At the same time there are no indications of the existence of two groups (slow and fast) of sodium conductivity.

The existence of two channels of potassium conductivity is shown for the fibers of transversely striated muscles. One of the channels (the second according to Noble's terminology) was studied relatively long ago. It is responsible for the phenomenon of "delayed rectification," its conductivity grows with the decrease of the membrane potential. The other channel clearly appears in special conditions, for example, in an isotonic solution KCl [63], it is responsible for the phenomenon of *abnormal rectification* its conductivity falls with the decrease of the membrane potential. The phenomenon of abnormal rectification appears in the muscle and under conditions of natural ionic concentration, but only in the presence of hyperpolarizing influences, since during depolarization it will be masked by the behavior of the second potassium channel. There is reason to assume that the properties of the first potassium channel are determined by the behavior of the membrane of the endoplasmic reticulum [2]; possibly that also takes place in the heart cells, while here this channel, in contrast to the muscles, plays a significant role in the determination of the shape of the action potential. L. M. Chaylakhyan [25, 26] has shown that with the increase of the membrane potential from a resting level of 70–80 mV there was observed an increase of membrane conductivity by 2 to 2.5 times, and that this process develops in a fraction of a millisecond. Extrapolating the obtained curve to the side of depolarization, one may assume that with lowering of the membrane potential from a resting level to 70–80 mV, the process described should lead to the lowering of the membrane conductivity by 5–10 times. Possibly this process is revealed in the Purkinje fibers during depolarization, as was shown in the experiments of Noble and Hutter, only because the phenomenon, dependent on the second potassium channel, develops in these fibers 100 times more slowly than in the muscle fibers. Thus, the properties of the membrane of the Purkinje cells, according to Noble's hypothesis, are essentially the same as those belonging to a series of other excitable cells. The differences are not major, and they make Noble's hypothesis very plausible and most probable. Experiments in which it is shown that the usual type of electrical activity can change into a "heart" activity point to the similarity of the mechanisms of generation in the Purkinje fibers and other excitable preparations. With an injection of tetraethylammonium into the giant axon of the squid, its action potential becomes amazingly similar to that of the heart. This similarity is not superficial since the impedance of the membrane in these conditions bears the same character as that of the heart.

3.2. THE CELLS OF THE FUNCTIONING MYOCARDIUM

The shape of the action potential of the cells of the functioning myocardium is similar to that belonging to the cells of the Purkinje fibers. Therefore, it is natural to assume that there are no principal differences between the electrical properties of the membranes of these two types of cells. However, in the literature there are a number of differences of opinions over the most common and principal points concerning the properties of the cells of the functioning myocardium.

True, for the rest potential, a number of authors show [17, 19, 98] that it is determined by the concentration gradient of potassium in the same way as in the Purkinje fibers. Moreover, the resting potential is always lower than the potassium equilibrium potential; a proof of that is, for example, the hyperpolarizing effect of acetylcholine on the heart muscle. In a series of papers [19, 45, 91] it is shown that acetylcholine increases the permeability for K^+, at least in some types of myocardial fibers, as a result of which their membrane potential approximates the potassium potential of equilibrium. As is to be expected, such a displacement of the potential is most significant in the region of low extracellular concentration of potassium.

The question about the mechanism of generation of the action potential of the myocardial cells is rather deeply complicated. Naturally the use here of the method of voltage fixation is to a great degree more difficult then for the Purkinje fibers. A comparison of factual material according to the effect on the electrical activity of the ion fluid with the results of experiments on the measurement of the membrane permeability in different phases of the cycle of excitation leads to considerable contradictions with Noble's hypothesis, contradictions which rather well explain the electrical activity of the Purkinje fibers. These contradictions concern not only the repolarization but also the depolarization phase. We will dwell briefly on these questions.

Coraboeuf and Otsuka [27] and Délèze [34] showed on the fibers of a porpoise's ventricle that the rate of the growth phase of the action potential depends linearly on the external concentration of ions of sodium. This fact indicates the accuracy of the sodium hypothesis for the phase of fast depolarization. But in addition to that, the same authors point out that the amplitude of overshoot depends much less on the concentration of sodium than is predicted by this hypothesis. However, Woodbury [105] presents records of the change of the overshoot and the growth rate of the action potential of a frog's ventricle which satisfactorily conform to the sodium hypothesis. Apparently one may assume that the ascending phase is guaranteed by the large sodium current inside the cells determined by the sharp and selective increase of sodium permeability. The results of an analysis of the changes of input resistance of fibers of the functioning myocardium in the first phase of the action potential proved to be all the more unexpected.

On the basis of their experiments with double-trunk intracellular electrodes performed on a rabbit's ventricle, Johnson, Robertson, and Tille [56, 59] conclude that at the moment of the maximum action potential the input resistance of the cells of the functioning myocardium is the same as that during diastole. Moreover, the presence of an overshoot 15 to 25 mV shows that the permeability of the membrane of these fibers for ions of sodium should by more than an order exceed the permeability for ions of potassium. In order to overcome the contradiction arising during examination of these two groups of facts, Johnson, Tille, and others proposed a hypothesis according to which there is an abrupt increase of the sodium permeability only during the fastest stage of depolarization, and then there takes place an abrupt lowering of conductivity both for ions of Na^+ and for ions of K^+, but with the preservation of the ratio of conductivities which remains greater for sodium [61]. By the moment of the peak of the action potential the sodium conductivity is equal to the potassium during the period of diastole, and the potassium is much lower. This rather artificial hypothesis overcomes the contradiction indicated. Woodbury [105] pointed out that if at the peak of the action potential, g_{Na} becomes comparable with g_{Cl}, then during the lessening of the concentration of sodium there should not be observed corresponding changes of overshoot. This observation makes it possible to make a decisive check of the Johnson and Tille hypothesis. The experiments of Hutter and Noble [46] and Carmeliet [22] on the measurement of the input resistance of the Purkinje fibers during the substitution of ions of Cl for the nonpenetrating anions showed that 30 percent of the current belonging to the surrendering membrane is transferred by the ions of chlorine. Similar experiments by Noble on the papillary muscle of the cat and the rabbit provide the basis to assume that the share of chlorine conductivity belonging to the cellular membrane of the functioning myocardium, if not greater, in any case, is not less (see below). Thus, according to the hypothesis of Johnson and Tille, at the moment of the peak of the action potential, g_{Na} exceeds in all a little over two times g_{Cl}. If we assume, according to the data of the same work of Noble, that $E_{Cl} = -50$ mV, then it is clear that in conditions of natural concentrations the ions of chlorine will lower the amplitude of overshoot some tens of millivolts, and consequently, if the hypothesis of Johnson and Tille is correct, during the substitution of chlorine for the nonpenetrating anions the overshoot should abruptly increase. A similar form of experiment was used in the already cited work of Hutter and Noble [46] on a cat's papillary muscle, during which the ions of chlorine were completely substituted for ions of NO_3 or ions of methyl sulfate. No change in the maximum size of the amplitude of the action potential was observed, that is, the hypothesis of Johnson and Tille is not proved correct and the contradiction cited above holds good.[7]

[7] We could have redeemed the situation had we assumed that in the first phase of the action potential there occurs a temporary lowering of conductivity for

The second significant contradiction with Noble's hypothesis involves the mechanism of the repolarization phase. In a series of works carried out on the myocardial cells of the auricle and ventricle of different animals, there was obtained a majority of the phenomena briefly listed in the beginning of this section (see pp. 108–109) in connection with the examination of the Purkinje fibers. We shall add to this that the substitution of ions of chlorine for nonpenetrating anions sharply prolongs the plateau; this points to the fact that the ions of chlorine are an important hyperpolarizing factor significantly accelerating the attainment of that size of the membrane potential during which the process of regenerative repolarization[8] develops. This phenomenon, as all the ones examined earlier, satisfactorily fits Noble's hypothesis and thereby indirectly speaks in favor of the hypothesis since the actions influencing the slow repolarization hardly change the character of the fast repolarization, which should behave rather independently because of the regeneration. The contradiction involves the fact that the strongest argument of Noble's hypothesis—the presence of the "all or nothing" repolarization phenomenon for the cells of the functioning myocardium—is called in question. True, Cranefield and Hoffman [30] observed on the papillary muscles of a cat and dog a phenomenon similar to the effect of "all or nothing" repolarization in conditions of polarization by a large extracellular electrode. However, later Johnson and Tille [57, 58], in experiments on the myocardial cells of a rabbit's ventricle in conditions of intracellular polarization, did not detect this phenomenon. And what is more, they pointed to the strictly linear behavior of the membrane of the myocardial fibers during significant polarizations (up to 100 mV). These results led the authors to conclude that ion conductivity during plateau does not depend on the membrane potential; it is only a function of time. In connection with this they quantitatively develop for the cells of the functioning myocardium a hypothesis [60] the principles of which were already briefly examined by us on p. 111, hypothesis 4.

---

ions of chlorine. However, this is a very powerful assumption since a similar process is not described for any of the excitable tissues.

[8] Another indication is the disappearance of the "peak" at the overshoot of the action potential of the papillary muscles during substitution of the ions of chlorine for nonpenetrating anions [46].

Briefly summing up what has already been said, it may be noted that two facts seriously contradict the attempt to explain the mechanism of generation of the action potential of the fibers of the functioning myocardium in the same way as for the Purkinje fibers: (1) the input resistance at the moment of the peak does not substantially differ from the input resistance during the period of diastole; (2) the "all or nothing" phenomenon is absent during intracellular hyperpolarization.

We will now try to show on the basis of the results of Noble's theoretical analysis and the analysis carried out by us, and on the basis of our experimental data, that the indicated contradictions are most probably connected not with the character of the surface membranes of the cells of the functioning myocardium, but with the special features of the geometry of the syncytial structures.

During the discussion of the conditions for the emergence of the "all or nothing" phenomenon in the Purkinje fibers it was pointed out that during point hyperpolarization it is more difficult to arouse this phenomenon than during hyperpolarization of a considerable region of membrane [78], which is explained by the lessening of the density of current in proportion to the removal from the polarizing electrode. It is clear that for the syncytial structures this effect is expressed to a still greater degree. In the first section of the article we showed that the fall of the potential from the source point of current in similar structures occurs several times faster than in a homogeneous fiber. For a quantitative evaluation of the effect of point polarization the problem is essentially formulated as, how does the displacement of the potential in the region of the polarizing electrode depend on the current passed through it for the given structure (a homogeneous cable in the case of the Purkinje fibers and a syncytium in the case of the fibers of the functioning myocardium), if we allow for the volt-ampere characteristic of the membrane and for the fact that at various distances from the electrode its resistance will be different. The solution of this problem may be obtained if we find the spatial distribution of the potential on the given structure with an estimate of the heterogeneity of the resistance of the membrane.

In the solution of this problem for the cells of the functioning myocardium, Noble [77] used the plane model of Woodbury and Crill [106] and the volt-ampere characteristic for the membrane was selected the same way as in the Purkinje fibers for the different moments of the plateau [76, 78]. The analysis showed that for the plane model the displacement of the potential on the membrane depends on the current passed almost linearly, as was also observed in the experiments of Johnson and Tille [57, 58]. Noble proposes that the presence of the "all or nothing" repolarization phenomenon in the experiments of Hoffman and Cranefield [30] is determined by the fact that they caused repolarization not by an intracellular electrode but by a comparatively large external one, and thereby established relatively homogeneous polarization in a rather large part of the myocardium. It seems interesting to us to conduct an analysis similar to that which Noble performed for our syncytial model of the myocardium, which apparently is closer to the real structure than the plane model of Woodbury and Crill. The results of Noble's studies have fundamental significance; they clearly show that there is no basis to assume that the mechanism of repolarization of the cells of the functioning myocardium is in principle different than in the Purkinje fibers.

The question about the behavior of the input resistance (R_{in}) of the functioning myocardium in different phases of the action potential was experimentally explored by us on a frog's ventricle for which the parameters of the syncytial model [14, 15] (section 1 of this chapter) were calculated earlier. The method described above was used: two intracellular electrodes with a distance between the tips of from 2 to 5 μ, across one of which hyperpolarizing shocks were delivered, and by the other the displacement of the membrane potential was measured. Since this displacement is directly proportional to the size of the input resistance, it was sufficient to compare the displacement of the potential during different phases of the action potential with its value during diastole. For the determination of the input resistance during the peak of the action potential there was done yet a somewhat different form of experiment: on an artificially stopped ventricle a depolarizing shock, causing excitation, was delivered across a current microelectrode. One may appraise the displacement of the potential in the state of rest by the amplitude of the step which precedes the ascent of the action potential, and compare it with the size of the corresponding displacement at the moment of the action potential peak. The results of these experiments coincide with the results obtained by the method of hyperpolarizing shocks.

In Fig. 8 two types of action potential are diagrammatically represented, having the shape which most often is found in the fibers of the ventricle of a frog's heart. The mean value of the changes of the R_{in} in percentages are presented in the diagram. The input resistance during the second half of the ascent phase and of the peak of the action potential falls 30–50 percent, and during the phase of fast repolarization approximately 13 percent. Statistical analysis showed that the changes of R_{in} in our experiments during the plateau are uncertain. On the basis of the analysis of passive electrical properties of the myocardium carried out in section 1 of this chapter, it is possible by the changes of the R_{in} to determine the changes of R_m. For this it is necessary, having used the graph in Fig. 3, to find the relation between these sizes for the hexagonal syncytium with the parameters determined earlier ($1 = 100\ \mu$; $R_{in} = 2$ megohms; $R_m = 2500$ ohm·cm^2). In Fig. 9 the dependence obtained is graphically expressed. From this graph it follows that the membrane resistance of the ascent of the action potential falls 12–16 times, and during the phase of fast repolarization, 2–3 times.[9] Thus significant changes of the R_m correspond to small changes of R_{in}, since for our syncytial model the dependence of R_{in} on R_m is weaker than for a nonbranching cable. It is obvious from the graph that for large R_m this dependence is weaker. So, for example, the decrease of R_m by two times from the initial value at rest leads to a much stronger change of R_{in} than with the same increase of R_m. Possibly the fact that we failed to establish a reliable increase of R_{in} during the plateau phase is connected with this phenomenon.

The computed fall of the membrane resistance (12–16 times) gives for R_m of the cells of the ventricular myocardium on the ascent phase the

[9] Using the results of previous experiments presented in footnote 4 by an analogous method we obtain a fall of the membrane resistance for the ascent phase of the action potential approximately 40 times, and for the fast repolarization phase approximately 7 times. These numbers are closer to Weidmann's classical results [102] obtained on the Purkinje fibers. However, they apparently require additional checking and therefore we do not use them here.

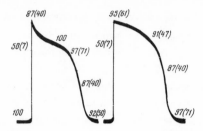

Fig. 8. Diagrammatic representation of two types of action potential of the ventricular fibers. The sizes of the input resistance (in percentages of the diastolic) are shown by the numbers. The numbers in parentheses are the corresponding values of the membrane resistance.

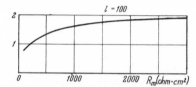

Fig. 9. Graph of the dependence of the input resistance of a hexagonal syncytium with a node spacing of 100 μ on the membrane resistance of the cable elements.

value 150–200 ohm·cm². It is interesting that these figures rather closely coincide with the values for sodium conductivity during the ascent phase of cells of a frog's ventricle, obtained with entirely different initial premises. Brady and Woodbury [18], on the basis of measurements of the maximum rate of depolarization during generation of the action potential of the membrane of cells of a frog's ventricle, assuming that the capacitance of the membrane is 30 mF/cm² and all the depolarizing current is carried by ions of Na⁺, figured out that the maximum size of sodium conductivity constitutes 10 millohm/cm², which corresponds to 100 ohm·cm².

The results obtained on the change of the membrane resistance of cells of a frog's ventricle during the growth phase of the action potential show that the mechanism of depolarization belonging to those cells does not differ fundamentally from that belonging to the Purkinje fibers or other excitable cells and the fall of sodium conductivity begins only after the action potential peak. At this time, according to Noble's hypothesis, the conductivity decreases even for the ions of potassium (the first potassium channel according to Noble's terminology) and, apparently, by an important hyperpolarizing factor become ions of chlorine. With the substitution of ions of chlorine for nonpenetrating anions, the "small peak" of the action potential vanishes and the plateau abruptly lengthens [46]. Incidentally, on the basis of the graph in Fig. 9 it is possible to estimate quantitatively the portion of chlorine conductivity by the change of the input resistance in Noble's experiments on the papillary muscle of a cat and dog[10] during the substitution of chlorine for nonpenetrating ions of CH_3SO_4 and others. In these experiments it was shown that the R_{in} increases on an average by 11 percent. According to our graph, this corresponds to a lessening of conductivity of the membrane up to 60 percent, that is, chlorine conductivity forms approximately 40 percent for the quiescent membrane of the cells of the functioning myocardium. Thus, the chlorine conductivity of the membrane of the cells of a functioning myocardium is similar to that of the Purkinje fibers. The fall of the resistance of the membrane 1.5 to 2 times during the fast repolarization phase is most likely connected with the rise of potassium conductivity determined both by reduction of the conductivity of the first potassium channel and the increased conductivity of the second potassium channel (according to Noble's terminology). We point out, by the way, that the impedance of the membrane of the functioning myocardium during generation of the action potential, according to our data, is very similar to its behavior in the membrane of the giant axon of the squid during generation of prolonged action potentials caused by an intracellular injection of tetraethyl ammonium.

The attempts to examine the changes of electrical conductance of the cells of the functioning myocardium during excitation were also made in a series of other works [29, 52]. However, the authors of these works do not present quantitative evaluations. Thus, the work of Johnson and Tille [59] is, apparently, the only one with which we can compare our experimental material, and even then only for the phase of growth and of

[10] In such an analysis the assumption is made that our model is applicable to these formations and its parameters differ rather slightly from the parameters obtained for a frog's ventricle.

the peak of the action potential. For other stages their analysis treats the so-called P fibers occupying probably the interstitial position between the Purkinje fibers and the fibers of the functioning myocardium [57][11]. This comparison shows that in contrast to our experiments, where a clearly registered although relatively slight decrease of R_{in} is observed during excitation, in the experiments of Johnson and Tille no changes of the R_{in} during excitation are generally observed. At present it is difficult to point to the cause of this divergence of results; it is scarcely probable that it is linked with the difference of the preparations used. For an explanation of the cause of the divergence of the data it apparently is expedient to analyze the electrical structure of the myocardium of the ventricle of a rabbit's heart by methods described in the present work since these methods appear more direct for such a task.

REFERENCES

1.
R. H. Adrian, *J. Physiol.* 1960:151, 154.
2.
———— and W. H. Freygang, *J. Physiol.* 1962:163, 61.
3.
Yu. I. Arshavskiy, M. B. Berkinblit, and S. A. Kovalev, *Biofizika* 1962a:7, 449.
4.
————, M. B. Berkinblit, and S. A. Kovalev, *Biofizika* 1962b:7, 619.
5.
————, M. B. Berkinblit, S. A. Kovalev, and L. M. Chaylakhyan, *Biofizika* 1964a:9, 365.
6.
————, M. B. Berkinblit, S. A. Kovalev, and L. M. Chaylakhyan, *Biofizika* 1964b:9, 634.

[11] It is necessary to note that in the analysis of the changes of electrical conductance of the membrane for the "P fibers" of a rabbit's ventricle in the slow repolarization phase Johnson and Tille proceeded from the inaccurate premise that the input resistance at the peak of the action potential equals zero. A similar hypothesis is tacitly assumed in the graphic blending at one point of the peaks of the action potential obtained in different degrees of hyperpolarization [58].

7.
————, M. B. Berkinblit, S. A. Kovalev, V. V. Smolyaninov, and L. M. Chaylakhyan, Chapter 2.
8.
Ye. B. Babskiy, V. S. Salmanovich, and Ye. A. Donskikh, *Dokl. AN, SSSR* 1963:153, 966.
9.
J. T. De Bell, J. del Castillo, and V. Sanchez, *J. Cell. Comp. Physiol.* 1963:62, 159.
10.
M. V. Z. Bennett, E. Alyure, Y. Nakajama, and G. D. Parras, *Science* 1963:141, 262.
11.
M. B. Berkinblit, Chapter 5.
12.
————, S. A. Kovalev, V. V. Smolyaninov, and L. M. Chaylakhyan, *Dokl. AN, SSSR* 1965:163, 741.
13.
————, S. A. Kovalev, V. V. Smolyaninov, and L. M. Chaylakhyan, *Biofizika* 1967, in press.
14.
————, S. A. Kovalev, V. V. Smolyaninov, and L. M. Chaylakhyan, *Biofizika* 1965:10, no. 2, p. 309, no. 5, p. 883.
15.
————, S. A. Kovalev, V. V. Smolyaninov, and L. M. Chaylakhyan, *Biofizika* 1965:10, 861.
16.
————, S. A. Kovalev, V. V. Smolyaninov, and L. M. Chaylakhyan, *Biofizika* 1967, in press.
17.
A. Brady and J. W. Woodbury, *Ann. N.Y. Acad. Sci.* 1957:65, 687.
18.
———— and J. W. Woodbury, *J. Physiol.* 1960:154, 385.
19.
A. S. V. Burgen and K. G. Terroux, *J. Physiol.* 1953a:119, 139.
20.
———— and K. G. Terroux, *J. Physiol.* 1953b:120, 449.
21.
E. Carmeliet, *Arch. Int. Physiol.* 1955:63, 126.
22.
————, *J. Physiol.* 1961:156, 375.

23.
——— and L. Lacquet, *Arch. Int. Physiol.* 1956:64, 513.
24.
L. M. Chaylakhyan, *Biofizika* 1962:5, 639.
25.
———, collection (sb) *Fiziko-Khimicheskie osnovy proiskhozhdeniya biopotentsialov* (*Physicochemical Fundamentals of the Origins of Biopotentials*), Trudy MOIP (Moscow Society of Naturalists) 1964:9, 120.
26.
———, collection (sb) *Biofizika Ketki* (*The Biophysics of the Cell*), Moscow: *Nauka* 1965, p. 175.
27.
E. Coraboeuf and M. Otsuka, *C. R. Acad. Sci.* 1956:243, 441.
28.
——— and S. Weidmann, *Helv. Physiol. et Pharmacol. Acta* 1954:12, 32.
29.
———, F. Zacouto, M. Gargouil, and J. Laplaud, *C. R. Acad. Sci.* 1958:246, 2934.
30.
P. F. Cranefield and B. F. Hoffman, *J. Gen. Physiol.* 1958:41, 633.
31.
———, B. F. Hoffman, and A. Paes de Carvalho, *Circulation Res.* 1959:7, 19.
32.
W. E. Crill, R. E. Rumery, and J. W. Woodbury, *Amer. J. Physiol.* 1959:197, 733.
33.
N. S. Daue and V. A. Shidlovskiy, collection *Materialy po eksperimental'-noklinicheskoy elektrokardiografii* (*Data on Experimental and Clinical Electrocardiography*), Moscow: Izd-vo AMN, SSSR (Publishing House of the Academy of Medical Sciences of the USSR) 1953, p. 18.
34.
J. Délèze, *Circulation Res.* 1959:7, 461.
35.
M. H. Draper and S. Weidmann, *J. Physiol.* 1951:115, 74.
36.
P. Fatt and B. Katz, *J. Physiol.* 1951:115, 320.
37.
D. W. Fawcett and C. C. Selby, *J. Biophys. Biochem. Cyctol.* 1958:4, 63.

38.
K. Frank and M. G. F. Fuortes, *J. Physiol.* 1956:134, 451.
39.
I. M. Gelfand, S. A. Kovalev, and L. M. Chaylakhyan, *Materialy II Zakavkazskoy Konf. Patofiziologov* (*Material of II Transcaucasian Conference of Pathophysiologists*), Yerevan, 1962, p. 437.
40.
———, S. A. Kovalev, and L. M. Chaylakhyan, *Dokl. AN, SSSR* 1963: 148, 973.
41.
——— and M. L. Tsetlin, *Dokl. AN, SSSR* 1960:131, 1242.
42.
J. P. Generick, *J. Cell Comp. Physiol.* 1953:42, 427.
43.
E. P. George, *Aust. J. Exptl. Biol.* 1961:39, 267.
44.
D. J. Glomset and A. T. A. Glomset, *Am. Heart J.* 1940:20, 677 (cited by S. Weidmann, 1956).
45.
E. J. Harris and O. F. Hutter, *J. Physiol.* 1956:133, 58pp.
46.
O. F. Hutter and D. Noble, *J. Physiol.* 1961:157, 335.
47.
A. L. Hodgkin and A. F. Huxley, *J. Physiol.* 1952a:116, 449.
48.
——— and A. F. Huxley, *J. Physiol.* 1952b:116, 473.
49.
——— and A. F. Huxley, *J. Physiol.* 1952c:116, 497.
50.
——— and A. F. Huxley, *J. Physiol.* 1952d:117, 500.
51.
———, A. F. Huxley, and B. Katz, *J. Physiol.* 1952:116, 424.
52.
B. E. Hoffman and P. Cranefield, *Electrophysiology of the Heart*, New York: McGraw-Hill, 1960.
53.
——— and E. E. Suckling, *Amer. J. Physiol.* 1953:173, 312.
54.
T. Hoshiko, N. Sperelakis, and R. M. Berne, *Proc. Soc. Exp. Biol. Med.* 1959:101, 602.

55.
O. F. Hutter and D. Noble, *Nature* 1960:188, 495.
56.
E. A. Johnson, P. A. Robertson, and J. Tille, *Nature* 1958:182, 1161.
57.
——— and J. Tille, *Austral. J. Exp. Biol.* 1960:38, 509.
58.
——— and J. Tille, *Nature* 1961:192, 663.
59.
——— and J. Tille, *J. Gen. Physiol.* 1961:44, 443.
60.
——— and J. Tille, *Biophys. J.* 1964:4, 387.
61.
———, J. Tille, L. Wilson, and E. P. George, *Biophysics of Physiological and Pharmacological Actions*, Washington, D.C.: American Association for the Advancement of Science, 1961:69, 525.
62.
B. Katz, *Proc. Roy. Soc. B.* 1948:135, 506.
63.
———, *Arch Sci. Physiol.* 1949:3, 285.
64.
S. A. Kovalev, in collection *Fizika -khimicheskie osnovy proiskhozhdeniya biopotensialov* (*Physicochemical Foundations of the Origin of Biopotentials*), Trudy MOIP (Moscow Society of Naturalists), 1964:9, 105.
65.
N. A. Kelareva and A. Nazarova, *Nauchn. Dokl. Vyssh. Shkoly, Biol. Nauki* 1963: no. 1, p. 61.
66.
M. V. Kirzon and S. A. Chepurnov, in collection *Fiziko-khimicheskie osnovy proiskhozhdeniya biopotensialov* (*Physicochemical Foundations of the Origin of Biopotentials*), Trudy MOIP (Moscow Society of Naturalists), 1964:9, 212.
67.
T. Lewis, *Clinical Disorders of the Heart Beat. A Handbook for Practitioners and Students*, London, 1912.
68.
W. R. Loewenstein and Y. Kanno, *J. Cell. Biol.* 1964:22, 565.
69.
K. Matsuda, T. Hoshi, and S. Kameyama, *Tohoku J. Exp. Med.* 1958a: 68, 16.
70.
———, T. Hoshi, and S. Kameyama, *Tohoku J. Exp. Med.* 1958b:69, 8.

71.
J. G. Mönckeberg, *Ergeb. Allgen. Path. Anat.* 1921:19, II 328 (cited by Hoffman and Cranefield, 1962).
72.
D. H. Moore and H. Ruska, *J. Biophys. Biochem. Cytol.* 1957:3, 261.
73.
A. R. Muir, *J. Anat.* 1957a:91, 251.
74.
———, *J. Biophys. Biochem. Cytol.* 1957b:3, 193.
75.
T. Nagai and C. L. Prosser, *Am. J. Physiol.* 1963:204, 910.
76.
D. Noble, *J. Physiol.* 1962:160, 317.
77.
———, *Biophys. J.* 1962:2, 381.
78.
——— and A. E. Hall, *Biophys. J.* 1963:3, 261.
79.
A. F. Samojloff, *Pflüg, Arch. Ges. Physiol.* 1929:222, 516.
80.
V. A. Shidlovskiy and N. S. Daue, in collection *Materialy po eksperimental'no-kinicheskoy elektrokardiografii* (*Data on Experimental and Clinical Electrocardiography*), Moscow: Izd-vo AMN, SSSR (Publishing House of the Academy of Medical Sciences of the USSR), 1953a, p. 5.
81.
——— and N. S. Daue, in collection *Materialy po eksperimental'no-klinicheskoy elektrokardiografii* (*Data on Experimental and Clinical Electrocardiography*), Moscow: Izd-vo AMN, SSSR (Publishing House of the Academy of Medical Sciences of the USSR), 1953b, p. 24.
82.
——— and N. S. Daue, in collection *Materialy po eksperimental'no-klinicheskoy eiektrokardiografii* (*Data on Experimental and Clinical Electrocardiography*), Moscow: Izd-vo AMN, SSSR (Publishing House of the Academy of Medical Sciences of the USSR), 1953c, p. 53.
83.
F. Sjöstrand and E. Andersson, *Experientia* 1959:10, 369.
84.
V. V. Smolyaninov, Chapter 4.
85.
N. S. Sperelakis, T. Hoshiko, and R. M. Berne, *Amer. J. Physiol.* 1960:198, 531.

86.
———, T. Hoshiko, Jr., R. F. Keller, and R. M. Berne, *Amer. J. Physiol.* 1960:198, 135.
87.
——— and D. Lehmkuhl, *J. Gen. Physiol.* 1964:47, 895.
88.
I. Tasaki and S. Hagiwara, *Amer. J. Physiol.* 1957:188, 423.
89.
S. Tawara, *Das Reizleitungsstem des Säugetierherzens*, Jena, 1906 (cited by Tigerstedt, 1921).
90.
W. Trautwein and J. Dudel, *Pflüg. Arch.* 1954:260, 24.
91.
——— and J. Dudel, *Pflüg. Arch.* 1958:266, 324.
92.
———, S. W. Kuffler, and C. Edwards, *J. Gen. Physiol.* 1956:40, 135.
93.
M. G. Udelnov, *Vestnik Mosk. Gos Un-ta (Journal of Moscow State University)* 1950, no. 6, p. 117.
94.
——— and A. P. Popova, in collection *Materialy po eksperimental'no-klinicheskoy elektrokardiografii (Data on Experimental and Clinical Electrocardiography)*, Moscow: Izd-vo AMN, SSSR (Publishing House of the Academy of Medical Sciences of the USSR), 1953a, p. 43.
95.
——— and A. P. Popova, in collection *Materialy po eksperimental'no-klinicheskoy elektrokardiografii (Data on Experimental and Clinical Electrocardiography)*, Moscow, Izd-vo AMN, SSSR (Publishing House of the Academy of Medical Sciences of the USSR), 1953b, p. 32.
96.
W. S. Wilde, *Ann. N.Y. Acad. Sci.* 1957:65, 693.
97.
——— and J. M. O'Brien, *Abstr. 19 Int. Physiol. Congr.* 1953, p. 889.
98.
E. M. V. Williams, *J. Physiol.* 1959:146, 411.
99.
S. Weidmann, *J. Physiol.* 1951:115, 227.
100.
———, *J. Physiol.* 1952:118, 348.
101.
———, *J. Physiol.* 1955:127, 213.

102.

————, *Elektrophysiologie der Herzmuskelfaser*, Bern: Huber, 1956.

103.

————, *J. Physiol.* 1960:153, 32pp.

104.

N. Wiener and A. Rosenblueth, *Arch. Inst. Cardiologia de Mexico* 1946: 15, 205.

105.

J. W. Woodbury, *Biophysics of Physiological and Pharmacological Actions*, Washington, D.C.: American Association for the Advancement of Science, 1961:69.

106.

———— and W. E. Crill, *Proceedings of the International Symposium "Nervous Inhibitions,"* New York: Pergamon, 1961, p. 129.

107.

J. W. Woodbury, *Handbook of Physiology, Circulation* 1962:2, 237.

108.

————, H. H. Hecht, and A. R. Christopherson, *Amer. J. Physiol.* 1951: 164, 307.

4

The Problem of the Electrical   V. V. Smolyaninov
Properties of Syncytia

Certain problems of electrophysiology [1, 3, 6–9] lead to the necessity of defining the electrical properties of systems of branching fibers or a large complex of electrotonically connected cells. Among these developments the following two categories naturally stand out. One constitutes closed networks—the closed syncytia or the syncytial media; the other, those that are not closed, or the open syncytia—the treelike structures. Typical examples of the first category are the heart and smooth muscles; of the second, the dendrites of the nerve cells.

In this chapter, methods for the calculation of passive electrical properties of homogeneous and for the most part closed syncytial networks will be described. The sizes of the input resistances and the law of the distribution of potential (drop) in the system from a fixed series of sources are related in the first place to the passive electrical characteristics of the fiber or system of branching fibers. For several reasons there is a very considerable opportunity to calculate the indicated passive characteristics. In the first place, one may sometimes determine them experimentally for real preparations, and, consequently, use theoretical relations for the calculation of the quantities which are not directly measured in the experiment, such as, for instance, the specific resistance of the membrane. On the other hand, knowledge of the passive properties of the syncytial networks is essential for study of the laws of the spread of excitation in them.

Calculations of the electrical properties of the branching treelike structures in connection with the determination of the properties of the dendrites of nerve cells were carried out by Rall [7, 8]. Besides stationary solutions, Rall also examined the transients of a specific class of trees "equivalent to a cylinder with a fixed diameter."

Recently, interest in the closed syncytial formations has grown and the question arises: How do the open and closed syncytia differ according to their electrical characteristics? George [3] tried to solve the problem. However, instead of a calculation of an infinite network, entirely constructed of pieces of fiber, he calculated the input resistance of a node of a small section of a dichotomous network not yet forming closed cells. Then, in order to imitate a closed network, he immediately, after the first closure, replaced the network with a "two-dimensional continuous fiber." Such an artificial substitution was offered because he used for the computations the "recurrent method," which has proved to be a good method in the calculation of the treelike formations.

A precise solution of the problem of the electrical properties of homogeneous closed syncytia is given in reference [1], and in addition, a comparison is presented there of hexagonal, square, triangular (two-dimensional) and cubical networks. However, the method for calculation used in [1] is based on ideas ensuing from unidimensional "chain" concepts. In this chapter a simpler and more versatile method is described based on the solution of a system of algebraic equations which represent equations of the state of the system. These equations describe the effect on the system of an arbitrary set of sources. Consequently, one may interpret the method described here for calculation of the syncytia as a distinct analog of the well-known method for calculation of continual systems based on the structure of Green's function.

1. THE FIBER

This section is introductory. Here, mainly on the example of a fiber of constant diameter, basic methods of calculating the input resistance and voltage are described.

1.1 THE CYLINDRICAL FIBER

The simplest model of a fiber is the cylindrical cable immersed in an ideal conducting medium. The outer covering of the fiber is a membrane characterized by a specific resistance R_m (ohm·cm²) and a specific capacitance C_m (F·cm²); and the inner medium is protoplasm of a specific resistance R_i (ohm·cm). Moreover, between the external and internal media of the fiber there is a constant difference of potential E_m. In Fig. 1A an equivalent diagram is presented of a piece of fiber dx long with radius r, and the basic relations are written down for the currents and voltages occurring in sections x and $x + dx$, distinguishing a given section of fiber. From these relations follows the well-known system of differential equations describing the distribution of voltage $u(x,t)$ and current $i(x,t)$ in a cylindrical

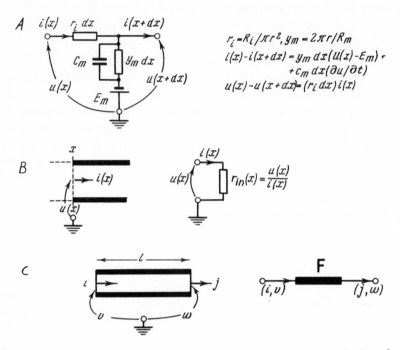

Fig. 1. Equivalent diagram of a piece of fiber dx in length, the determination of the input resistance, and the element F.

fiber:

$$\frac{\partial u}{\partial x} = r_1 i, \qquad \frac{\partial i}{\partial x} = -Y_m\left(u + \frac{\tau \partial u}{\partial t} - E_m\right) \tag{1}$$

where $\tau = R_m C_m$.

In this article we will ignore the membrane capacitance, that is, we will suppose $\tau = 0$. Under such conditions, solution (1) is well known and is represented as the sum of the exponentials

$$u(x) = ae^{x/\lambda} + be^{-x/\lambda}, \qquad i(x) = -\sigma(ae^{x/\lambda} - be^{-x/\lambda}), \tag{2}$$

where $\lambda = (r_m/r_1)^{1/2}$ and $\sigma = (r_m r_1)^{-1/2}$ are the characteristic length and conductivity, respectively; a and b are arbitrary constants.

If for certain sections of fiber the difference of the potentials $u(x)$ between the external and internal medium, and the current $i(x)$ flowing across this cross-section, for example clockwise (Fig. 1B) are known, then the ratio $u(x)/i(x)$ is the input resistance of the section of the fiber on the right and one may consider the ratio $u(x + l)/i(x + l)$ as the resistance of the load for the piece of fiber formed by the cross-sections x and $x + l$. Suppose that x is fixed, and that for simplicity $x = 0$; we will introduce the notation: $u(o) = u$, $i(o) = i$, $u(l) = w$, $i(l) = j$. Then $y_{in} = i/u$ is the input conductivity, and $y_L = j/w$ is the conductivity of the load for a piece of cylindrical fiber l long; such an "element" of fiber we will label F, and on the diagrams we will represent it by a length of heavy line (Fig. 1C on the right). (In electrical engineering such elements are called *triple poles with distributed parameters*.)

Using equations (2) it is not difficult to be convinced that the input and output currents and the voltage are related by the following system of equations:

$$i = \alpha j + \beta w, \qquad u = \gamma j + \delta w, \tag{3}$$

where $\alpha = \cosh(l/\lambda)$, $\beta = \sigma \sinh(l/\lambda)$, $\gamma = \rho \sinh(l/\lambda)$, $\delta = \alpha$ and $\rho = 1/\sigma$ is the characteristic resistance. It is customary to call matrix A, composed of the coefficients of system (3), the *transmission matrix* of F, its determinant $\det A = \alpha\delta - \beta\gamma = 1$:

$$A = \begin{pmatrix} \alpha & \beta \\ \gamma & \delta \end{pmatrix}$$

Then, from (3) we have

$$y_{in} = \frac{\alpha y_L + \beta}{\gamma y_L + \delta} = \chi(y_L), \tag{4}$$

that is, χ here stands for the fractional-linear function with the help of which one may determine the input conductivity of F with the fixed load y_L. This transform always has one static current: $\chi(\sigma) = \sigma$, which also defines in a given case the characteristic conductivity. With the change of y_L from zero to infinity, y_{in} changes from $\chi(\sigma) = \sigma \tanh(l/\lambda)$ to $\chi(\infty) = \sigma \operatorname{ctnh}(l/\lambda)$.

Instead of (3) the equations solved with regard to currents are often used:

$$i = \xi v - \eta w, \qquad j = \eta v - \xi w. \tag{5}$$

The coefficients of this system, $\xi = \alpha/\gamma$, $\zeta = \delta/\gamma$, and $\eta = 1/\gamma$, have the dimensions of the conductivities. Therefore, the matrix Y is called the *conductivity matrix* of element F:

$$Y = \begin{pmatrix} \xi & - \eta \\ \eta & - \zeta \end{pmatrix}. \tag{6}$$

1.2. THE COMPOUND FIBER.

The simplest generalization of a fiber of a fixed diameter is a fiber the diameter of which is constant within the bounds of certain sections—the *compound fiber* or the *chain*. We will label a section of fiber of length l_k, and diameter dk by F_k. In general the compound fiber is a successive ("chain") connection of elements and it is conveniently labeled $F_1, F_2, F_3, \ldots$. We will note that in this notation F^n is a fiber of constant diameter and length nl.

Interest in the compound fiber $F_1, F_2, F_3, \ldots$ arises chiefly from the fact that it represents a degenerated case of trees—a *monochotomous tree*, and of closed networks—a *unidimensional network* (see 2 and 3). Therefore, in the example $F_1, F_2, F_3, \ldots$ it is convenient to examine both of the methods applicable only to the systems of the chain and tree types, and the more universal method which permits one to compute just the networks apart from everything else.

THE CHAIN METHOD.   Let a chain $F_1, F_2, \ldots, F_n$ be loaded with conductivity y_L. The transmission matrix of this chain A is equal to the product of the transmission matrices of the sections $F_k, A = A_1, A_2, \ldots, A_n$. Carrying out all the multiplications, the elements $\alpha, \beta, \gamma, \delta$ are found for A, and the input conductivity of $F_1, F_2, \ldots, F$ is calculated according to (4): $y_{in} = \chi(y_L)$.

If in addition to y_L the chain $F_1, \ldots, F_n$ has the loads y_{Lk} in all the intermediate nodes, $k = 1, 2, \ldots, n - 1$ (see Fig. 2A), then the scheme for calculating y_{in} is the same. One may examine the inclusion of y_{Lk} between F_k and F_{k+1} as the inclusion of an element, the transmission matrix of which, B_k, consists of the components $b_{11} = b_{22} = 1$, $b_{12} = y_{Lk}$, $b_{21} = 0$.

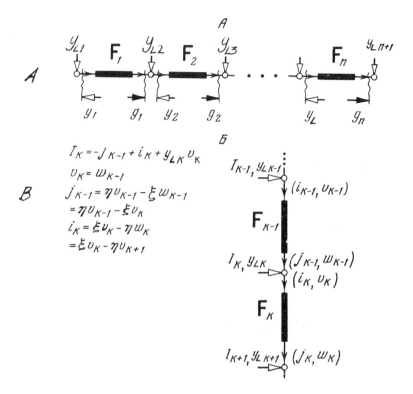

Fig. 2. Diagrams for the two ways of calculating the input resistances of a chain composed of sections of fiber. (A) recursive method (see text); (B) matrix method.

Then the transmission matrix of the whole chain is $A = C_1, C_2, \ldots, C_n$, where $C_k = A_k, B_k$ and $y_{Ln} = y_L$.

Now suppose we have to determine the input conductivity of the kth node, that is, the size of $y_{in,k}$. Obviously, two branches of the chain come together at this node: $F_k, \ldots, F_n$ and $F_{k-1}, \ldots, F_1$. Therefore the input conductivity of the node is defined as the result of the connection in parallel of the input conductivities of these branches and y_{Lk}, that is, $y_{in,k} = y_k + g_{k-1} + y_{L,k}$ (Fig. 2A). Naturally for each node the whole procedure of calculation is done anew, beginning with the determination of the transmission matrices of the corresponding branches.

It is easy to see that this method for calculation of the trees is impracticable.

THE RECURSION METHOD. This method, based on the in series use of dependence (4), is applicable for trees and is the most common in practical calculations.

Beginning with the penultimate node there are figured out in series the input conductivities $y_k = \chi_k(y_{k+1} + y_{Lk})$ of the chains F_n, $F_{n-1}F_n$, $F_{n-2}F_{n-1}F_n$, etc. In an analogous way, starting with the second node, there are figured out the reverse input conductivities $g_k = \chi_{k-1}(y_{n,k-1} + g_{k-1})$ of the chains F_1, F_2F_1, $F_3F_2F_1$, etc.

THE MATRIX METHOD. For the formation of suitable equations we use a method which resembles an experimental way of measuring the input resistances and the potential distribution along a fiber.

We examine the chain $F_1, F_2, \ldots, F_{n-1}$, in all the nodes of which "point" sources of current are included: $I_1, I_2, \ldots, I_n$ (Fig 2B), some of these currents can equal zero. Using the symbols and relations presented in Fig. 2B, the latter written on the basis of (5) for current I_k, we obtain the following term:

$$I_k = -\eta_{k-1}v_{k-1} + (\xi_{k-1} + \xi_k)v_k - \eta_k v_{k+1} + y_{Lk}v_k.$$

The aggregate of these equations, when $k = 1, 2, \ldots, n$, gives the desired system of equations describing the electrical properties of $F_1, \cdots, F_{n-1}$. In matrix notation, this system is written

$$I = (Y + Y_L)v \tag{7}$$

where the vectors $I = (I_1, \ldots, I_n)$, $v = (v_1, \ldots, v_n)$, the matrix of the conductivities of loads Y_L is diagonal with the elements $Y_{L1}, \ldots, Y_{Ln}$, and Y is a Jacobian matrix:

$$Y = \begin{pmatrix} \xi_1 & -\eta_1 & 0 & 0 & \ldots & 0 \\ -\eta_1 & \xi_1 + \xi_2 & -\eta_2 & 0 & \ldots & 0 \\ 0 & -\eta_2 & \xi_2 + \xi_3 & -\eta_3 & \ldots & 0 \\ 0 & 0 & -\eta_3 & \xi_3 + \xi_4 & \ldots & 0 \\ \cdot & \cdot & \cdot & \cdot & & \cdot \\ 0 & 0 & 0 & 0 & & \xi_n \end{pmatrix} \tag{8}$$

Solving system (7) with regard to v, we obtain

$$v = (Y + Y_L)^{-1}I = RI. \tag{9}$$

It is not difficult to be persuaded that the diagonal elements of matrix R determine the input resistances of the corresponding nodes $F_1, \ldots, F_{n-1}$. Indeed, suppose only that current I_k in (9) is different from zero, and all the rest are equal to zero. Then the voltage in the kth node will be equal to

$$v_k = r_{kk}I_k \quad \text{or} \quad r_{kk} = v_k/I_k. \tag{10}$$

The second equation is a definition of the input resistance of the kth node.

If the matrix R is calculated then equation (9) also defines the entire picture of the distribution of the potentials along the nodes of $F_1, F_2, \ldots, F_n$ for the arbitrary set of active sources of current. Thus, system (9) completely solves the problem of the electrical properties of the chain $F_1, F_2, \ldots, F_{n-1}$ with arbitrary loads at the nodes.

We will make some observations regarding the system of equation (7) and the ways of correlating the chain and recurrent methods in the case of the networks.

1. The matrix Y has a Jacobian structure because of which the elements of F_k in the chain under consideration connect only the adjoining nodes: for node k nodes $k + 1$ and $k - 1$ are adjacent. Therefore, in Y the elements different from zero fill the left diagonal, that is, $Y = P_{-1} + P_0 + P_1$. (We label as P_k the matrix which has nonzero elements located only on the kth parallel to the left diagonal. If $k > 0$, then the parallel is higher than the diagonal; if $k < 0$, then it is lower than the diagonal.) One may theorize a definition of the neighborhood: the nodes $k + j$ are neighbors of the jth order of the node k. That is to say the nodes of an ordinary chain (Fig. 3A) have neighbors only of the first order, and nodes of the chain represented in Fig. 3B have neighbors of the first and second orders. Therefore, for the last chain, $Y = P_{-2} + P_{-1} + P_0 + P_1 + P_2$. For the chain to which the neighbors of the jth order are connected, the matrix Y contains P_{-j} and P_j. Thus the existence of nonzero elements on those or

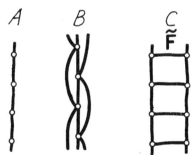

Fig. 3.

other parallels testify to the character of the combined state of the nodes of the chain.

2. Chains were examined above composed of the pieces F, the properties of such elements are described by the (2×2) matrices A or Y.

However, one may also form chains out of more complex "multidimensional" elements. For example, one may take the "aggregate" depicted in Fig. 3C as a compound element of a chain. This aggregate $\tilde{F}$ is constructed of pieces of F and represents a part of a square network. We will consider the left vertical series of nodes of $\tilde{F}$ as its entrances and the right as its exits. If we write a system of equations relating the input and output currents and the voltage of $\tilde{F}$, then it will look like

$$i = Aj + Bw, \qquad v = \Gamma j + Dw$$

where vectors $i = (i_1, \ldots, i_n)$ and $j = (j_1, \ldots, j_n)$ are the inflowing and outflowing currents for the corresponding entrances and exits $v = (v_1, \ldots, v_n)$ and $w = (w_1, \ldots, w_n)$; and $(n \times n)$ matrices A, B, Γ, D are blocks in the transmission matrix for $\tilde{F}$:

$$A = \begin{pmatrix} A & B \\ \Gamma & D \end{pmatrix}.$$

The transmission matrix A of the chain $\tilde{F}_1, \tilde{F}_2, \ldots, \tilde{F}_n$ is obviously defined as the product $A_1, A_2, \ldots, A_{n-1}$. If the load of this chain is determined by the matrix Y_L, then its input conductivity is

$$Y = (AY_L + B(\Gamma Y_L + D)^{-1}.$$

Changing the form of the chosen "aggregates" $\tilde{F}$, one may obtain a rather wide class of systems. In particular, for the presented sample $\tilde{F}$ (Fig. 3C), a homogeneous chain of such elements provides a part of the square network $(\tilde{F})^n$.

3. One may define the characteristic conductivity of F as the input conductivity of the chain $(F)^\infty$, or as the conductivity satisfying the equation $\chi(\sigma) = \sigma$.

By an analogous way one may determine the matrix of characteristic conductivity for the multidimensional element $\tilde{F}$; this is the input conductivity of the chain $(\tilde{F})^\infty$ or the matrix G satisfying the equation

$$G\Gamma G + GD - AG = B.$$

1.3. THE HOMOGENEOUS CHAIN.

We will describe the results of the use of the matrix method for the simplest case when the compound elements of the chain $F_1, F_2, \ldots, F_{n-1}$ are the same, which corresponds to the examination of a fiber with a constant diameter. The final results are obvious and here we dwell on this case only for the sake of the convenience of carrying out later on "by analogy" generalizations on the networks (see section 3).

This, given the chain F^{n-1}, the loads of which to all the inside nodes for simplicity equal zero, and the loads to the outside nodes equal ξ (i.e., the same elements of F are switched onto the edges, and only their free ends are shortcircuited). For such a chain system (8) has the altogether simple form:

$$
Y = \begin{pmatrix}
2\xi & -\eta & 0 & 0 & \ldots & 0 \\
-\eta & 2\xi & -\eta & 0 & \ldots & 0 \\
0 & -\eta & 2\xi & -\eta & \ldots & 0 \\
0 & 0 & -\eta & 2\xi & \ldots & 0 \\
\multicolumn{6}{c}{\cdots\cdots\cdots\cdots\cdots\cdots} \\
0 & 0 & 0 & 0 & & 2\xi
\end{pmatrix}.
\tag{11}
$$

It is easy to calculate the matrix R, the reciprocal of (11), by reducing Y to the canonical form [2], $Y = T \Lambda T^{-1}$, where $\Lambda = \mathrm{diag} \times (\lambda_1, \ldots, \lambda_n)$, the diagonal matrix of the eigenvalues, and $T = (t_{pq})_1{}^n$ is the transforming matrix.

$$
\lambda_k = 2\xi - 2\eta \cos \frac{k\pi}{n+1}, \qquad t_{pq} = \sqrt{\frac{2}{n}} \sin \frac{pq\pi}{n+1}.
$$

Therefore, the elements r_{pq} of the matrix R have the following form:

$$
r_{pq} = \sum_{k=1}^{n} \frac{t_{pq} t_{kq}}{\lambda_k} = \frac{1}{n} \sum_{k=1}^{n} \frac{\sin pq\pi/n + 1 \cdot \sin qk\pi/n + 1}{\xi - \eta \cos k\pi/n + 1}.
\tag{12}
$$

Hence, according to (10) the input resistance of the pth node of our chain F^{n-1} equals

$$
r_{pp} = \frac{1}{n} \sum_{k=1}^{n} \left(\sin^2 \frac{pk\pi}{n+1} \right) \Big/ \left(\xi - \eta \cos \frac{k\pi}{n+1} \right).
\tag{13}
$$

Suppose we measure the input resistance r_{in} in some inside node of the chain, then the increase of the length of the chain on both sides from this node leads to the increase of the input resistance of the node. However, in the end with the unrestricted growth of the length of the chain this resistance will apparently tend to a size equal to half the characteristic resistance, that is, $r_{in} \to \rho/2$ with $n \to \infty$. Carrying out the appropriate limiting transformation in the right part of (13) we obtain

$$
r_{in} = \int_0^1 \frac{dx}{\lambda(x)} = \frac{1}{2} \int_0^1 \frac{dx}{\xi - \eta \cos \pi x}.
\tag{14}
$$

This integral is figured out in quadratures and gives

$$r_{in} = (\tfrac{1}{2}) \times (\xi^2 - \eta^2)^{-1/2} = \tfrac{1}{2}\sigma = \rho/2.$$

If we examine infinite chains, then it makes sense to renumber the nodes choosing one of them for the starting one—the zero one. For such a chain the matrix R has an infinite order and its elements

$$r_{pq} = \int_0^1 \frac{\cos{(p - q)\pi x}}{\lambda(x)} \, dx \qquad (15)$$

where $p,q = 0, \pm 1, \pm 2, \ldots$, $\lambda(x) = 2(\xi + \eta \cos \pi x)$. As is clear from the preceding, the term (15) with $p = 2$ is the input resistance of the node and turns into (14). The integral (15) is also calculated in quadratures, therefore

$$r_{pq} = \frac{\rho}{2} \exp{(-|p - q|l/\lambda)}. \qquad (16)$$

Now one can find, using (16) and (9), the distribution of the potential along the chain for an arbitrary series of sources.

2. THE TREES

Here we will briefly examine the applicability of the methods described to treelike fibers that do not form closed contours. Such fibers are called *trees*. The initial branch of the tree or the base of the tree is called the branch of the zero order, the branches produced by the initial branch are called the branches of the first order; beyond the branches of the first order come the branches of the second order, etc. The order of the tree is set by the maximum order of its branches.

If each branch of a lower order (except the terminal ones) always produce exactly k branches of the next order, then the tree is called k-*chotomous*. A k-chotomous tree of the order n will be called a (k,n) tree. If a (k,n) tree has all the branches of all the orders assumed, such a tree is called *complete*. A complete (k,n) tree has k^m branches of the mth order, and the whole tree has $(1 + k + k^2 + \ldots + k^n)$ branches.

If all its branches of a single order are identical, then the tree may be called *normal*. (Another, more broad definition of normal trees, since they are constructed of pieces of fiber, can be as follows: for a source active in the base of the tree, all the nodes of one order are equipotential.) Obviously, a section of fiber represents a $(0,1)$ tree, and the chain $F_1, \ldots, F_{n+1}$ is equivalent to a $(1,n)$ tree; the dendrites of the nerve cells generally are incomplete $(2,n)$ trees.

2.1. THE BRANCHING FIBER

The simplest variant of a fiber with a single node of branching (a "T-joint" or $(2,1)$ tree) is shown in Fig. 4. The input conductivity of the central node, on the basis of (1.4), obviously is defined as the sum $\chi(y_{L0}) + y_{L1} + \chi(y_{L2}) + \chi(y_{L3})$. The input conductivity of one of the outer nodes, for example the zero one, can be figured out by the recursive method:

$$y_{\text{in}0} = y_{L0} + \chi_1(y_{L1} + \chi_2(y_{L2}) + \chi_3(y_{L3})).$$

We will now determine the input resistance of the nodes of the T-joint by the matrix method, compiling for it a system of equation of the (1.11) type. According to the symbols presented in Fig. 4, we have $I_0 = i_1 + i_2 + i_3$ and $I_k - j_k + y_{Lk}w_k$, where $k = 1,2,3$. Substituting in these

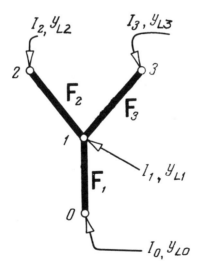

Fig. 4. Diagram of a node of branching.

equations the currents i_k, j_k, with the voltages in the nodes, we obtain

$$I = (Y + Y_L)v \tag{17}$$

where $I = (I_0, I_1, I_2, I_3)$, $v = v_0, w_1, w_2, w_3)$, Y_L is the diagonal matrix of the 4th order, $Y_L = \text{diag}(0, y_{L1}, y_{L2}, y_{L3})$, and

$$Y = \begin{pmatrix} \xi_1 + \xi_2 + \xi_3 & -\eta_1 & -\eta_2 & -\eta_3 \\ -\eta_1 & \xi_1 & 0 & 0 \\ -\eta_2 & 0 & \xi_2 & 0 \\ -\eta_3 & 0 & 0 & \xi_3 \end{pmatrix}.$$

We will now solve system (17) regarding the voltages, that is, we will determine the coefficients r_{pq} ($p, q = 0,1,2,3$) for the system

$$v = (Y + Y_L)^{-1}I = R \cdot I.$$

After simple computations we obtain

$$R = \frac{\xi_1 \xi_2 \xi_3}{\Delta} \begin{pmatrix} 1 < \Delta_{23} & a_1 & a_1 a_2 & a_1 a_2 \\ a_1 & 1 & a_2 & a_3 \\ a_1 a_2 & a_2 & 1 + \Delta_{13} & a_2 a_3 \\ a_1 a_3 & a_3 & a_2 a_3 & 1 + \Delta_{12} \end{pmatrix}.$$

Here

$$\Delta = \det Y = \sigma_1^2 \xi_2 \xi_3 + \xi_1 \sigma_2^2 \xi_3 + \xi_1 \xi_2 \sigma_2^2,$$
$$\Delta_{ij} = (\xi_i \sigma_j^2 + \sigma_i^2 \xi_j)/\xi_1 \xi_2 \xi_3, \qquad a_k = \eta_k/\xi_k.$$

2.2. THE ARBITRARY TREES

In order to compute the input conductivity of the base of a $(2,n)$ tree, y_0 (for simplicity of symbols we examine only dichotomous trees), it is necessary to know the input conductivities y_{11} and y_{12} of the primary subtrees $y_0 = \chi(y_{11} + y_{12})$, and for the computation of y_{11} and y_{12} it is necessary to know the values of y_{21}, y_{22}, and y_{23}, y_{24}—the input conductivities of the secondary subtrees, etc. Thus one may compute the input conductivity of a tree by a one-sided recurrent movement "downward":

$$y_{k,s} = \chi_{ks}(y_{k+1,2s-1} + y_{k+1,2s}),$$

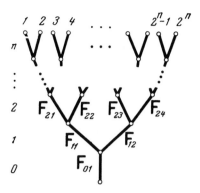

Fig. 5. Dichotomous tree of the nth order.

where $k = n, n - 1, \ldots, 1, 0$; $s = 1, 2, 3, \ldots, 2^k$ (Fig. 5); and the conductivities $y_{n+1,s}$ are the loads of the terminal, top branches.

Essential for the determination of the input conductivities of the inner nodes of a $(2, n)$ tree are the additional calculations of the "reverse" input conductivities $g_{k,s}$, which represents the input conductivities of the distal ends of the branches approaching the given node of branching (analogous values were determined earlier for the chain—a *monochotomous* tree, see Fig. 2A). For example, $g_{k,2s-1}$ and $g_{k,2s}$ are the reverse input conductivities of the two branches $F_{k,2s-1}$ and $F_{k,2s}$ coming from one node ($k - 1, s$). Therefore they are computed by a reverse movement "downward" according to the recursive formula

$$g_{k,2s-1} = \chi(g_{k-1,s} + y_{k,2s}), \qquad g_{k,2s} = \chi(g_{k-1,s} + y_{k2s-1}).$$

In a normal tree the input conductivities of the bases of all the subtrees of one order are also identical: $y_{k1} = y_{k2} \ldots y_{k,2}k$. In the computation of the input resistance and the pattern of the distribution of the potentials for the source, active in the base, one may change the normal tree connecting all the nodes of one order into an equivalent chain $F_1, F_2, \ldots, F_n$.

In some cases an examination of an equivalent chain (an "equivalent cylinder," [7, 8]) is extremely helpful, for example, in the calculation of the input conductivity of a tree.

As in the case of a single node of branching (see section 2.1), one may look for the input conductivities of nodes of a $(2,n)$ tree by the matrix method. For a complete $(2,n)$ tree the matrix Y is of the order 2^{n+1} and has the following structure:

$$\begin{pmatrix} \xi_{01} & -\eta_{01} & 0 & 0 & 0 & 0 & 0 & 0 & \cdots \\ -\eta_{01} & y_{22} & -\eta_{11} & -\eta_{12} & 0 & 0 & 0 & 0 & \cdots \\ 0 & -\eta_{11} & y_{33} & 0 & -\eta_{21} & -\eta_{22} & 0 & 0 & \cdots \\ 0 & -\eta_{12} & 0 & y_{44} & 0 & 0 & -\eta_{23} & -\eta_{24} & \cdots \\ 0 & 0 & -\eta_{21} & 0 & y_{55} & 0 & 0 & 0 & \cdots \\ 0 & 0 & -\eta_{22} & 0 & 0 & y_{66} & 0 & 0 & \cdots \\ \cdots & \cdots & \cdots & \cdots & \cdots & \cdots & \cdots & \cdots & \cdots \end{pmatrix}$$

where $y_{22} = \xi_{01} + \xi_{11} + \xi_{12}, y_{33} = \xi_{11} + \xi_{21} + \xi_{22}, y_{44} = \xi_{12} + \xi_{23} + \xi_{24}$, etc.

As we have already pointed out, the matrix method of calculation is more universal in comparison with the recursive one because it provides the opportunity to obtain at once a complete description of the electrical properties of the whole tree. However, on the bulk of the computing work the recursive method in a given case is not inferior to the matrix.

2.3. THE OPEN SYNCYTIUM

Especially interesting [3] are the homogeneous (k,∞) trees all the branches of which are identical. All the subtrees of a homogeneous tree are also homogeneous and coincide with the tree itself. We will find the input conductivity of a (k,∞) tree which, obviously, coincides with the input conductivity of all the subtrees. This means the load of the first and each following branch is the conductivity

$$y_L = k y_{in}. \tag{18}$$

Therefore for the determination of the input conductivity we obtain the equation

$$y_{in} = \chi(k y_{in}) \quad \text{or} \quad y_{in}^2 - \frac{k-1}{k} \xi y_{in} - \frac{\sigma^2}{k} = 0.$$

Hence

$$y_{in} = \frac{k-1}{2k} \xi + \sqrt{\left(\frac{k+1}{2k} \xi\right)^2 + \frac{\sigma^2}{k}}. \tag{19}$$

Condition (18) is valid not for all branches branching out from some node of a homogeneous (k,∞) tree but only for branches which can serve as bases of similar trees. If by means of $k + 1$ homogeneous (k,∞) trees we form homogeneous syncytia, connecting these trees by the bases, then all the nodes and branches become equivalent. Apparently the common input conductivity of a node of such a homogeneous open syncytium is equal to $(k + 1)y_{in}$, where y_{in} is from (19).

We also note that the homogeneous (k,∞) tree is equivalent to the infinite chain $F_0, F_1, F_2, \ldots, F_m \ldots$, the characteristic conductivity of the elements of which changes (with the growth of m) according to the law of geometric progression with the common ratio k. Actually, the mth element of the chain is the result of the parallel connection of k^m branches of F_0. Therefore, its characteristic conductivity is equal to $k^m \sigma$.

3. THE NETWORKS

We will call *networks* lattice structures the ribbing of which are sections of fiber F. Each node of an n-dimensional network is marked by a number $(x_1, x_2, \ldots, x_n)$, where each of the numbers of x_k, if the network is infinite, can take on any whole number value: $x_k = 0, \pm 1, \pm 2, \ldots$; for a finite network with the dimensions $m_1 \times m_2 \times \ldots \times m_n$ it is convenient to assume that $x_k = 1, 2, \ldots, m_k$.

A network all the nodes of which (except the border ones for a finite network) contain the same number of r branches, can be called, using the terminology of the theory of graph, a network of the r degree; an n-dimensional network of the r degree will be called a (r,n) network.

Of the two-dimensional networks we will examine the most simple—a homogeneous square network, that is, a (4.2) network (Fig. 6), and we will cite the results for an n-dimensional cube network.

The applicability of the matrix method (see section 1.2) to the different systems does not depend on the sort of structural features they have as the presence or absence of closed contours. Therefore, with the extension of this method to systems of the type of the networks, no difficulties of a fundamental nature arise; basically, only the complexity and quantity of calculations increase.

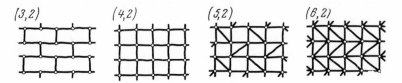

Fig. 6. Examples of two-dimensional networks of the degree $r = 3,4,5,6$.

3.1. THE TWO-DIMENSIONAL NETWORKS

Suppose as in the examination of the chain $F_1, F_2, \ldots, F_n$ (see section 1.2), that the sources of current I_{pq} act in the nodes pq of a certain two-dimensional network. If the network has the dimensions $m_1 \times m_2$, n_1 nodes along the vertical and m_2 along the horizontal, then $p = 1, \ldots, m_1$; $q = 1, \ldots, m_2$. Generally speaking, I_{pq} form the $(m_1 \times m_2)$ matrix, but we will examine the aggregate of these values as the vector (recording them by the lexicographical method);

$$I = (I_{11}, \ldots, I_{m_11}; I_{12}, \ldots, I_{m_12}; I_{1m_2}, \ldots, I_{m_1m_2}).$$

In an analogous vectoral form one may represent the aggregate of the value defining the potentials v_{pq} in all the nodes of the network:

$$v = (v_{11}, v_{m_11}; \ldots; v_{1m_2}, \ldots, v_{m_1m_2}).$$

On the basis of Kirchhoff's law and equation (15), the equation relating I and v can be written:

$$I = Yv; \tag{20}$$

moreover, the conductivity matrix Y will have the order (m_1m_2) and one may examine it as the matrix containing $m_2{}^2$ blocks of Y_{pq} each of which has the order m_1:

$$Y = (Y_{pq})_1{}^{m_1}, \qquad Y_{pq} = (Y_{ij,pq})_1{}^{m_2}. \tag{21}$$

Thus, the elements of Y have two pairs of indexes $y_{ij,pq}$. The first pair ij indicates the affiliation to the (i,j) block, and the second pair p,q indicates the layout inside the block, that is, $i,j = 1,2, \ldots, m_1$; $p,q = 1,2, \ldots, m_2$. The element $y_{ij,pq}$ is diagonal if $i = j$ and $p = q$.

In exactly the same manner as earlier (see section 1.2) the diagonal elements of the matrix

$$R = Y^{-1} \tag{22}$$

define the input resistances of the nodes of the network under consideration.

We will now examine a "square" (4.2) network, assuming for simplicity that all the sections of fiber constituting this network are identical. Using the relations for current and voltages of the (p,q) node presented in Fig. 7, we obtain the system of equations

$$I_{pq} = -\eta v_{p-1,q} + 4\xi v_{pq} - \eta v_{p+1,q} - \eta v_{p,q-1} - \eta v_{p,q+1} \tag{23}$$

After the analogy of an unidimensional network (see section 1.3) one may, in addition, load along the edges so that the matrix Y in system (20) will become homogeneous in the make-up of the elements, that is, will have the form

$$Y = \begin{pmatrix} Y_0 & -Y_1 & O & O & \dots & O \\ -Y_1 & Y_0 & -Y_1 & O & \dots & O \\ O & -Y_1 & Y_0 & -Y_1 & \dots & O \\ O & O & -Y_1 & Y_0 & \dots & O \\ \cdot & \cdot & \cdot & \cdot & \cdot & \cdot \\ O & O & O & O & \dots & Y_0 \end{pmatrix}. \tag{24}$$

Here the elements Y_0 and Y_1 are the matrices of order m_1, and the remaining elements are zero matrices of the same order. In conformance

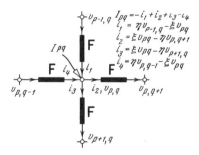

Fig. 7. Formation of an equation for the input current I_{pq} of a network.

with (23), Y_1 is a diagonal matrix with elements η, Y_0 is a Jacobian matrix and differs from (11) only by the fact that on its diagonal is the value 4ξ and not 2ξ. Thus, for a square network the matrix Y with regard to its blocks has a Jacobian structure.

The elements of the matrix of the resistances $R = Y^{-1}$ (computed in an analogous manner, see section 1.3, by the reduction of Y to canonical form) have the following form:

$$r_{ij,pq} = \sum_{s=1}^{m_1} \sum_{k=1}^{m_2} \lambda_{sk}^{-1} t_{is,pk} t_{sj,kq}$$

where

$$\lambda_{sk} = 2\xi - \eta \cos \frac{s\pi}{m_1 + 1} - \eta \cos \frac{k\pi}{m_2 + 1},$$

$$t_{is,pk} = \frac{2}{\sqrt{m_1 m_2}} \sin \frac{is\pi}{m_2 + 1} \sin \frac{pk\pi}{m_1 + 1}.$$

Calculating the limit with $m_1, m_2 \to \infty$ for the diagonal element of the matrix R, we find the input resistance of the node of an infinite square lattice

$$r_{in} = \frac{1}{2} \int_0^1 \int_0^1 \frac{dx\,dy}{2\xi - \eta \cos \pi x - \eta \cos \pi y}. \tag{25}$$

One may integrate this expression once by one of the variables, then

$$r_{in} = \frac{1}{2} \int_0^1 \frac{dx}{\sqrt{(2\xi - \eta \cos \pi x)^2 - \eta^2}}. \tag{26}$$

OBSERVATION: The idea of the calculation of the input resistance of an infinite network described in reference [1] was as follows. First was examined an infinite strip of a (4,2) network having a finite number of nodes along the vertical and an infinite number along the horizontal. One may consider such a strip the infinite chain $\tilde{F}^\infty$, the elements $\tilde{F}$ of which are the elementary vertical strips (see observation 2 in section 1.2 and Fig. 3C). It is easy to write the system of equations relating the input and output currents and the voltage of such an element and the equations relating the input conductivity matrix and the load matrix: $Y_{in} = \chi(Y_L)$. Now one may find the input conductivity matrix of the chain $\tilde{F}^\infty$ by solving the matrix equation $G = \chi(G)$. Such a matrix of input conductivity $Y_{in} = G$ is called the characteristic matrix of the conductivities. If we then change the matrix G and proceed to a network infinite both along the horizontal and the vertical, then for the input resistances of the node we obtain expression (26) and not (25). An analogous connection exists between the integral in equation (16) and the characteristic resistance ρ, and it applies to a n-dimensional case. Thus, the chain examination of a system allows us to reduce the order of the matrices to one measurement (to reduce the examination of an n-dimensional case to a $(n - 1)$ dimensional one) and is equivalent to a single integration of the matrix of the resistances R. But it is more natural to use an electronic computer for the calculation of the integrals of form (26).

To find the rest of the elements of matrix R for an infinite (4,2) network, we renumber accordingly the nodes of the lattice, choosing one of them for the first. Making the limiting transition, we obtain

$$r_{ks,pq} = \frac{1}{2} \int_0^1 \int_0^1 \frac{\cos(k - s)\pi x \cos(p - q)\pi y}{\lambda(x,y)} \, dx \, dy \qquad (27)$$

where $k,s,p,q = 0, \pm 1, \pm 2, \ldots$ and $\lambda(x,y) = 2\xi - \eta \cos \pi x - \eta \cos \pi y$. When $k = s$ and $p = q$ we have the diagonal element (25). With a single source active in the node (0,0) the distribution of the potential along the network, that is, the voltage in the arbitrary node (p,q) is identified, obviously, by the expression

$$v_{pq} = \frac{1}{2} I_{00} \int_0^1 \int_0^1 \frac{\cos p\pi x \cos q\pi y}{\lambda(x,y)} \, dx \, dy. \qquad (28)$$

The integral in the right part is (p,q)—the coefficient of the expansion of the even (in both variables) function $\lambda^{-1}(x,y)$ into a double Fourier series.

EXAMPLE. For the problem of the spread of excitation in syncytial media, examinations of the distinct configurations of the active sources simulative of a wave front are of interest. Suppose that at the initial moment of time, node (0,0) is excited, and that within time t the excitation moves to the first nodes $(0, \pm 1) (\pm 1,0)$. Then in time Nt the "wave front" will have the shape of a rectangular ring schematically depicted in Fig. 8. The coordinates of the excited nodes in the first square are thus: $(N - k,k)$ where $k = 0,1,\ldots, N - 1$. We will assume that the excited node differs from all the unexcited ones by the presence of an active source of current I, and that all these sources are identical. We will determine the potential in the arbitrary node (p,q). For $N = 0$ the solution of the problem gives expression (28). Assume now $N = 0$, $N = 1,2,\ldots$, then the solution of the

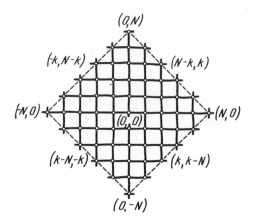

Fig. 8. Diagram of the spread of excitation from the node (0,0) of a square network (the nodes are joined by a dotted line for the moment N = 5); (N,0), (N − k,k), etc., are the present coordinates of the excited nodes $k = 0,1,\ldots,$ N − 1.

problem is the calculation of the sum

$$v_{pq} = I \cdot \sum_{s} \sum_{t} r_{pk_s,qk_t},$$

where $(k_s k_t)$ are the nodes in which the sources act. Representing the co-ordinates of the excited nodes as shown in Fig. 8, we can substitute a double sum for v_{pq} by a sum along one index k; introducing the sum under the symbol of integral (27) we have

$$\sum_{k=0}^{N-1} [\cos(p - N + k)\pi x \cdot \cos(q - k)\pi y$$
$$+ \cos(p + k)\pi x \cos(q - N + k)\pi y$$
$$+ \cos(p + N - k)\pi x \cos(q + k)\pi y$$
$$+ \cos(p - k)\pi x \cos(q + N - k)\pi y].$$

After simple transformations, we obtain

$$v_{pq} = I \int_0^1 \int_0^1 [F_{N+1}(x,y) - F_{N-1}(x,y)] \frac{\cos p\pi x \cos q\pi y}{\lambda(x,y)} \, dx dy,$$

where

$$F_N(x,y) = (\cos N\pi x - \cos N\pi y)/(\cos \pi x - \cos \pi y).$$

3.2. THE MULTIDIMENSIONAL NETWORKS

The nodes of a three-dimensional network are numbered as the set pqr. If the network has the dimensions $m_1 \times m_2 \times m_3$, then $p = 1, \ldots, m_1$;

$q = 1, \ldots, m_2; r = 1, \ldots, m_3$. Recording the currents I_{pqr} and the voltages of the nodes in a lexicographical way, we obtain the vectors I and v of the order m_1, m_2, m_3. The conductivity matrix Y of the equation which I and v satisfy consists of the m_3^2 blocks Y_{rt}, each of which can be broken into the m_2^2 subblocks $Y_{qj,rt}$ of the order m_1. Therefore, in the given case, the elements of Y have three pairs of indexes: $Y_{pk,qj,rt}$.

For a cubic network, the matrix Y with regard to the blocks Y_{rt} has a Jacobian structure. It is easy to be persuaded that the Jacobian structure with regard to the basic blocks Y_{rt} occurs also in the general case of a n-dimensional cubic network. And what is more, for a n-dimensional cubic network all the subblocks of matrix Y will have a Jacobian structure. This circumstance significantly simplifies all the computations connected with the handling of Y. Omitting the intermediate computations in view of their bulk, we present only the final result. For the input resistance of a node of an infinite n-dimensional "anisotropic" cubic network we have

$$r_{\text{in}} = \frac{1}{2} \int_0^1 \int_0^1 \cdots \int_0^1 \frac{dx_1 dx_2 \cdots dx_n}{\sum_{k=1}^{n} (\xi_k - \eta_k \cos \pi x_k)} \cdot$$

Hence for a homogeneous net ($\xi_1 = \cdots = \xi_n, \eta_1 = \cdots = \eta_n$ is obtained:

$$r_{\text{in}} = \frac{1}{2} \int_0^1 \int_0^1 \cdots \int_0^1 \frac{dx_1 dx_2 \cdots dx_n}{n\xi - \eta(\cos \pi x_1 + \cos \pi x_2 + \cdots + \cos \pi x_n)} \cdot$$

The networks, the conductivity matrix of which has a Jacobian structure with regard to all its subblocks, should satisfy the following condition (this results from observation 1 in section 1.2, where it was explained how the structure of Y is determined by the extent of the connections of each node of the network): each node of the network is connected only with the nearest (along the index) neighboring node. Examples of two-dimensional networks of the degree $r = 3,4,5$, and 6, satisfying this condition are presented in Fig. 6. The already examined cubic network, the degree of which is $2n$, provides the simplest example of a n-dimensional network of the Jacobian type. Obviously, the minimum degree of a spatial network ($n \geqslant 2$) is three. If in a cubic network we restore all the diagonals of all the cells, then we will obtain a Jacobian network of the maximum degree

$n^3 - 1$. Networks of intermediate degrees result with the removal of those or other diagonals or ribs in each cell of a network of the maximum degree.

In the example of the family of cubic networks we tried to describe a general scheme for computation of the passive electrical characteristics of closed syncytia. The calculations of networks of other types do not differ in principle from those described. The results of the tabulation of integral relations for the input resistances of the syncytia are presented in reference [1].

REFERENCES

1.
M. B. Berkinblit, S. A. Kovalev, V. V. Smolyaninov, and L. M. Chaylakhyan, *Biofizika* 1965:10, 309.
2.
F. R. Gantmakher and M. G. Kreyn, *Ostsillyatsionnye matritsy i yadra i malye kolebaniya mekhanicheskikh sistem* (*Oscillatory Matrices and Nuclei and Small Fluctuations of Mechanical Systems*), Moscow: GITTL (State Publishing House of Telegraph-Telephone Literature).
3.
E. P. George, *Aust. J. Exp. Biol.* 1961:39, 267.
4.
A. L. Hodgkin, A. F. Huxley, and B. Katz, *J. Physiol.* 1952:116, 424.
5.
——— and W. A. H. Rushton, *Proc. Roy. Soc.* [*Biol.*] 1946:B133, 444.
6.
C. L. Prosser, G. Burnstock, and J. Jakn, *Amer. J. Physiol.* 1960:119.
7.
W. Rall, *Exp. Neurol.* 1959:1, 491.
8.
———, *Biophys. J.* 1962:2, 145.
9.
J. W. Woodbury and W. E. Crill, *Proceedings of the International Symposium "Nervous Inhibitions,"* New York: Pergamon, 1961, p. 129.

5

The Periodic Blocking of Impulses   M. B. Berkinblit
in Excitable Tissues

The problem concerning the behavior of excitable tissues during action on them of a periodic wave of stimuli arose in connection with the study of two types of biological processes: first, in connection with the study of the rhythmically functioning formations of, for example, the heart or electric organ nerves of fish; second, in connection with the study of the reprocessing of the input signals by neurons.

An analysis of corresponding phenomena was carried out both on physiological preparations, for example, single nerve fibers, the nodes of Ranvier, and neurons, and on various models: on physical models of neurons, on a digital electronic computer, and for the simplest models by means of analytical description.

1. ANALYTICAL MODELS OF THE WORKING OF THE SIMPLEST EXCITABLE ELEMENTS

Let us examine the behavior of the simplest excitable element—an arrangement with one input and one output, and possessing a threshold and refractoriness. Suppose that during excitation the threshold changes as shown in Fig. 1C, that is, that at the moment of excitation it increases instantaneously, whereupon the element during time R_a, the *absolute refractory period*, is unexcitable, and then during time R_0, it linearly falls from some value h_1 to its value at rest h_0.

We will say that at the output of the element, the rhythm $n:(n - m)$ is realized if during continuous stimulation by each of the successive n stimuli the element responds as follows: from the first $n - m$ stimuli it is excited, from the following m stimuli it is not excited, etc. We will assume also that the change of the threshold after the kth stimulus occurs in exactly the same manner as after the first, that is, that fatigue processes are absent.

At first we will examine the behavior of an excitable element during action on it of a recurrent wave of supraliminal stimuli having a very small duration (the width of the impulse, θ, is substantially shorter than R and the period of stimulation T). Suppose that the first incoming impulse

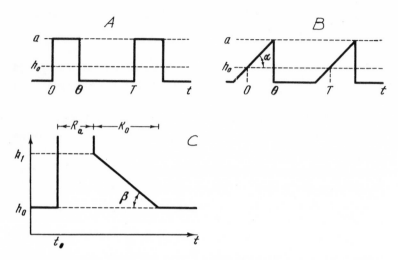

Fig. 1. Series of rectangular (A) and triangular (B) stimuli; (C) the change of the threshold of the element after excitation.

causes excitation. The second incoming impulse either causes excitation or arrives (depending on the frequency of stimulation) at that moment when the threshold is too high and does not cause excitation. If the second impulse excites the element, then all the following ones will also excite it (since there is no "fatigue"). If the second impulse does not cause excitation, then we will obtain omissions of impulses of the 2:1 type, or even more frequent ones. Thus, during stimulation of the element by very short impulses, the only possible transformation of rhythm is the *frequency division* type, that is, a rhythm of the n:1 type.

The frequency during which omissions of impulses begin depends on the amplitude of the stimuli (a). The greater is the amplitude, the higher will be the critical frequency. The time interval during which an impulse of a given amplitude cannot excite an element, or the time after excitation during which the threshold is higher than the amplitude of the given input signal, is called the *functional refractory period* (R_F) [32]. The functional refractory period differs from the *relative refractory period* by its instability and its connection (according to definition) with the amplitude of the stimulus; the bigger it is, the smaller is the amplitude. The functional refractory period is, so to say, the absolute refractory period for stimuli of a given amplitude. Obviously, a condition of frequency division will be observed when T is less than R_F.

It is interesting that the electroreceptors of certain fish operate on the principle of frequency division [16]. Their electrical organs generate a signal of high frequency so that the receptors do not respond to every stimulus. With the appearance in the environment of any object the amplitude of the signal is changed and accordingly the conditions of frequency division also changes, for example, not every third but every fourth stimulus begins to be skipped and accordingly the frequency of the impulses transmitted from the receptors changes.

If the stimulus has a finite duration $\theta \neq 0$, then other routines of the element are possible besides frequency division. Their character depends in particular upon the shape of the stimulus.

We will examine the excitation of an element by a recurrent wave of rectangular impulses (Fig. 1A). The idea of this model was proposed by Rosenblueth (33) for an explanation of the periodic blocking of impulses in the heart (see section 4). For brevity we will subsequently call it the R-model.

If the line representing the threshold, as it drops after the preceding excitation, intersects the resting level before the arrival of the next stimulus or falls on its leading edge, then no impulse will be skipped and all the impulses except the first will emerge in the same stage of refractoriness: the later the stage, the lower the stimulation frequency (stable conditions), as shown in Fig. 2A. This behavior occurs when $T > R_F$. If after the first excitation the threshold is not able to drop to a by the end of the next stimulus, then the element will operate in a frequency division regime. Obviously, this will occur when $T < R_F - \theta$. Finally, if the line representing the threshold which is dropping after the preceding excitation intersects the horizontal part (the plateau) of the following stimulus (Fig. 2B), then the element operates in a $n:(n-1)$ regime, that is, an omission of every nth impulse is observed. In order for a recurrent impulse dropout of that type to take place, the period of stimulation must be in an interval from R_F to $\frac{2}{3}R_F$ or even still narrower. We will examine this routine in somewhat greater detail.

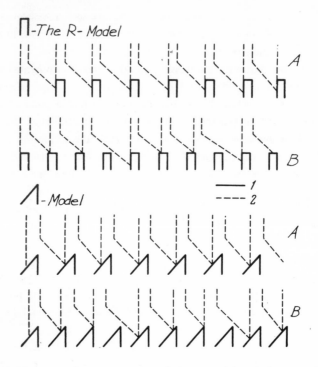

Fig. 2. Diagrammatic representation of the R-model and the ∧-model. (A) stable mode; (B) periodic omissions; (1) stimuli; (2) threshold changes in time.

When the period of stimulation becomes shorter than R_F, a stimulus does not cause excitation by its leading edge. However, due to the finite duration of the stimulus, its maximal amplitude lasts for some time during which the threshold can drop rather sharply in order for an excitation to arise. The stimulus "waits" while the threshold drops sufficiently. On account of this fact the second impulse arises with some delay in comparison with the first. The third stimulus finds the element in a condition with a higher threshold than the second found it, since the third stimulus arrived within the same time after the second as the second after the first; but the excitation response to the second stimulus arose with a delay and therefore the third stimulus causes a response with a still greater delay.

As is apparent from Fig. 2B, each following response impulse will arise still later after the arrival of the next stimulus until finally some stimulus is not able to be completed, for it does not cause an excitation.

We will note that the output frequency in this behavior is determined only by the functional refractory period of the element. Each of the following impulses arises as soon as the line of the threshold intersects the plateau of the new stimulus, that is, as soon as the functional refractory period ends. Therefore, the spacing between the output impulses is equal to R_F. In the R-model the nature of the threshold decrease generally is of no importance since only one point, the threshold value, which is equal to the amplitude of the stimulus, is "effective."

We will examine now how the delays in successive responses of the element change. If τ_i is the delay in the response to the ith stimulus (Fig. 3), then the period between the ith and the $(i + 1)$th responses of the element is $T_i = T + \tau_{i+1} - \tau_i$. But as already said, $T_i = R_F$. Hence, $\tau_{i+1} = \tau_i + (R_F - T)$, and since $\tau_1 = 0$, we derive that the delay in the response to the ith stimulus is

$$\tau_i = (i - 1)(R_F - T). \tag{1}$$

Thus the delay increases linearly with the number of the stimulus. The dropping threshold moves, in successive responses, along the plateau of a stimulus with equal "steps," such that the length of one step equals $R_F - T$.

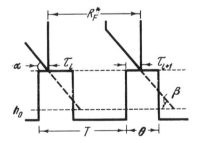

Fig. 3. Change of the delays of successive responses to rectangular stimuli.

Hence, it follows that the length of the cycle[1] or the number of the impulses skipped is determined by the fact of how many such "steps" fit on the plateau. For the cycle length we obtain

$$n = 1 + \left[\frac{\theta}{R_F - T}\right], \tag{2}$$

where the brackets are indication of the whole part.

Now it is clear that for the rise of a cycle with the duration n the line representing the threshold should intersect the plateau in the interval $\theta/n, \theta/(n-1)$, the width of which $\theta/(n-1) - \theta/n = \theta/[n(n-1)]$ decreases with the number n (roughly, as n^2). Therefore, the longer the cycles the narrower the range of frequencies in which they can be received. We will illustrate this phenomenon in somewhat greater detail. Assume, for example, that the second response arises at the moment when almost half of the period of the second stimulus has already passed. Then the third impulse will still be rising at the moment of the very completion of the period of the third stimulus, and the fourth impulse will be skipped. If the second response arises when the first third of the second stimulus is completed, then the third impulse arises when somewhat more than two-thirds of the next stimulus is passing, but the fourth impulse will again be skipped. Thus, in order to obtain omissions of every fourth impulse, it is necessary that the second impulse arises after a third impulse passes, but before half of the duration of the second stimulus passes. In exactly this way we can show that for omissions of every fifth impulse the second impulse should arise after a fourth finishes, but before one-third of the duration of the second stimulus finishes, that is, the range of the period of stimulation in which the omission of every fifth impulse will be observed is narrower than during the omission of every fourth impulse.

It is easy to note that during a period of stimulation shorter than R_F, only for the time of the duration of the stimulus will there be observed

---

[1] For precision of terminology we will subsequently use the word *cycle* for designation of the time from one impulse omission to another, and in the case of the heart (see section 4), for designation of the time from one ventricular systole omission to another. We keep the term *period* for the interval between two stimuli, and in the case of the heart, for the interval between two successive systoles of the auricles.

omissions of every second impulse, that is, the range of the periods in which it is possible to obtain all the possible behaviors of the $n:(n - 1)$ type is equal to the width of the stimulus. The shorter is the stimulus, the more exactly must the frequency of stimulation be maintained.

We can show that the omission of the nth impulse will be systematically repeated if the width of the stimulus is small in comparison with the period of stimulation. Otherwise, the length of each cycle can vary. (If in the first cycle the nth impulse was skipped, then the behavior will be stable, that is, all the cycles will have the same length when the "gap" between successive stimuli is greater than $1/n$ of the period.)

I. M. Gelfand and M. L. Tsetlin proposed using for the examination of biological questions a model with stimuli of finite duration and arbitrary form of the leading edge. The basis of this model was the method used by Teodorchik [42] for examination of the processes of synchronization of relaxation generators. In the work of Arshavskiy, Berkinblit, and Kovalev [2] the behavior of an excitable element during a triangular-shaped stimulus was examined in detail. This method of examination was used also in the work of Ivanov and Telesnin [20]. Harmon [17] and Wilson [45] employed a similar method.

The model with a triangular form of the stimulus (see Fig. 1B) in the future will be called a ∧-model. At the time that the rectangular stimulus instantaneously increases to the maximum, and then waits until the fiber is not excited because of the decrease of the threshold, the triangular stimulus gradually increases and during the decrease of the threshold continues to grow. Descriptively speaking, in the R-model the stimulus "waits" for the decreasing threshold, but in the ∧-model it "goes to meet it."

In reference [2] it is shown that with a low frequency of stimulation, when the period of stimulation is longer than R_F, a stable condition (see Fig. 2A) is established during which all the impulses arise in the same stage of refractoriness. During the critical period of stimulation, equal to R_F, the omissions of impulses begin. The mechanism of emergence of the omissions is next (see Fig. 2B). The first stimulus after an omission causes a response with a low value of the threshold. The following stimulus arrives when the threshold is still high, and the stimulus must grow to a larger size than the first in order to cause an excitation, which arises with a delay in time. Because of this delay the third stimulus cannot cause an excitation with the same threshold value as did the second since the third stimulus arrives at an earlier stage of refractoriness. Therefore, the third stimulus causes triggering with a higher threshold value and with a still greater delay in time, and so forth until an omission occurs.

In reference [2] it is shown that for the rise of rhythm of the $n:(n - 1)$ type, the period of stimulation should be at least in the interval from R_F to $\frac{2}{3}R_F$ or still narrower for stimuli increasing sharply. The range in which the period must lie in order that cycles of n length arise equals

$$\Delta T = \theta \frac{(p - 1)^2 p^{n-2}}{(p^{n-2} - 1)(p^{n-1} - 1)} \tag{3}$$

where

$$p = 1 + \left| \frac{tg\ \alpha}{tg\ \beta} \right|.$$

The angles α and β are indicated in Fig. 1. An analysis of this expression shows that the smaller period is ΔT, and that the greater is n.

During stimulation of an excitable element by triangular impulses, the delay, that is, the time between the moment when the stimulus attains the value of the initial threshold and the response of the element, increases nonlinearly with the number of the impulse [2]:

$$\tau_n \sim 1 - \frac{1}{p^{n-1}}. \tag{4}$$

The output wave of impulses in this instance is not recurrent, and hence the spacing between adjacent impulses of the cycle gradually decreases.

One of the questions which arise in the examination of a period of the $n:(n - 1)$ type concerning the stability of the emerging type of transformation, that is, whether the pattern of the omission of the nth impulse will be regularly repeated. (This, generally speaking, is far from obvious, since the first cycle, for example, begins under different conditions than the following ones.) As was shown by Smolyaninov [38] for stimuli with an arbitrary form of the leading edge, the length of the cycles can change by no more than a unit; that is, if in the first cycle the nth impulse was skipped, then in the next cycle either the same nth or the $n - 1$th impulse will be skipped. Later on the order of cycles having the lengths n and $n - 1$ will be paired. It is interesting that in physiological experiments, periodic fluctuations of the length of the cycle by one impulse were observed. Thus, on the axons of the earthworm periodic alternation of cycles was observed 3:2, 2:1, 3:2, 2:1, etc. (Arshavskiy, Berkinblit, and Kovalev in unpublished data). Wilson [45] observed in the ganglion of a grasshopper a change of cycles 3:2, 3:2, 2:1, 3:2, 3:2, 2:1. An analogous pattern was observed during transformation of rhythm in the model studied on the digital electronic computer {8].

In summing up this examination, we will note the following points. (1) During stimulation with rectangular and triangular stimuli, the excitable element can operate in a frequency division mode of periodic omissions of impulses, and in a stable mode. (A mixed mode is also possible during frequency division when some stimuli remain without response because of the mechanism described above.) (2) A fixed mode sets in provided that $T < R_F$; during stimulation by rectangular stimuli it is established from the second stimulus, and during stimulation by triangular stimuli, it occurs after some transitional process. During the stable mode all the impulses arise in the same stage of refractoriness; the earlier it occurs the higher is the frequency of the stimulation. (3) The cycles $n:(n - 1)$ set in with $\frac{2}{3}R_F < T < R_F$ or in a still narrower range of the frequencies of stimulation. (4) The larger the number of impulses in the cycle, the narrower the range of changes of the period of stimulation during which a cycle of a given length can be obtained. (5) In the case of triangular stimuli, the delay in each cycle increases linearly, the output impulses follow periodically with the period R_F, and all the impulses arise in the same stage of refractoriness.

In the case of triangular stimuli the delays in the cycle increase nonlinearly and the output impulses arise nonperiodically, gradually converging, each of the following impulses of the cycle arising in an earlier stage of refractoriness.

2. THE ROLE OF TISSUE INHOMOGENEITY IN THE ORIGIN OF PERIODIC BLOCKING OF IMPULSES

The models examined above essentially describe the processes at the point of stimulation. In reference [13] is shown that in a homogeneous fiber all the dropouts of impulses can arise only at this point. We will examine the spread of successive impulses which arise at some point of the fiber at equal intervals and consequently during the same stage of refractoriness. The very first impulse will spread along the normal nonrefractory fiber at a high rate. The second impulse spreads along the refractory tissue. During refractoriness the spread rate of the impulses is lowered; therefore, the second impulse will fall behind the first and fall into a later stage of refractoriness on account of which its rate will increase. Thus, the second impulse goes on accelerating, encountering better conditions all the time. The second impulse spends more time for conducting than the first, therefore, at the end of the fiber the difference of the times of arrivals of these impulses can be much greater than the period of stimulation. The third impulse spreads after the second, which proceeded under worse conditions than the first; therefore, the rate of the third impulse turns out lower than the second. Moreover, the third impulse also goes on accelerating, falling all the time into later stages of refractoriness. The third impulse spends more time for spreading along the fiber than the second, but the difference of the times of arrival at the end of the fiber of the third and second impulses is less than that of the second and first. In references [13] it is shown that within limits all the impulses will spread along the fiber at a steady rate in the same stage of refractoriness and in precisely that one which they meet at the first point. In these conditions the difference of the times of arrival of the impulses at the end of the fiber becomes equal to the period of stimulation.

From this examination we may draw the following conclusions. (1) In a homogeneous fiber the impulses which pass through without dropouts the region adjoining the site of stimulation cannot fall out at other points since they do not encounter at any point a stage of refractoriness earlier than at the first point of the fiber. Dropouts not at the point of stimulation are possible only in tissues containing heterogeneity. (2) Impulses arising at the beginning of the fiber during the same stage of refractoriness arrive at the end of it in irregular intervals: the difference of times of the arrival of the second and first impulses is the greatest; after that, this difference decreases aiming at the period of stimulation.

We will examine now what sort of character of heterogeneity there must be in order for it to lead to a dropout of an impulse. Obviously, a dropout of an impulse can be caused by only that sort of heterogeneity which leads to a decrease of the safety factor, that is, a decrease of the ratio of the amplitude of the action potential to the threshold potential. We will examine, therefore, the heterogeneity of which paramenters of the excitable tissues may lead to the decrease of the factor of safety.

First of all, it is obvious that if there is a tissue consisting of two sections which differ in threshold, then at the junction of these sections an increase of period is possible. A single impulse may pass through a section of heterogeneity, but in the refractory period the threshold of the second section of the tissue rises still more and according to calculations some impulses may be blocked. In exactly the same manner the dropout of impulses is possible at the junction of sections differing in refractoriness. At some moment of time the threshold of stimulation in the section of fiber with a short refractoriness manages to decrease while the threshold of stimulation in the section with long refractoriness is still high (this question is examined in detail in [2] and [3]). The presence of heterogeneity in the amplitude of the action potential leads to the same result as did the differences in the refractoriness or threshold. The equivalence of heterogeneity in the refractoriness and amplitude of the action potential is clear from the fact that in the increase of a period the functional refractory period, which is greater the lower the amplitude of the action potential, plays a decisive role. Finally, the heterogeneity of the geometric structure, for example, fiber enlargement (see Chapters 2 and 3 for greater detail), may also lead to a decrease of the safety factor. Thus, in a heterogeneous tissue, consisting of two sections, the impulses can fall out, if in the second section the threshold is higher, the amplitude of the action potential is lower, the refractoriness is longer, or the diameter of the fibers forming the tissue is larger.

In physiological experiments a periodic blocking of impulses was observed near the stimulating electrodes (in a homogeneous fiber, in a strip of the myocardium) or in some heterogeneity of the medium (in the region of local cooling or local action of chemical substances, at the site of the widening of the fiber, for example of the passing of the axon into the body of the cell, on the border of the auricle and of the conducting system in the upper part of the atrioventricular node). We will examine some experiments of that sort.

3. THE PERIODIC BLOCKING OF IMPULSES IN SINGLE NERVE FIBERS

A vast amount of published material is devoted to the study of responses of excitable tissues to rhythmic stimulation, in particular, of the periodic transformation of rhythm of the $n:(n-1)$ type described in Latmanizova's work [26] and in the works of Kirzon and Chepurnov [10, 22].

In this section we will examine only the results of the experiments performed by Arashavskiy, Berkinblit, and Kovalev [2, 3] at the suggestion of Gelfand and Tsetlin for the verification of the theoretical ideas developed in section 1. The experiments were conducted on separate nerve fibers of the earthworm and frog. A section of fiber was cooled to 1 to 2° C; at the junction of this section with the warm section, dropouts were observed. The action potentials spreading along the warm section of fiber to the site of cooling were the activating stimuli. The warm end of the fiber was rhythmically stimulated.

In a series of experiments there was observed the periodic dropout of a single impulse beyond the region of cooling during a constant frequency of the impulses approaching the chilled section. Moreover, a latent period, that is, the interval from the moment of stimulation to the recording of the impulse in each cycle, gradually increased (Fig. 4A) which corresponds

Fig. 4. Examples of the periodic dropout of impulses obtained on the axon of an earthworm. (A) example of an individual cycle. The growth of a latent period in the cycle is apparent. (B) series of successive cycles. The number of impulses in the cycle gradually decreases. Note that there are only two impulse cycles, while the 4-impulse cycles continue for a prolonged time, indicating the greater stability of the short cycles. The top recording shows the absence of a dropout till a region of cooling. The time mark in (B) is 50 Hz.

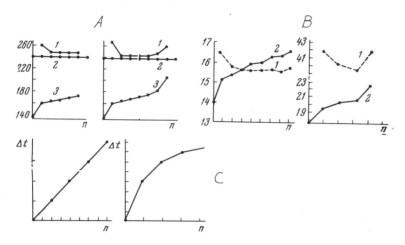

Fig. 5. Nature of the change of the AV delay and the latent period in the cycles. (A) growth of the AV delay [33]; (1) time between ventricular impulses; (2) time between auricular impulses; (3) atrioventricular delay. (B)'growth of the latent period (1); (1) time between successive impulses; (2) latent period. Along the axis of the abscissa in each instance is the impulse number; for description of (C) see text.

to the growth of the delay in the model described in point 1. The latent period fluctuated the most widely for the second impulse in the cycle. Direct measurements showed that the basic cause of increase of the latent period is the decrease of the rate of the impulses spreading along the refracting fiber. The maximum growth of the latent period for the second impulse is explained by the sharp decrease of its rate in comparison with the first impulse. In many experiments a more abrupt growth of the latent period was also observed for the final impulse in the cycle (see Fig. 5). We can explain this phenomenon thus. The final impulse before a dropout falls into a very early stage of refractoriness, when the rate is very sharply lowered; therefore, the latent period for it increases abruptly. This means that the different impulses of the cycle arise during different stages of refractoriness, a fact which corresponds to the model with triangular stimuli closer in form to the action potential.

In section 1 it was pointed out that the nature of the blocking is determined by the relation of the stimulation period and the functional refractory period. Therefore we may observe different types of blocking by changing either the frequency of stimulation or the duration of the refractory period. We know that during rhythmic stimulation the duration of the refractory period gradually increases, if it continues without interruption for a long enough time. During such prolonged stimulation with a frequency 20 to 40 times per second we could observe the following pattern. At first some steady regime was established when all the impulses had the same amplitude and the same latent period; then there began the periodic dropout of impulses during which the first cycles were rather long but were retained for a short period (see section 1, conclusion 4); after that the cycles became still shorter and lasted still longer (Fig. 4B). A period of the type 5:4, 4:3 and 3:2 was most characteristic in the experiments. Thus the experimental results obtained on the whole corresponded well with the theoretical ideas stated above.

The periodic dropout of individual impulses was also observed by Wilson [45] in the nerve cells of a grasshopper and a cockroach, and also in the motor axon of a grasshopper. In the explanation of the rising period Wilson used two models: one similar to the one examined above, but with an exponentially increasing leading edge of the impulse (Wilson used an electronic model instead of analytical examination for the verification of his ideas); and the other, in which the refractoriness does not remain unchanged but increases after each stimulus, which leads to a dropout of some impulses. The latter version is similar to Samojloff's model [35] for the heart (see below).

4. THE PERIODIC BLOCKING OF IMPULSES IN THE HEART (THE WENCKEBACH PERIOD)

4.1. CHARACTERISTIC FEATURES OF THE WENCKEBACH PERIOD

One of the types of a partial atrioventricular block is the so-called Wenckebach period, discovered by him in 1899 [43, 44]. Later on the Wenckebach period was described in a series of clinical papers and was repeatedly produced in experiments [6, 11, 19, 33–36].

The characteristic feature of this type of heart pathology is the recurrent dropout of ventricular systoles during the unchanged rhythm of auricular impulses. After each such omission the atrioventricular (henceforth the

AV) delay, that is, the passage of time of the impulse from the auricle to the ventricle, for successive heart impulses gradually increases until a new omission of the ventricular systole sets in.

The basic characteristics of this behavior of the heart were basically revealed under the conditions of a physiological experiment with artificial stimulation of the auricles.

1. First characteristic. An increase of the stimulation frequency of the auricles leads to the lengthening of the AV delay, which can last a long time, and not lead to a dropout of impulses. The higher the stimulation frequency, the longer is the delay. A critical frequency exists during which the Wenckebach period arises.

2. Second characteristic. The cycles arising during frequencies similar to the critical frequency have a greater length. With the further increase of the stimulation frequency the length of the cycles quickly decreases down to the omission of every second impulse. In addition, the range of stimulation frequencies during which all the possible instances of Wenckebach cycles are observed is very narrow. Thus, in Rosenblueth's experiments [33] Wenckebach cycles were observed during variation of a period of stimulation from 233 milliseconds (the critical frequency for the start of omissions) to 210 milliseconds (the omission of every second impulse begins).

3. Third characteristic. The longer the cycles, the less frequently are they encountered in the clinic and the less stable are they in the experiment; that is, very small changes of the period of stimulation are sufficient to change the number of impulses in a long cycle. The most common in the clinic and easiest of all to produce in experiments are cycles of the 3:2 or 4:3 type. However, the 6:5 cycles can also be observed for minutes. In a shorter time in the experiment even 20:19 cycles are observed.

4. Fourth characteristic. The increase of the AV delay is different for successive impulses of a cycle. The maximum lengthening of the delay is observed during passage of the second impulse, after which this growth becomes less and less for subsequent impulses. Sometimes this growth again appears greater for the final impulse of the cycle. In a series of cases the growth of the AV delay for all impulses, beginning with the third, proves to be the same (see Fig. 5). The reasons for the diverse character

growth of the AV delay in different experiments at the present time are not clear.

4.2. BASIC HYPOTHESES SUGGESTED FOR AN EXPLANATION OF THE WENCKEBACH PERIOD

Already the simple comparison of the characteristics of the Wenckebach period, enumerated at the end of the preceding section, with the characteristics of an exciting element, enumerated at the end of section 1.1, shows their exceedingly great similarity. This forces us to assume that any additional assumptions, except the ordinary attributes inherent to all excitable tissues, are not required for an explanation of the Wenckebach period. From the examinations carried out, it follows that in any tissue possessing a threshold and refractoriness, periodic omissions of impulses can be obtained. Model experiments on individual nerve fibers fully confirmed this hypothesis ([2] and section 3). Therefore, we assume that the mechanism of the Wenckebach period does not differ from the one described above. For brevity we will call this model the *model of temporary displacement* and explain the period by the fact that a stimulus, arriving during refractoriness, evokes a response with a lag on account of which the next stimulus falls into an earlier phase of refractoriness, which still more increases the delay of the response, etc.

In this model the changes of the threshold are absolutely the same after each impulse (Fig. 2). The model does not presume any changes of duration of refractoriness for the duration of the cycle. It is important to emphasize this in connection with misunderstandings which sometimes arise. Babskiy, Salmanovich, and Donskikh [6] write thus about the origin of the Wenckebach period: "the majority of authors connected it with the changes of the refractory period (Samojloff, Rosenblueth, Arshavskiy and others)." Actually, a change of the refractory period is assumed only in Samojloff's model [35]. Rosenblueth, who for the first time had proposed for an explanation of the Wenckebach period a model of temporary displacement and had examined a variation of it with rectangular impulses (R-model), does not use the change of refractoriness for an explanation of the dropouts; the same applies to Arshavskiy, Berkinblit, and Kovalev [2].

The authors of another group of hypotheses suggested for an explanation of the Wenckebach period assume that during the time of the cycle, the condition of the AV node changes, an indicator of this in particular being the growth of the AV delay. Wenckebach assumed that during cardiac disturbances or with a high frequency of cardiac rhythm the conducting system of the heart does not manage to rest during a diastole, and its fatigue gradually builds up to the impossibility of the conduction of excitation. During the omission of an impulse, the conducting system manages to rest and proves able for the conduction of impulses of the next. cycle. This hypothesis is attractive in its simplicity and has already survived nearly three-quarters of a century since Wenckebach's work [43]. Unfortunately, an experimental identification of the "law of the buildup of fatigue" is lacking, a circumstance which hampers a comparison of this hypothesis with factual material. On the other hand, the very facts of fatigue, the gradual lengthening of the refractory period, and the growth of the threshold during rhythmic stimulation, especially with a high frequency, are unquestionable.

Thus it is important to evaluate the relative role of these processes, to ascertain whether they are key factors in the rise of the Wenckebach period or whether they determine only certain accompanying phenomena. At the present time we can present only some considerations which, as it seems to us, indicate the secondary role of the fatigue process.

First of all, during long cycles, "fatigue" would have had to accumulate during for example twenty impulses, but rest would have had to take place during only one omission, which seems scarcely possible. The prolonged retention of the period, for example, of the 6:5 type or the systematically recurrent fluctuations in the length of the cycles (3:2, 2:1, 3:2, 2:1), also testify to the fact that for the extent of several cycles the condition of the tissue does not essentially change.

Ideas about the role of gradual fatigue are argued, in particular, with the gradual growth of the AV delay in the individual cycle. In this condition it is interesting to note that during an auricular extrasystole the early appearance of a systole always is accompanied by the growth of the AV delay during the conducting of this impulse. Moreover, the longer is the AV delay, the earlier the stage of diastole at which an extrasystole occurs. Obviously, in this case there is no indication of fatigue. The growth of the delay in these conditions can be explained only by refractoriness after the preceding impulse. The gradualness of the growth of the delay during the Wenckebach period is scarcely a sufficient reason to assume a different mechanism of its lengthening in this case.

The model of the Wenckebach period proposed by Samojloff [35] is one of the concrete models in which some process of accumulation is employed. Samojloff proposed that an impulse arising in the refractory phase leaves a longer "trail" after itself, that is to say, the condition of the tissue changes. The next impulse, arriving in an equal period T after the second, as the second after the first, finds the tissue in a condition with a higher threshold; from this it leaves a still longer "trail" after itself; etc. After each of the following impulses there remains a still longer trail with a higher threshold until an omission of the usual impulse sets in. Samojloff assumes that the duration of the refractory period after a given impulse depends on the phase of refractoriness from the preceding impulse in which it arose. It is possible that under actual conditions during the Wenckebach period the mechanism proposed by Samojloff plays a certain role since there are indications [19] that fibers of the upper part of the ÁV node possibly behave as Samojloff assumed. However, the Wenckebach period can be obtained also on a strip of the myocardium, in fibers of which, as is known, the impulse, having arisen during a refractory phase, leaves not a longer but on the contrary a shorter trail after itself.

Another explanation of the period also connected with the processes of accumulation was proposed by Babskiy, Salmanovich, and Donskikh [6]. The Wenckebach period was observed by them in a strip of the myocardium during threshold stimulation. A dropout of impulses occurred in the region of the stimulating electrodes. The authors assume that the excitability of the tissue gradually fell owing to the summation of positive afterpotentials which led to the gradual growth of the AV delay and the blocking of the impulse. What are the facts presented in support of this viewpoint?

In the first place, with registration of the potential in the region of stimulation, the authors observed a gradual lengthening of the latent period of response, a phenomenon which indicates, in their opinion, the fall of excitability. As was shown in section 1, during stimulation of any excitable element with refractoriness by impulses of a finite duration, the delay of the response to successive stimuli should increase (formulas 1 and 4) during a fixed duration of the refractory period.

In the second place, the time of spread of successive impulses of the cycle gradually increased, but the average rate for these impulses dropped. Such an effect should be observed for periodically rising impulses spreading along a refractory medium (section 2) and cannot indicate a decrease of excitability. The same effects are observed, for example, during calculation of the spread of impulses on the model of a fiber with refractoriness on the digital electronic computer when the processes of fatigue or summation are not taken into consideration.

In the third place, the Wenckebach period, which arose with a weak intensity of stimulation, disappeared with an increase of the intensity of the stimulus, although the frequency of stimulation did not change; but the rate of the conducting of impulses became even lower than during weak stimulation. As was pointed out in section 1, a condition for the appearance of periodic omission of impulses is that a period of stimulation becomes shorter than R_F. Meanwhile, the longer is the R_F, the shorter is the amplitude of the stimulus. Therefore, the frequency essential for the rise of the Wenckebach period should be different for a weak stimulus and for a strong stimulus. During that frequency in which a weak stimulus causes an increase of the period, a stronger stimulus should lead to the appearance of the stable behavior which was observed in experiments. The stimuli in the experiments of Babskiy, Salmanovich, and Donskikh were weaker than the natural stimuli of the myocardium. This results from the fact that during high intensity stimulation all the impulses spread along the fiber without dropouts with a lower velocity, but as a consequence, in a deeper stage of refractoriness. The absence of dropouts of the spreading impulses shows that the functional refractory period was relatively short for natural stimuli, and consequently their amplitude was higher than that of the stimulating ones.

Finally, the authors of the article note that they observed the Wenckebach period during frequencies of stimulation so low that the refractoriness of the tissue should by that time have ended. However, the indispensable condition for the appearance of the period was the lowered rate of the conduction of impulses and their smaller amplitude; consequently the conduction was realized while the threshold of the tissues was increased, which, however, managed to drop noticeably during an omission. It is doubtful that there are grounds to contrast such a resulting rise of the threshold with the refractory period.

There is only one experimental fact which can be immediately cited in defense of the hypothesis concerning the role of fatigue. During prolonged stimulation of a tissue with a certain frequency no omissions of impulses were observed, but then the omissions rose; with constant parameters of stimulation the length of the cycles appearing during a long stimulation gradually decreases (see Fig. 4). Babskiy, Salmanovich, and Donskikh observed such dynamics during stimulation of a strip of the myocardium; Arshavskiy, Berkinblit, and Kovalev [2] during stimulation of a single nerve fiber, finally managed to observe occasionally in the clinic the same sequence of events.

During attacks of auricular tachycardia all the impulses at first can pass to the ventricle; then the Wenckebach period sets in with cycles which progressively shorten, whereupon the ventricle begins to respond to every second or even only to every third auricular impulse. Surely it is impossible to explain such a change of pattern by anything except fatigue. However, the processes described above usually develop rather slowly in comparison with the time of the individual cycle, and they cannot by themselves serve as proof of the decisive role of fatigue in the periodic blocking of impulses. It is possible that these slow processes cause only a change of one type of cycle to another and not the dropout of an individual impulse at the end of the cycle and the lengthening of the delay in it.

Thus, at present there are no experimental facts which would indicate a significant or more predominant role of the process of fatigue in the origin of periodic blocking. This does not exclude, of course, the possibility that such facts will be obtained in the future.

Another explanation of the Wenckebach period is that it is determined by the interaction of two generators (the sinus and AV node) of a beating type. The model of that sort is proposed by Grant [15, see also 9]. A hypothesis of the existence of two generators with different periods makes possible an easy explanation of the periodicity of the omissions; the gradual change of the AV delay is explained in this model by the phase difference of the generators. This hypothesis can come in for criticism on many sides. We will note only one, from our point of view the weakest point of this hypothesis. It assumes that one generator, the sinus node, periodically sends impulses; and the second generator, the AV node, guarantees periodic fluctuations of the threshold of the fibers, but with a different period owing to which some one of the impulses arriving from the auricle is subliminal.

In this hypothesis it is assumed that the impulse passing from the auricle through the AV node does not have an influence on the rhythmic fluctuations of the threshold in the latter, that is, it does not disrupt the phase of its own node generator. After transmission of an auricular impulse a further change of the condition of the AV node should proceed, since if there is no auricular impulse, periodic omissions do not otherwise develop. Of course, one can devise mechanisms of functioning of the node which guarantee such characteristics. For example, one may propose that the threshold of the fibers of the AV node is regulated by nerve elements, themselves not subject to the effect of the passing impulses, and that maintain the phase.

However, at the present time there are no experimental reasons to adopt this or some analogous assumption. On the contrary, the Wenckebach period rises in the sort of preparation (a strip of myocardium, a nerve fiber) where it is impossible to assume the presence of a complex mechanism of that kind.

We will note, by the way, that the conditions supposed by this hypothesis result, apparently, during an extremely interesting disturbance of rhythm discovered by Babskiy and Ulyaninskiy [7]. These investigators, stimulating the heart during respiratory arrhythmia, observed a periodic blocking of impulses. The auricular impulses moreover followed rhythmically with a frequency determined by the stimulator, and the condition of the AV node also changed periodically under the influence of the vagus, but with the frequency of respiration. During such behavior of the heart, they sometimes managed to see a dropout of not one but two or three impulses in succession.

In section 1, two variations of the model of temporary displacement were examined—with rectangular stimuli (the R-model) and with triangular (the ∧-model). Both these models well explain not only the fact itself of periodic blocking of the impulses but also a series of more minute details of the period: the presence of a critical frequency during which the period begins; the dependence of the character of dropouts of the omissions on the frequency; the greater stability of short cycles; etc. Strictly speaking, the basic difference of these models is that in one of them the plateau of the stimulus plays the chief role and in the other, its leading edge.

This difference leads to certain peculiarities in the explanation of the details of the period. Therefore, the question naturally arises of which of the variations of the model more precisely corresponds to one or another preparation. For example, is it not possible to assume that the variation with rectangular stimuli is better applied to the heart, the action potential of which has a plateau, and the variation with triangular stimuli to a nerve fiber?

Unfortunately, the majority of the experiments which can be offered for an answer to this question are methodically very complicated. Thus, for example, different variations of the model predict a rather varied character of change of the range of frequencies in which cycles of different length should be observed. However, for verification of these predictions there is required a relationship of the duration of the period of stimulation and the functional refractory period maintained with a sort of exactness which exceeds the resources of the experimenter, if we are to take into consideration the processes of fatigue and the change of the condition of the preparation.

We would like, however, to cite several facts and considerations which quickly speak in favor of the model with a triangular form of stimulus for the blocking of impulses in the heart. First of all, we will examine in somewhat greater detail the question about the formation of the delay of successive impulses of the cycle.

As was pointed out in section 4.1, characteristic 4, the AV delay sharply increases for the second impulse of the cycle, for subsequent impulses its growth is either constant or gradually decreases, and in a series of cases this growth sharply increases again for the final impulses of the cycle (Fig. 5).

At the same time the R-model predicts a linear growth of the delay for all impulses of the cycle (see Fig. 5C, left) while the ∧-model predicts in accordance with the facts the maximum growth of the delay for the second impulse (Fig. 5, right). In order to explain the abrupt growth of the delay for the second impulse of the cycle, Rosenblueth proposed that, at the same time, the dependence of the refractory period of the cardiac tissue on the frequency of the succession of impulses plays an essential role. (We stress that Rosenblueth required this dependence of the duration of the refractory period on the frequency not for an explanation of the period, but for explanation of one of its details. Moreover, the hypothesis proposed by Rosenblueth by itself hampers an explanation of the period since it assumes that with the growth of the frequency the refractory period decreases.)

Rosenblueth assumes that the first impulse after an omission leaves after itself a longer refractoriness since the effective frequency for it is lower and therefore the delay for the second impulse increases very abruptly. However, the growth of the latent period in separate nerve fibers where there is no dependence of the refractory period on the frequency bears exactly the same character as in experiments on the heart (Fig. 5). Therefore, the growth of the delay for the second impulse apparently has a different cause than that supposed by Rosenblueth.

The atrioventricular delay consists of two components: the time of the delay at the site of stimulation, that is, at the upper border of the AV node, and the time of conduction along the more distantly situated tissue in a state of refractoriness. At the site of stimulation, according to the R-model, the delay should increase linearly, in like manner for all the impulses of the cycle. All the impulses should rise at the same stage of refractoriness. For such impulses [see 3] the maximum increase of time of the conduction will be for the second impulse, and then it should decrease. Thus the overall delay for successive impulses of the cycle should increase more slowly than according to a linear law; therefore, Rosenblueth's experimental data concerning the linearity of the growth of the delay, generally speaking, still does not indicate the accuracy of the R-model.

According to the ∧-model, successive impulses arise in different stages of refractoriness. The growth of the delay along this model is not linear, it should decrease for successive impulses of the cycle. In addition to that, the time of conduction for successive impulses of the cycle can increase to a different degree, determined by the nature of the dependence of the rate of conducting on the refractoriness. If the rate of conducting depends slightly on the phase of refractoriness, then the delay should increase more slowly than according to a linear law; with a stronger dependence, the lengthening of the time of conducting can compensate for the slow growth of the delay at the site of blocking and we will obtain an approximately linear overall growth of the delay; finally, if the rate drops very sharply during the early phases of refractoriness, then this can explain the often observed vigorous growth of the delay of final impulses of the cycle. It is not possible apparently to explain the last fact from the viewpoint of the R-model.

Thus, the nature of the change of the AV delay, observed in experiments, more quickly testifies in favor of the ∧-model.

If we record the action potential in that region where the blocking sets in, then in the R-model all the impulses should arise at the same stage of refractoriness, while in the ∧-model the impulses should all arise in earlier stages of refractoriness. Intracellular recording from the fibers of the AV node shows that successively arising impulses of one cycle have all a lower amplitude which indicates their rise during early stages of refractoriness [19] and speaks in favor of the ∧-model. However, with the interpretation of this result certain caution is also needed; if the site of recording is in fact a certain distance from the site of blocking, then the successive impulses can arrive at the recording site in different stages of refractoriness even if they rose in the same stage [see 3 and 13].

We think that the results of the model experiments on separate nerve fibers [2, 3] also testifies in favor of the ∧-model. The period obtained in these experiments was very similar to the Wenckebach period. In Fig. 5 there is presented data on the change of the AV delay from Rosenblueth's paper [33] and data on the change of the latent period from reference [2]. A very great similarity of results is apparent. Thus, in nerve fibers, where there are no singularities of structure inherent to the heart, a specific form of the action potential, and a dependence of the refractory period on the frequency (of the type which belongs to the heart), the Wenckebach period in fact arises with all its characteristic features. Probably this shows that the leading edge of the impulse plays a decisive role in the origin of a period of that type and not its plateau which in nerve impulses is absent.

Apparently, the lengthening of the time of conduction along the refractory tissue constitutes the basic part of the increase of the delay. In the model experiments on separate nerve fibers [2, 3] this was shown by direct measurements. An analysis of the formation of the delay allows us to understand one of the variations of the AV blockade during which a periodic dropout of individual impulses is observed, but without a lengthening of the AV delay. Such a regime should rise in the small dimensions of the region where the refracting period is lengthened. At that time in this region the Wenckebach period will rise, but the impulse will spread along the remaining tissue at the usual rate, and therefore the increase of the AV delay in this instance is imperceptible. Clinical material [36], apparently, tallies with such a viewpoint. Actually, with this type of period a complete blockade of conducting often rises, which clinic physicians connect with an organic disease of some part of the conducting system. This indicates the presence of a sharp heterogeneity of the properties of the tissue. At the same time a relatively small duration of the AV delay with this type of blockade shows that the dimensions of this region of heterogeneity are small.

4.3. THE ATRIOVENTRICULAR DELAY AND THE BLOCKING OF IMPULSES

In electrophysiological research it is shown that the dropout of impulses arises in the upper part of the AV node. This is to be expected since in precisely this region there is discovered a longer refractoriness and a small amplitude of the action potential [19]—just those characteristics which can lead to a blocking.

Generally speaking, the reduction of the safety factor in this region of the heart is to be already expected from the facts ascertained in the study of the mechanism of the AV delay. It is shown that the AV delay is connected with an extremely low rate for the distance of a relatively short path

Such a low rate (several centimeters per second) can be explained either by the very small diameter of the fibers of this region of the heart, or by the small safety factor. It is difficult to assume that such low rates of conducting are ensured only by the small diameter of the fibers, since for this result extremely fine fibers would be required with a very large ratio of surface area to volume. This would seriously hamper maintaining a stability of the inner medium of such cells for a stretch of uninterrupted long rhythmic work. We would have to assume that the fibers of this region of the heart possesses a membrane with particularly high-grade ionic pumps.

Obviously, it is more natural to expect that the low rate of conduction is assured by some deterioration of the properties of the membrane with a not too small fiber diameter. As we have already said, in fibers of the upper portion of the AV node there has actually been discovered a lower safety factor than in fibers of other regions of the heart. It is quite probable that the reduction of the safety factor in this region of the heart is determined not only by the properties of the membrane of its cells but also by special features of geometry of the branching of the fibers of this region. This question is examined in greater detail in Chapter 3. Here we would like to emphasize only that the guarantee of the decrease of the rate of conducting by means of geometric factors is also inevitably connected with the reduction of the safety factor.

For solution of the problem of hemodynamics a temporary succession of systoles of the auricles and ventricles is important, but in the realization of that sort of succession the AV delay plays a significant role. Such a delay is assured by reduction of the rate of the conducting at the junction of the auricle and ventricle, which is connected with the reduction of the safety factor for the spreading excitation. At the same time the heterogeneity of the excitable tissue, according to the safety factor, inevitably leads during deterioration of conditions in this region or during an increased frequency of succession of impulses, to the blocking of these impulses because of the mechanisms described above. Thus, the possibility of a dropout of impulses at the junction of the ventricle and auricle, the possibility of the appearance of blockades is the distinctive price for the presence of the atrioventricular delay.

5. SOME MORE COMPLEX MODELS

We examined the simplest models of excitable elements having a threshold and refractoriness, and some physiological data which can basically be explained by these simplest models. However, in the investigation of real biological preparations the experimenters encounter other phenomena for the explanation of which a slight modification of the models described above is sometimes sufficient, but sometimes models of an essentially different type are required.

We will examine one example. In Kitamura's work [23] carried out on a single node of Ranvier of the fiber of a toad, the responses of the node to rhythmic stimuli were studied. With the use of subliminal stimuli and rather high frequencies of stimulation, the node began to respond to every third, every fifth, or even every eighth stimulus. A node which does not respond to individual stimuli of a given strength begins to respond to every nth stimulus, because of summation. Outwardly this behavior is similar to frequency division, but in its mechanism it is close to a periodic blocking, when the successive impulses find a fiber with an increasingly higher threshold; but in this case there is a progressively smaller difference between the potential shift on the membrane of the node and its threshold. The pattern arising is, so to say, a mirror image of a period of the type $n:(n-1)$, and it can be examined in an analogous manner. It is necessary only that the excitable element, beside a threshold and refractoriness, should possess the capability of summation.

Another result by Kitamura [23] is that with high frequencies of stimulation the node responds only once. This result emphasizes the limitations of too simple formal models.

5.1. THE BEHAVIOR OF ELECTRONIC MODELS OF EXCITABLE ELEMENTS DURING RHYTHMIC STIMULATION

The properties of elements approximating in their complexity real preparations were studied in corresponding electronic models and with the use of programs for the digital electronic computer. The great complexity of these models hampers an interpretation of the results obtained with their help. However, in such models the experimenters can theoretically change at their own pleasure one or another of the parameters of the model, making clear thereby the roles of these parameters in the observed pattern. Moreover, in such experiments the experimenter can be sure that the properties of the model did not change during the experiment. All this makes the study of such models in certain instances easier than a study of real preparations.

An electronic model having a threshold, refractoriness, accommodation, summation capability, and rhythmic responses was studied by Harmon [17]. During rhythmic stimulation Harmon observed the frequency division and the rise of a period of the $n:(n - 1)$ type. The qualitative explanation of a period of this type offered by Harmon and given in radio engineering terms by him essentially agrees with what is presented above: a growth of the delay leading to an omission of an impulse. Harmon observed behaviors of the type $5:3, 16:5, 19:30$ etc. He explained such modes in the general form as a combination of the effect of refractoriness and the capability of summation. It is wholly possible that at least some of these periods are explained by the fact that the model was working in a frequency division mode responding only to every kth stimulus, and besides that, due to the mechanism described in the first section, not responding from these stimuli to the nth stimulus.

Using a program for the digital electronic computer which stimulated the excitation of a membrane with the properties set forth by the Hodgkin and Huxley equations [18], Jenik [21] studied rhythmic processes under the effect of triangular stimuli of current with a duration of 0.2 millisecond. He also studied the transformation of rhythms in an electronic model imitating the behavior of the membrane of a motor neuron (among them after hyperpolarization, hyperexcitable phase, etc.). The results of these works were presented in the form of the diagrams proposed by Kupfmuller and Jenik [25]. On these diagrams, along the axes of the coordinates were plotted the strength and frequency of the stimulus, and then on this plane were drawn lines which separate the regions corresponding to the constant ratio of the mean output frequency to the input frequency. Thus, for example, in the region of low frequencies of stimulation, when summation and refractoriness are of no importance, there will be only one line separating the regions with the ratio of the frequencies equal to zero (the subliminal strength of the stimulus) and those equal to one (the strength of the impulse is higher than the threshold and the model responds to every stimulus). In the model there were obtained a division of frequencies, periods of the $n:(n-1)$ type, and also the rise of responses because of the summation of subliminal stimuli.

We would like to consider here in somewhat greater detail the possibility of excitation of fibers and tissues by frequent subliminal stimuli. It is well known that in numerous instances the stimulation of the same biological object by stimuli of the same amplitude, but of different frequency, leads to completely different results. So, for example, the stimulation of the cortex of the cerebellum with a frequency of 300 Hz leads to an inhibiting of the activity of some cells of a netlike structure, and stimulation with a frequency of 20 Hz, on the other hand, increases their activity [28]. The localization charts for the motor cortex, plotted along different frequencies of stimulation, also differ noticeably. It is entirely possible that one of the causes of these phenomenon is the drawing into activity with high frequencies that sort of neuron or fiber for which the stimuli are subliminal and, on the other hand, the inactivation of elements on which strong and frequent stimuli are acting. Thus, for example, Lewis [27], who also studied the properties of an electronic model based on the Hodgkin and Huxley equations, obtained during high frequency of stimulation only one or several responses after which the model ceased to respond to stimuli —an effect similar to Wedensky inhibition. As pointed out above, Kitamura observed an analogous effect on the node of Ranvier [23]. Apparently, the stimulus arriving during a refractory period is not indifferent for an excitable object. Although it does not cause a response, it can prolong the duration of the refractory period, for example by inactivating the sodium carriers in the membrane. Then the next impulse also does not cause a response, but in its turn lengthens the refractoriness, etc. With certain parameters of the stimulus—high frequency, great strength—and with a certain condition of the membrane, this behavior can set in. For example, a long preliminary stimulation is conducive to such behavior.

5.2. MODELS WITH SEVERAL INPUTS, WITH ACTIVE ELEMENTS AND CHAINS OF EXCITABLE ELEMENTS

Elaborations of the models examined above develop in several directions. We will examine several examples.

1. Models with several inputs. The behavior of an element with two

stimulating inputs was examined by Jenik [21], The element possessed summation capacity and triggered only on that occasion when the stimuli along both inputs were arranged in time no greater than a certain interval. In this instance the inflow along both inputs of periodic successive impulses of different frequencies results in the effects of a beating type. Depending on the parameters of the system, namely, the time of summation, the difference of the frequencies of the input waves, and the amplitude of the stimuli (a larger stimulus leaves a longer trace), the element either functioned with an alternating frequency or produced at the output groups of several impulses or separate impulses separated by long intervals.

The behavior of models with several inputs, and instances when an output signal of the same element was fed into one of the inputs with a certain delay, were examined in the work of Ganzen and Granovskaya [12].

2. An interesting direction of research on excitable elements was outlined by Perkel et al. [30], who examined a spontaneously active element, operating periodically, and the effect on its functioning of both periodic excitatory and inhibitory stimuli. Moreover, the excitatory stimuli are subliminal and only shorten the time of the element's own triggering, while the inhibitory stimuli prolong it. A corresponding program was written for the digital electronic computer in which the behavior of the element was studied. It was shown that with the action of the input stimuli after a transition process, a new steady working frequency of the element is established. The authors emphasize that the presence of refractoriness can lead to paradoxical effects. For example, the increase of the frequency of the inhibitory input stimuli can in this way displace the natural working frequency of the element, that a part of the inhibitory stimuli begins to fall into the refractory phase. In this instance the output frequency of the element increases because of the increase of frequency of the inhibitory stimuli. The conclusions obtained as a result of the calculations were checked by the authors in experiments conducted on the neurons of Aplysia and the receptor neuron of a crayfish. A good confirmation of the theory developed was obtained.

3. Chains of excitable elements. With a combination of several elements there can arise very complex results. However, these problems more

directly concern the theory of nerve nets and we will not examine them here with any amount of detail. The only question which we will examine here is the properties of several successively joined elements, the *chains*. Moreover, we will assume that the output impulse of each element has the same amplitude and duration as the input stimulus.

In Smolyaninov's work [37] it was shown that the simplest system of two, differing in the refractoriness of the elements, can produce at the output a rather complex nonmonotonic change of frequency during a monotonic increase of the input frequency. The resulting effects are similar to the paradoxical behavior described above for an active element. If the chain is made up of m elements, differing in refractoriness, then a rhythm of the $n:(n - m)$ type can be obtained in it.

For simplicity we will examine a chain of elements producing rectangular pulses at the output. We will recall that in this instance each element gives out impulses with a period equal to its R_F and the number of the responses of the element until an omission is equal to the width of the input impulse divided into the difference of the input and output frequency, that is, in a given instance, into the difference of the functional refractory periods of the two neighboring elements. Obviously, in the last two elements of such a chain it is possible to select the sort of gradient of refractoriness so that the last element would not respond to a $n - m$th impulse, the gradient of refractoriness between $m - 1$ and $m - 2$ elements must be smaller so that they do not respond to the following impulses, etc. The first element should not respond to the nth stimulus.

Thus in a chain it is possible to obtain a group of impulses separated by a group of omissions. For this during a given input frequency there should be a fixed gradient of refractoriness along the length of the fiber. As shown above, in a chain of elements with rectangular impulses the refractoriness should not increase linearly; the gradient should be greater at the end of the chain. Moreover, each of the following impulses fall out at a different place on the chain and the rate of the dropout shifts toward the passing impulses. The question of the behavior of chains of excitable elements with an arbitrary form of impulses was examined in detail by Smolyaninov [39].

An extreme case of a chain of excitable elements is a continuous fiber with gradually changing properties. The study of the spread of impulses in a fiber with a gradient of refractoriness was conducted by Arshavskiy, Berkinblit, Kovalev, and Chaylakhian [4] on the giant axon of the earth-worm. The gradient was established by nonuniform cooling of the fiber. In such a fiber there was observed the periodic recurrence of cycles consisting of groups of impulses separated by a group of omissions. Multi-channel recording showed that the first skipped impulses fall out in colder sections of the fiber, and the subsequent ones in the warmer sections. In some cases, more complex modes arose which are combinations of group omissions and periodic omissions of separate impulses [5]. Recently Paintal [29] proposed that an analogous mechanism is responsible for the origin of a block of a group of impulses in fibers of the vagus during the local cooling of it.

The range of phenomena examined are closely related and in several cases they imperceptibly blend with a whole series of other biological processes. Thus, the behavior analyzed of the simplest excitable elements are a particular case of the processing of the impulse by neurons. The citing of the phenomena of summation, inhibitory inputs, and the combination of several elements greatly expand the sphere of possible periodic behavior [see, for example, 31 and 45], and lead to the examination of the simplest nerve networks. On the other hand, the study of chains of excitable elements directly borders upon the examination of the spread of impulses in continuous linear excitable media [1, 8, 13, 14, 41, 42]. Likewise the transition to an examination of the periodic blocking in two-dimensional excitable media is also natural and also the examination of the role of such blocking in the origin of behaviors analogous to flutter and fibrillation of the myocardium. Such an approach was used in the work of S. S. Balakhovskiy (*Biophysics*, 1965:10, 1063) and V. I. Krinskiy [24].

REFERENCES

1.
Yu. I. Arshavskiy, M. B. Berkinblit, and V. L. Dunin-Barkovskiy, *Biofizika* 1965:10, 1048.
2.
————, M. B. Berkinblit, and S. A. Kovalev, *Biofizika* 1962:7, 449.
3.
————, *Biofizika* 1962:7, 619.
4.
————, M. B. Berkinblit, S. A. Kovalev, and L. M. Chaylakhyan, *Biofizika* 1964:9, 365.
5.
————, *Biofizika* 1964:9, 634.
6.
Ye. B. Babskiy, V. S. Salmanoivh, and Ye. A. Donskikh, *Dokl. AN, SSSR* 1963:153, 966.
7.
———— and L. S. Ulyaninskiy, *Dokl. AN, SSSR* 1960:133, 716.
8.
M. B. Berkinblit, S. V. Fomin, and A. V. Kholopov, *Biofizika* 1966:11, 329.
9.
A. Bethe, *Pflüg. Arch. Ges. Physiol.* 1951:254, 1.
10.
S. A. Chepurnov, Mestnaya i rasprostranyayushchayasya elektricheskaya reaktsiya v odinochnykh nervnykh voloknakh pri formirovanii ritmicheskogo otveta (Local and Spreading Electrical Responses in Individual Nerve Fibers During Formation of a Rhythmic Response), doctoral dissertation, Moscow, 1963.
11.
I. A. Chernogorov, *Narusheniya ritma serdtsa* (*Disturbance of the Cardiac Rhythm*), Moscow, 1962.
12.
V. A. Ganzen and R. M. Granovskaya, *Vestnik LGU* 1965: no. 3 issue (vyp) 1, 142.
13.
I. M. Gelfand and M. L. Tsetlin, *Dokl. AN, SSSR* 1960:131, 1242.
14.
S. I. Gelfand and D. A. Kazhdan, *Dokl. AN, SSSR* 1961:141, 527.

15.
R. P. Grant, *Amer. J. Med.* 1956:20, 334.
16.
S. Hagiwara and H. Morita, *J. Neurophysiol.* 1963:26, 551.
17.
L. D. Harmon, *Kybernetik* 1961:1, 89.
18.
A. L. Hodgkin and A. F. Huxley, *J. Physiol.* 1952:117, 500.
19.
B. F. Hoffman and P. F. Cranefield, *Electrophysiology of the Heart*, New York, McGraw-Hill, 1960.
20.
A. F. Ivanov and V. R. Telesnin, *Izvestiya vysshikh uchebnykh zavedeniy, Radiofizika* (*News of the Higher Educational Institutions, Radiophysics*) 1959:2, no. 1.
21.
F. Jenik, in *Neural Theory and Modeling*, ed. R. F. Reiss, Stanford: Stanford University Press, 1964.
22.
M. V. Kirzon and S. A. Chepurnov, Symposium *Fiziko-Khimicheskie osnovy proiskhozhdeniya biopotentsialov i ikh rol' v funktsiyakh* (*Physico-chemical Foundations of the Origin of Biopotentials and Their Role in the Functions*), Moscow, 1964.
23.
S. Kitamura, *Jap. J. Physiol.* 1961:11, N4.
24.
V. I. Krinskiy, *Biofizika* 1966:11, no. 5.
25.
K. Kupfmuller and F. Jenik, *Kybernetik* 1961:1, 1.
26.
L. V. Latimanizova, *Zakonomernosti vvedenskogo v elektricheskoy aktivnosti vozbudimykh edinits* (*The Wedensky Principles in the Electrical Activity of Excitable Units*), Leningrad State University, 1949.
27.
E. R. Lewis, in *Neural Theory and Modeling*, ed. R. F. Reiss, Stanford: Stanford University Press, 1964.
28.
E. Manni, G. B. Azzena, and R. S. Dow, *Exp. Neurol.* 1965:13, 252.
29.
J. Paintal, *J. Physiol.* 1965:180, 1.

30.
D. H. Perkel, J. H. Schulman, T. H. Bullock, G. P. Moore, and J. P. Secundo, *Science* 1964:145, no. 627, 61.

31.
R. F. Reiss, in *Neural Theory and Modeling*, Stanford: Stanford University Press, 1964.

32.
A. Rosenblueth, *Amer. J. Physiol.* 1958:194, 171.

33.
————, *Amer. J. Physiol.* 1958:194, 491.

34.
V. S. Salmanovich and M. G. Udelnov, sb *Voprosi patologii i fiziologii serdtsa* (*Collection Problems of the Pathology and Physiology of the Heart*), Moscow: Medgiz, 1955.

35.
A. F. Samojloff, *Pflüg. Arch. Ges. Physiol.* 1929:222, 516.

36.
A. M. Sigal, *Ritmy serdechnoy deyatel'nosti i ikh narusheniya* (*The Rhythms of Cardiac Activity and Their Disturbances*), Moscow, 1958.

37.
V. V. Smolyaninov, *Biofizika* 1964:9, 639.

38.
————, *Biofizika* 1966a:11, 337.

39.
————, *Biofizika* 1966b, in print.

40.
V. R. Telesnin, *Izvestiya vysshikh uchebnykh zavedeniy, Radiofizika* 1963:6, 624.

41.
————, *Izvestiya vysshikh uchebnykh zavedeniy, Radiofizika* 1965:8, 169.

42.
K. F. Teodorchik, *Avtokolebatel'nye sistemy* (*Self-oscillating Systems*). Gostekhizdat, 1952.

43.
K. F. Wenckebach, *Z. Klin. Med.* 1899:36. 181.

44.
————, *Z. Klin. Med.* 1899:37, 475.

45.
D. M. Wilson, *J. Exp. Biol.* 1964:41, 191.

II
The Organization of
Certain Parts of the
Central Nervous System
and the Modeling of
Their Operation

6

Characteristics of the   I. A. Keder-Stepanova
Respiratory Neurons of Different
Levels of the Central Nervous
System

Physiologists have at their disposal vast material concerning the adaptive responses of respiration. There is a great deal of data about the role of different structures of the central nervous system in the development of the adaptive respiratory responses. However, without a knowledge of the properties and functions of their individual elements, it is impossible to approach an understanding of the operation of these structures. But there is a shortage of such knowledge, since obviously each element's contribution in the work of the system is infinitesimal. All the diversity and subtleties of behavior of the control system are determined by the specific character of the intrasystem and extrasystem connections of its elements. Precisely this specific character and the vastness of the connections determine the functional importance of the different structures in the operation of the whole system controlling respiration.

There is now at the disposal of physiologists only one clear criterion for the assigning of one or another neuron to the system controlling respiration: function with rhythmic volleys synchronous either with the inhalation phase (the inspiratory neurons) or with the expiration phase (the expiratory neurons). It is customary to call such neurons respiratory. The existing definition of respiratory neurons undoubtedly narrows the study of nerve structures involved in the regulation of respiration; even now we know neurons that do not work in phase with respiration, but that have a direct relation to respiration regulation. However, just such a definition facilitates an approach to the study of the localization and function of the central structures controlling respiration.

The method of recording volleys of action potentials synchronous with phases of respiration make it possible to detect respiratory neurons on the level of the medulla oblongata, the pons, and also the cervical and thoracic regions, of the spinal cord. We now have available a range of data concerning the localization and the nature of activity of these neurons, and their behavior in different experimental conditions. But there still remain unknown those functional properties of the neurons and the nature of their interrelationship which determine, in the first place, the rise of rhythmic volleys in the system of the respiratory neurons of the medulla oblongata and, in the second place, the coordination of the neurons of all levels of the system ensuring the necessary volume of ventilation at a given moment.

Later on, experimental data will be presented on the localization, the nature of activity, and the functional properties of respiratory neurons, and existing concepts about the reasons for the appearance of rhythmic volleys and the interrelations of respiratory neurons of different levels.

1. THE LOCALIZATION OF RESPIRATORY NEURONS OF THE MEDULLA OBLONGATA

With the localization of the respiratory neurons in the medulla oblongata, physiologists encountered a surprising fact which has been emphasized only in a few works of recent years. By means of localized recording of the action potentials, respiratory neurons were found in the lateral regions of the reticular formation of the medulla oblongata, and not in the medial portion of this region of the brain which had been identified as the region of the respiratory center.

We will briefly recall the history of the question of the localization of the respiratory center. In 1885 N. A. Mislavskiy [90] summarized the results of earlier works of other authors and his own data of experiments with partial and complete transsections of the brain stem and definitively located the general boundaries of the respiratory center, that is, the region without which respiration achievement is impossible. Rhythmic respiration is maintained if the cellular groups located in the so-called "intermediate bundle" from the level of the roots of the hypoglossal nerve to the inferior angle of the calamus scriptorius remain unaffected. In subsequent, more subtle experimental conditions with transsections of the brain stem, it was shown that a rhythmic succession of inspiration and expiration is possible with the preservation of the lower two-thirds of the medulla oblongata, even with the absence of the cervical vagi [22–24, 57–59, 73, 81, 82, 101, 143].

The use of the method of localized stimulation with depth needle electrodes made it possible still more exactly to limit the zone of the respiratory center. Pitts and other authors [84, 106, 107, 109–111] produced the first charts of the distribution of the inspiratory and expiratory divisions of the respiratory center. Membership in one of these divisions was determined by what type of maximum response of respiration (that is, whether respiration ceased in inhalation or in expiration) the stimulation of a given point of the medulla oblongata caused. The authors especially demonstrated that during stimulation the neurons were involved, and not the conducting pathways. The neurons of the respiratory center, according to Pitts, occupy the caudal part of the floor of the IV ventricle between the auditory tubercles and a line passing approximately 1 to 2 mm below the level of the obex. They were found among the cells of the reticular formation on the two sides from the median line to the restiform bodies.

Pitts maintained that a strict segregation exists of the two types of neurons. The neurons causing inspiratory reactions are located in the ventromedial and caudal, and the neurons the stimulation of which leads to expiratory reactions, in the dorsomedial and more rostral parts of the reticular formation. Histological indentification of the structures stimulated and matching of pictures obtained with charts of the reticular formation of the medulla oblongata made it possible to establish that the first type of these neurons is found in the large part of the nucleus giganto cellularis (the ventral part) and in those regions of the ventral and medial part of the lateral nuclei from which the reticulospinal tract begins. The second type of neuron is found in the dorsal and rostral sections of the nucleus giganto cellularis [25, 139, 140]. It should be noted that after destruction of the anterior columns and the anterior part of the lateral columns at the C_1 level, Pitts [105] observed retrograde degeneration mainly in the ventral part of the n. reticularis inferior, that is, where he localized the inspiratory neurons by the method of stimulation. The results of other authors using stimulation for localization of the respiratory center differ from the Pitts data only in regard to the relative distribution of the inspiratory and expiratory divisions; the general boundaries of the center coincide [2–5, 26, 78, 103, 141, 147].

Thus, they localize the respiratory center in the medial part of the reticular formation of the medulla oblongata. However, in a large part of this region no one succeeded in detecting neurons which would function with rhythmic volleys synchronous with phases of respiration (i.e., respiratory neurons). In the medial region of the medulla oblongata the cells function continuously or by volleys, but their activity is not connected in a visible way with the respiratory cycle [9, 10, 80, 83, 123].

The respiratory neurons are found among the cells of different nuclei of the reticular formation in the lateral regions of the medulla oblongata approximately 2.5 to 5 mm from the median line, 3 mm above and 3 mm below the level of the obex. Some authors located inspiratory neurons mainly in the ventrolateral and rostral parts, and the expiratory in the dorsolateral and caudal parts [2, 11–13, 30, 55, 96, 97, 130, 132]. Other authors maintain that in the lateral regions the inspiratory and expiratory neurons are interspersed at random [44, 45, 48, 61, 120]. However, almost everyone agrees that the respiratory neurons do not form a nucleus and cannot be morphologically distinguished from the cells of the reticular formation. True, Batsel [9] recently reported that he discovered two masses of respiratory neurons. One of these masses, located near the nucleus ambiguus, the motor nucleus of the nucleus glossopharyngeus, and the nucleus vagus, included inspiratory and expiratory neurons. The second mass, containing only inspiratory neurons, is localized more ventrolaterally of the solitary tract and somewhat more rostrally of the obex. The author of this chapter thinks that such a mass of respiratory neurons is justified from the viewpoint of the evolution of the structures controlling respiration. It is known that these nuclei play a leading role in the regulation of respiration of, for example, fish.

Stimulation of the lateral regions of the medulla oblongata in which the respiratory neurons were discovered do not produce clearly pronounced and stable changes of respiration.

Oberholzer and Tofani [102], considering the question of the variability of the respiratory zones localized by the two methods (stimulation and removal), suggested that the respiratory neurons of the lateral region more probably belong to the afferent structures of the respiratory muscles than to the actual respiratory center. Meanwhile, still more data accumulate proving the inaccuracy of such a hypothesis. It has been demonstrated that the rise of rhythmic volleys in the respiratory neurons of the medulla oblongata is maintained after section of the vagi, section of the spinal cord at C_1, and finally after the curarization of the animal and the transfer to diffuse respiration, that is, after complete removal of all respiratory afferentation [10, 11, 14, 36, 55, 62, 96, 119]. Moreover, there are data that the neurons of the medial region, unlike the respiratory neurons of the lateral zone, do not react in a specific way to those factors so important for the volume determination of ventilation, as for example expansion of the lungs and the change of the gas composition of inhaled air [10, 68].

On the basis of the material presented we can assume that it is impossible to limit localization of the respiratory center to only the medial region of the medulla oblongata. Probably the respiratory center consists of two zones: lateral and medial. In the lateral zone there are situated respiratory neurons in the system of which rhythmic respiratory volleys originate, determining the rhythm of replacement of inhalation with expiration. The functional role of the neurons of the medial zone remains vague. However, we must keep in mind that only during stimulation of this zone is it possible to obtain maximum responses in respiration (Pitts and others). Moreover, there are data which show that even very brief stimulation of this zone (0.2–0.8 sec) shifts in an abrupt way the activity of the respiratory neurons of the lateral region and is able to change the working phase of the entire respiratory system [69]. Prolonged stimulation of certain sections of the medial zone causes emergence of activity in the respiratory neurons during their pause, while the discharge appearing has the character of a normal volley [10].

2. THE NATURE OF THE ACTIVITY AND THE FUNCTIONAL PROPERTIES OF THE RESPIRATORY NEURONS OF THE MEDULLA OBLONGATA

Respiratory neurons account for 10 to 15 percent of all active neurons of the lateral region of the medulla oblongata [13, 76, 122]. The remaining cells of this region function continuously or with volleys not apparently connected with respiration. Among the respiratory neurons we can distinguish inspiratory neurons, the volley activity of which is synchronous with the inhalation phase, and expiratory neurons working with volleys synchronous with the expiration phase. Many authors note that the respiratory neurons do not change the type of their action under conditions of prolonged observations during an acute experiment. As already mentioned, the inspiratory and expiratory neurons maintain their volley activity after vagotomy and curarization. Moreover, we can see that the number of impulses in the volley of inspiratory neurons increases, but in the expiratory ones it diminishes. In the lateral region, neurons also are found, the volley activity of which can begin during the inhalation period and end during expiration, or begin at the start of the inhalation phase and end at the start of the next inhalation. Such neurons belong to the vagal system since after section of the vagi, activity in them completely disappears [11, 36].

In most papers devoted to a description of the nature of the activity of the respiratory neurons, the authors attempt to single out several similar types of inspiratory and expiratory neurons. If we examine these papers, it becomes obvious that the difference in the resulting types depends on the quantity of the material and the methodical procedures and criteria used for sampling. Thus were singled out types of inspiratory and expiratory neurons according to the nature of the frequency change of the volley in relation to the phase of the respiratory cycle [36, 44, 61]. Gesell and others [44], and at the same time Dirken and Woldring [36] maintain that the volleys of different inspiratory neurons are very similar; the frequency in the volley slowly increases to the maximum and by the end of inhalation the volley abruptly stops. In expiratory neurons we can observe two types of volleys. In some neurons the frequency in the volley very rapidly increases to the maximum and then slowly falls toward the end of expiration. In other neurons the frequency of the impulses in the volley rapidly reaches a maximum; then during the whole of expiration remains constant and falls abruptly toward the end of expiration. Hukuchara and others [61], according to frequency characteristics, single out five types of inspiratory and five types of expiratory neurons.

Depending on the phase of the start and end of activity, inspiratory and expiratory neurons can also be classed in several types [8–11, 13, 36, 55, 96, 97]. There were singled out early inspiratory neurons the discharge of which begins within approximately 0.15 to 0.20 sec before the start of inspiration and ends either at the very start of expiration or at the end of inspiration. Moreover, late inspiratory neurons were recorded in which activity appears at different moments after the start of inspiration and ends at the end of inspiration or the start of expiration. Very seldom were neurons observed the activity of which was continuous; this activity rose abruptly in frequency during inspiration. Among the expiratory neurons one can find also early neurons in which the activity appears within approximately 0.15 sec before the start of expiration and ends by the start of inspiration. One can also find expiratory neurons with a late discharge. In them activity appears at different moments of time after the start of expiration and ceases by the start of inspiration. In each of these types of inspiratory and expiratory neurons, volleys with different frequency characteristics can be observed.

In the study of the reaction of respiratory neurons to expansion of the lungs, it was found that all inspiratory and expiratory neurons can be classed in two types. Baumgarten and Kanzow [14] discovered that inspiratory neurons having almost the same frequency characteristics as well as times of start and termination of volley can behave quite differently in response to the expansion of the lungs. The neurons of one type, called the R_α neurons by Baumgarten, completely cease to function simultaneously with the disappearance of activity in the diaphragm during inflation of the lungs. At that time, the neurons of the second type, the R_β neurons, accelerate discharge more rapidly and with longer volleys during expansion of the lungs and inhibition of the activity of the diaphragm. During the expiratory pause the inflation of the lungs causes intense activation of the R_β neurons with complete silence of the diaphragm neurons. The same phenomenon was also observed during succinylcholinic paralysis [12, 30].

Examining the relation of activities in the R_α and R_β neurons during expansion of the lungs, Baumgarten concluded that the R_β neurons inhibit the activity of the R_α neurons. The R_β neurons are apparently excited by the neurons of the vagus system and work similarly to the Renshaw cells.

Recently Nesland and Plum [98] described two types of expiratory neurons, in one of which, called by them α, occurred inhibition of activity during inflation of the lungs; in the other, β, continuous activity was observed during inflation.

Correlating their own experimental material and the data of other authors' works, Nesland and Plum [98] concluded that 7 types of respiratory neurons exist. They are (1) α inspiratory neurons, which begin to work at the start or somewhat before inspiration, the frequency of the impulses in their volley gradually increasing to a maximum and the volley stopping suddenly at the end of inhalation; with inflation of the lungs they are inhibited; (2) β inspiratory neurons, having approximately the same frequency and phase characteristics, but inflation of the lungs causes continuous activity in them; (3) late inspiratory neurons, with a very short volley which appears after the middle of inspiration and stops suddenly at the start of expiration; during inflation of the lungs they are inhibited; (4) early inspiratory neurons, the volley of which is likewise very short, also appearing at the end of expiration and stopping abruptly at the very start of inspiration; with inflation of the lungs they are inhibited; (5) early expiratory neurons, which begin to work either at the end of inspiration or at the start of expiration with maximum frequency and cease working at the end of expiration; inflation of the lungs causes activation of these neurons; (6) α late expiratory neurons, in which the volley arises after the start of expiration, but the frequency maximum is reached by the end of expiration; inflation of the lungs inhibits their activity; and (7) β late expiratory neurons, with the same frequency and phase characteristics of (6), but inflation of the lungs causes continuous activity in them. Nesland and Plum maintain that neurons of different types retain the nature of their volleys in different experimental conditions, for example, during vagotomy.

Not all authors are agreed that it is possible to separate definite groups of inspiratory and expiratory neurons. Salmoiraghi and Burns [120], for example, maintain that the examination of the volley character of individual neurons does not have particular meaning, although the same neuron for a prolonged time gives almost steady volleys. Among the inspiratory and expiratory neurons a multitude of variations are observed of (1) duration of the volley and of times of starting and ending of the volley in relation to the corresponding phases of respiration; (2) number of impulses per volley; and (3) point of achievement of maximum frequency of discharge in the volley. In spite of the infinite variety of volleys, it is possible to find the mean activity level of the inspiratory and expiratory neurons and the probability with which these neurons will discharge during different moments of the respiratory cycle. It is claimed that the greater the general activity of the inspiratory neurons, the less the activity of the expiratory ones. Moreover, the sum of probabilities of discharges of inspiratory and expiratory neurons is approximately the same in the course of the whole respiratory cycle. From this point of view the nature of the discharge of separate neurons does not indeed have functional significance.

At present there is very little knowledge about the properties of respiratory neurons. By intracellular electrodes the activity of 7 inspiratory neurons and one expiratory neuron was successfully recorded [12, 119]. The membrane potential of the respiratory neurons reaches on an average 45 mV (from 25 to 70 mV) during measurements in the interval between volleys. The action potential fluctuates from 22 to 55 mV. Probably these figures do not give the absolute sizes, since their great range as well as instability in the course of the experiment indicate, apparently, the very small dimensions of the cells and the easiness of damaging them with the microelectrode.

The authors observed in the inspiratory neurons depolarizing and hyperpolarizing synaptic potentials: a slow depolarization which precedes the action potential, and a slow hyperpolarization. Furthermore, during the development of the volley of impulses one could see a general depolarizing shift not exceeding 2 to 6 mV. In the expiratory neuron the same changes were observed. Unfortunately, these data do not permit us either to confirm or reject the hypothesis about the spontaneous activity of respiratory neurons; with intracellular recording, a slow depolarizing prepotential is distinctly manifested (characteristic for structures which work automatically); however, against its background synaptic potentials are seen.

The influence was also observed of anodic and cathodic polarization on the activity of single respiratory neurons [27]; a steady current was passed through the extracellular electrode which at the same time was also used for recording. During cathodal action (the tip of the electrode is negative) an increase of the number of impulses and of the length of the volleys occurred. Occasionally only an increase of the frequency of discharge could be seen. In some instances the cathode managed to evoke activity of a respiratory neuron which has not fired until then. During anodic action (the tip of the electrode is positive) the length of the volley lessened, the action potentials appeared later and disappeared earlier, and their amplitude sometimes increased. In some instances the neurons on the whole ceased their action. The change of discharging of one neuron did not affect the character of the diaphragm activity. The authors did not notice any differences in reaction of the inspiratory and expiratory neurons.

Do the respiratory neurons have spontaneous activity? This question long ago attracted the attention of physiologists. A vast number of papers exist that concern respiratory deafferentation of the medulla oblongata. The main result of these papers is that the respiratory neurons continue to function with rhythmic volleys with the complete removal of all respiratory afferentation. But it is possible that for the appearance of rhythmic volleys in the system of respiratory neurons, any nonspecific afferentation is essential [121]. For verification of this hypothesis experiments were conducted with complete deafferentation of the brain stem. It is necessary to note right away that in the two most subtly conducted experiments, exactly opposite results were obtained. In one of the papers, complete denervation of the pons and the medulla was carried out. The pons was cut along the border of the lamina quadrigemina. At the site of the sections, Novocain was applied [40, 41]. The authors note that such a brain is very "silent" but there was success in recording neurons in it with continuous activity which intensifies with the increase of the concentration of CO_2 and disappears with the increase of O_2. Occasionally under the action of CO_2 it is possible to discover neurons with rhythmic volley activity; this activity completely disappeared with introduction of pure O_2. In the same zones, the authors observed slow fluctuations of the potential in response to the change of the percentage of CO_2 in the blood. In another work [121] there was carried out a longitudinal separation of the brain stem into halves, one of which served as the control for the second, which was subjected to further cuts. Initially, in both halves, normal activity of the respiratory neurons was maintained and their frequency of occurrence was as usual. With subsequent cuts of one of the halves of the pons at its upper border and at the spinal cord at the C_1 level, a noticeable reduction of the frequency of occurrence of respiratory neurons is observed. Additional section of all the cranial nerves leads to the complete disappearance of any activity within the limits of the corresponding half of the brain stem. Neither in the first nor the second studies is it possible to detect any particular methodical procedures which could explain the differences in the results obtained. True, one should note that the removal method used by Burns and Salmoiraghi excluded the possibility of recording, for example, of slow rhythmic potentials which could remain in the isolated brain.

In the majority of cases it is impossible to determine whether the response of the respiratory neurons is a manifestation of individual properties of the neuron of a given type or the result of a change of activity of the whole system of neurons. Thus, for example, the characteristic reactions of inspiratory and expiratory neurons to the change of partial pressure of O_2 or CO_2 in the blood is described in detail. As is known, in ordinary conditions the expiratory neurons are encountered much more seldom than the inspiratory. But if we increase the partial pressure of O_2 in the blood, then the frequency of occurrence or expiratory neurons sharply increases, their volleys become longer, and the frequency of the impulses increases. In the inspiratory neurons there is observed a decrease of frequency of the impulses and a shortening of the volley down to its disappearance. During an insufficiency of O_2 the expiratory neurons generally do not function, and in the inspiratory neurons the frequency of the impulses grows, their volleys lengthen, and an involvement of supplementary neurons occurs [11]. It turned out that after the section of the vagosympathetic, aortal, and glossopharyngeal nerves all these changes disappear [42]. Hence, the reactions observed are not determined by the specific effect of O_2 on individual neurons.

We know that with the increase of CO_2 partial pressure or during hypercapnia the volume of ventilation of the lungs increases because of both the deepening and increase of frequency of respiration. With denervation of the peripheral chemoreceptive zones the increase of frequency disappears and the volume increases because of deepening of respiration during decrease frequency of its rhythm [49]. In conditions of hypercapnia the expiratory neurons almost disappear. It is difficult to locate them in the medulla oblongata and in the thoracic section of the spinal cord; expiratory activity disappears in the recurrent laryngeal nerve. In the inspiratory neurons the frequency of impulses in the volley increases and the volley lengthens; the supplementary inspiratory neurons and the diaphragmic motor neurons are involved [11, 19, 42, 49]. During hypocapnia abrupt changes in the activity of the respiratory neurons are also observed.

Hyperventilation of the animal with air leads to the appearance of apnea, the duration of which depends on the degree of hyperventilation, the presence or absence of vagi, and the general condition of the preparation [151]. It was established that during this apnea the thorax is in an expiratory position. During the period of such apnea in the medulla oblongata it is impossible to locate inspiratory neurons. The frequency of occurrence of expiratory neurons sharply increases. Moreover, in the diaphragmatic neurons activity is completely absent, but in the efferent fibers of the recurrent laryngeal nerve, continuous activity of the expiratory fibers is recorded [11, 19, 42, 156].

Before the reestablishment of rhythmic respiration following apnea resulting from hyperventilation, in the medulla oblongata continuously active inspiratory neurons appear, in the diaphragmatic nerve weak continuous activity also begins to be recorded which gradually becomes more frequent and intensifies; after reaching a certain level, it begins to inhibit rhythmically; volleys thus arise [27, 42, 151]. In this way, for the normal working relation of the inspiratory and expiratory neurons, the maintenance of a certain level of CO_2 in the blood is essential. We can assume that this necessity is determined by specific properties of the different types of respiratory neurons. However, on the basis of certain experiments, yet another hypothesis was suggested, namely, that CO_2 causes generation of the impulses in specific chemoreceptive systems, the cells of which are mixed in the reticular formation with the respiratory neurons and affect their activity [40–41].

During study of the properties of the respiratory neurons and their responses to different effects, experiments in the investigation of the action of specific respiratory and nonspecific afferentation provide great interest. The sum total of all this data will probably make it possible for us to come to an understanding of the interaction of the different types of respiratory neurons.

Up to now we do not know exactly how the vagus or the inflation of the lungs acts on the inspiratory and expiratory neurons. The existing data are singularly contradictory. It is known, for example, that during weak intensities and lower frequencies of stimulation [36, 94] or slight expansion of the lungs [62] there occurs an acceleration of the discharge of the inspiratory neurons, the maximum of frequency is reached earlier, and then a continuous impulse rises the frequency of which increases periodically. The expiratory volleys, in this connection, virtually do not change. During higher intensities and frequencies of stimulation or during pronounced lung expansion there is observed an inhibition of the activity of inspiratory neurons and an increase of frequency of the impulses in the volley of expiratory neurons which gradually develops into a continuous behavior [36, 62, 94].

The examination of a great quantity of experimental data permits us to say that during action of the vagi and during expansion of the lungs it is impossible to observe thoroughly well-defined responses of individual neurons of either the inspiratory or the expiratory types. Moreover, even though the responses of the inspiratory and expiratory neurons are not always reciprocal, it is almost never possible to observe an isolated reaction of just the inspiratory or just the expiratory neurons. In connection with these observations, the research of the behavior of respiratory neurons of the medulla oblongata during the reflexes of swallowing [62, 130, 131, 135], coughing, and sneezing [116] is interesting. These reflexes involve the systems of nuclei of the cranical nerves, in particular the n. vagi, n. glossopharyngei, and n. trigemini, which are closely connected with the neurons of the reticular formation.

During stimulation of different somatic nerves the expiratory neurons show only a small decrease in frequency of impulses in the volley. The inspiratory neurons abruptly fire more frequently and their volley lengthens (for example, during stimulation of a muscle nerve) but they never change the phase of their activity. If stimulation is given during the period of expiration, then the inspiratory neuron changes its volley, but this volley rises punctually in its own time even in that case when stimulation ends long before the rise of the next volley [76].

It is possible to use still another experimental approach to solve the question of the interaction, or rather the dependence, of activity of the neurons of the lateral zone. Although we achieve this phenomenon very seldom, nevertheless we do sometimes succeed in recording simultaneously for a prolonged time with one electrode the activity of two neurons. If the number of intervals between the impulses of two cells is sufficiently great, then the analysis of the histograms, plotted along these intervals, makes it possible for us to form an opinion about the character of the interaction between the two neurons. It appears possible to assert with confidence [70] that in a pair of two inspiratory neurons excitatory interaction occurs; and in a pair consisting of an inspiratory neuron and a neuron of the reticular formation which works continuously, there is inhibitory interaction. The interaction in pairs of inspiratory and expiratory neurons proved to be very interesting. In that instance when the expiratory neuron was included in the background of a still functioning inspiratory one, the histogram of the intervals between impulses of these neurons reliably showed that the expiratory neuron inhibits the inspiratory, but the inspiratory neuron has virtually no effect on the work of the expiratory. In the second pair, in which the inspiratory neuron was included in the background of the activity of the expiratory, the interactions were reversed.

Experimental data about the nature of the activity of respiratory neurons and their behavior in different conditions were examined. The respiration rhythm is undoubtedly determined by the rhythm of the rise of volley activity in the respiratory neurons. In order to change the length of the respiratory cycle, that is, the period of rhythmic respiration, it is essential to change the period of the volley activity of the respiratory neurons of the medulla oblongata. Hence, the mechanisms of the formation of rhythmically recurrent volleys in the system of the respiratory neurons also guarantee rhythmic respiration. In the present case we are talking only about the mechanisms of the origin and the regulation of rhythm of the replacement of inspiration with expiration; the mechanisms of interaction of the respiratory systems of the medulla oblongata and spinal cord are not considered.

3. HYPOTHESES FOR THE ORIGIN OF RHYTHMIC RESPIRATION

Let us examine several hypotheses concerning the origin of rhythmically recurrent volley activity in respiratory neurons of the medulla oblongata and of the rhythm respiration.

1. When there is a transsection of the brain stem below the pons and the vagi it is possible to observe the appearance of a prolonged tetanus of the inspiratory muscles or inspiratory apnea. Many authors considered this phenomenon as a manifestation of continuous spontaneous activity of inspiratory neurons of the medulla oblongata [85, 86, 106, 111] released from inhibitory effects. Continuous spontaneous activity is a specific characteristic of neurons similar to the characteristics of automatic structures of the heart; it can depend on the specific character of the metabolism and the particular sensitivity of the inspiratory neurons to the surrounding medium [109]. The rhythm of respiration is determined by the fact that the continuous activity of inspiratory neurons is normally inhibited by the expiratory neurons, which in their turn are rhythmically excited either by afferent impulses of the vagus or by neurons on the pneumotaxic center of the pons (see section 4, pp. 213–217). The mechanisms of the origin of inspiratory apnea are still vague, but it positively cannot be considered a manifestation of the initial continuous activity of inspiratory neurons. Moreover, it was shown that after section of the vagi and brain stem below the striae acusticae, rhythmic respiration is retained (see above).

However, the idea itself of the existence of the spontaneous continuous activity of inspiratory neurons seems very attractive. Actually respiration of newborns begins from protracted inspiration. After apnea from hyperventilation, caused by any means, rhythmic respiration appears only against a background of continuous activity of inspiratory neurons, which gradually reaches a certain level. How from these positions is the origin of rhythmic respiration explained?

Normally the inspiratory neurons, sending impulses to the diaphragm and the intercostal muscles, cause expansion of the lungs. At the same time there occurs activation of different receptors by the expansion of the lungs. Along the afferent fibers of the vagus in the medulla oblongata arrive impulses, part of which, acting immediately on the inspiratory neurons, excite them and increase inspiration. The impulse arriving along other afferent fibers, and impulses arriving during the still more powerful expansion of the lungs at the end of inspiration, arrive at the expiratory neurons and cause their activation. The expiratory neurons begin to inhibit the inspiratory ones; inspiration ceases, and the lungs collapse. During disappearance of excitatory influences from the vagus to the expiratory neurons, in the inspiratory neurons continuous spontaneous activity again develops [149, 151, 152]. Thus, the inclusion of the mechanical system (the expansion of the lungs) completely guarantees the delay, essential for the rise of rhythm, in the sequence of activation of inspiratory and expiratory neurons.

But at this point let the vagi be cut. The rhythm is maintained and the frequency of respiration becomes even more infrequent. Now the inspiratory neurons send impulses to the pneumotaxic center, which excites the expiratory neurons, and those in their turn inhibit the inspiratory ones, etc. At the same time respiration remains uniformly rhythmic with a smooth transition from inspiration to expiration. We will note that in these conditions, the delay, still longer than normal, should be guaranteed by synaptic switchings in the brain stem. This circumstance does not seem very likely.

After an additional cut below the pons, rhythmic respiration is maintained, but it becomes spasmodic and an abrupt transition from inspiration to expiration is observed. One can assume that the inspiratory neurons have an excitatory connection with the expiratory neurons. Working continuously, they activate the expiratory neurons up to a certain level, after which the latter, in their turn, begin to inhibit the activity of the inspiratory neurons, etc. Respiration at this time becomes more frequent, but all the same it is difficult to set up a delay which would guarantee an emerging rhythm. Moreover, it should be stressed that such a viewpoint requires that the inspiratory neurons always exert an excitating action on the expiratory ones, and that the expiratory neurons always inhibit the inspiratory ones [67, 153, 154]. We can show that a system in which one type of neuron virtually inhibits itself, with a certain delay, in the end emerges into a fixed behavior of continuous activity, and no rhythm arises in it.

2. The retention of rhythmic volleys in respiratory neurons after a cut of the vagi and removal of the pons led some authors to the hypothesis that the inspiratory and expiratory neurons work spontaneously with rhythmic volleys [57, 142, 144]. Within the systems of inspiratory and expiratory neurons there must be, apparently, synchronizing excitatory connections since in the reverse case it would be difficult to maintain the stability of the whole system of respiratory neurons in response to different effects. Among the inspiratory and expiratory neurons there should exist reciprocal interrelations.

It should be stressed that the utter absence of concrete facts in this viewpoint does not permit us to discuss it. Moreover, in the present case the difficulties in the understanding of the mechanism of the origin of rhythmic volleys are carried over into the area of the study of specific biochemical properties of the respiratory neurons.

3. At present the viewpoint expressed by Salmoiraghi and Burns [121] wins ever increasing attention. Using their own data, obtained during the complete isolation of half the brain stem, the authors maintain that the respiratory neurons do not possess spontaneous activity, and, what is more, in the medulla oblongata there are not any neurons which work spontaneously. Consequently, rhythmic volleys arise in a system of neurons having specific interneural connections. For the appearance of a rhythm neither the vagi nor the pons is necessary.

There are two systems of respiratory neurons—the inspiratory and the expiratory. The neurons of each of these systems have a rather different excitability and refractory period. Within each system the neurons are linked among themselves by extensive excitatory connections. The system of inspiratory neurons is connected with the system of expiratory neurons by inhibitory connections; they are found in reciprocal proportions. The appearance of activity in such systems depends on the working of another chain of neurons which do not have to be rhythmically active but which are able to exert a facilitatory effect on the respiratory neurons.

The inspiratory and expiratory neurons are scattered diffusely in the lateral zone of the medulla oblongata. Some of the inspiratory neurons are located very close to the nucleus of the solitary tract and to the nuclei of the sublingual nerve. It is highly probable that these neurons can receive impulses coming from different receptors or from neurons of the facilitatory reticular formation. If as a result of temporary summation of these impulses the membrane potential of any inspiratory neuron reaches the discharge level, then there arises in it the action potential which starts the operation of the whole inspiratory chain. It should be kept in mind that in this case the antagonistic expiratory chain should not exert inhibitory effects. Each of the members, for example, of the inspiratory chain will work asynchronously. Its inclusion or exclusion from activity depends on the level of synaptic bombardment which it receives not only from the neurons of its own population, but also from different afferent systems and central sources. We can explain by that the varied form of the discharges of the inspiratory and expiratory neurons. However, the whole system as a functional unit works relatively synchronously, a fact which is explained by the general character of its internal and external connections.

In this way any afferent impulses can start volley activity, for example, in an inspiratory chain. The self-exciting chain, having begun to work, will operate endlessly. What are those mechanisms which limit its work and stop the volley? One of the self-limiting mechanisms can lie in the involvement of inspiratory cells of the R_β type. The R_β cells tend to limit the activity of the inspiratory neurons but they more likely decrease the frequency of their volley; apparently they are able to maintain some mean frequency, but they are not able to stop all activity. The reciprocal expiratory chain can stop a volley, but it can be activated only after cessation or abrupt reduction of activity in the inspiratory chain. At this point the authors use data obtained during examination of the behavior of cells in conditions of intracellular recording [119]. During the development of a volley in an inspiratory neuron, the discharge level progressively increases, that is, as a result of some inner developments there arises a depolarizing shift of the membrane potential, and the threshold of the neuron increases by 2–6 mV. Consequently, the greater the discharge frequency, the sooner the threshold increase, and the greater should be the level of synaptic bombardment for the rise of the next action potential. Within the inspiratory chain temporary summation becomes still less effective, the discharge frequency drops, and this combination leads to the release of expiratory neurons which, being activated at once, completely inhibit the inspiratory neurons. Thus, the volleys arise.

During discussion at a symposium of Salmoiraghi's report on the regulation of respiration [118] two principal objections were expressed against the viewpoint set forth (Hoff and Wang). In the first place it was maintained that it is impossible to base the hypothesis expressed on very crude experiments with the isolation of half the brain stem. Second, the observation was made that from this viewpoint it is impossible to understand the roles of the pons, the vagi, and other factors in the origin of normal rhythmic respiration. It should be stressed that the authors themselves do not consider these questions. They are trying to give an explanation of the mechanisms of the origin of rhythmic volleys in respiratory neurons and of the rhythm of respiration, but not of normal, rhythmic respiration, which is impossible without the pons and connections with the spinal cord.

It is difficult briefly to state all the experimental data confirming this viewpoint or contradicting it. However, let us recall that from this point of view the chief cause of the emergence of rhythmic volleys lies in the rise of the threshold of individual neurons, that is, in the development of fatigue. It seems scarcely probable that the system of respiratory neurons works on the basis of fatigue.

On the basis of the hypothesis of Burns and Salmoiraghi and of Baumgarten it is still difficult for the present to construct a quantitative model of the working of the system of respiratory neurons. And in this instance the construction of the model is complicated by the use of specific properties of individual neurons.

4. Nesland and Plum [98] recently suggested dividing all the respiratory neurons of the medulla oblongata into 7 subtypes. They assume that each of these subtypes forms a functional group. Interaction among the functionally different groups guarantees both the rise of rhythmic volleys and the rhythm of the replacement of inhalation with expiration. Proceeding from this hypothesis and not considering the question about the mechanism of the initial emergence of activity in the system of respiratory neurons, they assume that (1) the activity of the neurons of one of the groups arises as a result of the action on it of active neurons of another

group; (2) the cessation of activity of neurons of one of the groups is the result of the inhibitory action of the active neurons of another group; (3) the system of the respiratory neurons is always active and is never quiet.

We have already described at length the types of respiratory neurons according to Nesland and Plum. The principal functional group consists of the α inspiratory neurons (the first type, pp. 199–200). These neurons activate the motor nuclei of the diaphragm and other muscles. They are inhibited during inflation of the lungs. This inhibition is guaranteed by the β inspiratory neurons, the second type, which are activated during expansion of the lungs. It should be stressed that activity in the neurons of the first and second types is maintained after vagotomy and succinylcholine paralysis. The function of the neurons of the third and fourth types, that is, early and late inspiratory neurons, apparently is that they ensure the transition from expiration to inspiration, and from inspiration to expiration. They either inhibit the already active group of neurons or facilitate the appearance of the discharge of impulses in a reciprocally quiet group of neurons. The early expiratory neurons, the fifth type, probably inhibit inhalation and start expiration. One can assume that between this type of neurons and the late expiratory neurons there exists a reciprocal inhibition since the activity of one group reaches the maximum at the moment when the activity of the other is minimal. Of the α and β late expiratory neurons, the sixth and seventh types, some probably maintain expiration, and others limit its duration.

The authors assume that disturbances of the rhythm of respiration are likely the result of changes in the activity of specific functional groups, and not of the whole population of inspiratory or expiratory neurons.

This latter hypothesis is of undoubted interest since it is based on an attempt to determine those conditions of interaction in the system of the types of respiratory neurons known to us which guarantee the rise of rhythmically recurrent volleys and the rhythm of respiration.

On the basis of our experimental material on the work of inspiratory and expiratory neurons [67–69], A. N. Chetaev (1965, unpublished data) suggested a mathematical model of the formation of rhythmic volleys in the lateral zone of the medulla oblongata. The inspiratory and expiratory neurons in this model are presented as combinations of several functional types of neurons. The special structure of interaction of the neurons of all types guarantees the rhythmic nature of their activity. The model can be checked on an electronic computer.

In the present collection there is another mathematical model (Keder-Stepanova, Rikko) which does not at all claim to be a modeling of the respiratory center but which yields only that sort of system of neuron the particular type of interaction of which guarantees the appearance of periodic volleys. The operation of this system appears resistant to random external stimulation.

The model of the origin of respiratory volleys must be very concretely defined. It is essential to know the importance of each functional group of respiratory neurons and their interrelation at any moment of the respiratory cycle. If in such a model the specific properties of individual neurons are used—for example, the spontaneity or fatiguability in the course of functioning—then we should scrupulously investigate these properties in different conditions. Only then will we say that a model of the system of the respiratory neurons exists when on the basis of a quantitative analysis of it we can predict the behavior of the system and its elements under specific conditions.

4. THE NATURE OF THE ACTIVITY AND CERTAIN DATA ON THE
FUNCTIONAL SIGNIFICANCE OF THE RESPIRATORY NEURONS OF THE PONS

Transsection of the pons lead to pronounced disturbances of respiration [22, 23, 56, 58, 64, 73, 81, 82, 85–87, 136, 138, 142, 144]. By the method of sectioning it was established that in this region of the brain stem two very important zones for respiration exist: the so-called pneumotaxic and apneic centers. By procedures of localized lesions and stimulation, it has been shown that the pneumotaxic center occupies the locus coeruleus of the pons [17, 30, 31, 56, 71, 136, 137, 143]. The precise localization of the zone of the apneic center still has not been established.

The respiratory neurons of the pons do not form accumulations but are scattered among the cells of the reticular formation. In different regions volleys of impulses are recorded, synchronous with the phase of inhalation or expiration, and also continuous activity which becomes more frequent during either the inspiration phase or the expiration phase. There is reason to assume that the periodic respiratory activity of the neurons of the pons depends on, and possibly is initiated by, the respiratory neurons of the medulla oblongata. After isolation of half of the pons and the disappearance in it of rhythmic volleys, neurons with nonperiodic activity [30, 31] are found in the same zones.

Examination of the nature of the activity of respiratory neurons of the pons makes it possible to show a certain similarity of these neurons to the neurons of the medulla oblongata, as well as important differences between them. We have considered respiratory only those neurons which maintain some type of periodicity synchronous with respiration after sectioning of the vagi and introduction of curare-like substances [31].

All respiratory neurons can be divided into 2 classes: (1) noncontinuous, that is, having a period of silence and producing a volley during a specific phase of the respiratory cycle; and (2) continuous, working all the time, but exhibiting an increase of frequency of impulses during some phase of respiration. Among the neurons of the first class we can single out those neurons which produce a volley either during the inspiration phase (inspiratory neurons) or during the expiration phase (expiratory ones). Yet, besides that, there are neurons which begin to work during one phase of the cycle (for example, in inspiration or expiration) and cease discharging in another phase of the cycle (correspondingly, in expiration or inspiration). As a rule, the purely inspiratory neurons of the pons, unlike the neurons of the medulla oblongata, discharge during the entire phase of inspiration; only very seldom are neurons encountered the volleys of which occupy only part of the inspiratory phase. The very same thing can be said also about purely expiratory neurons.

The majority of the inspiratory-expiratory neurons begin to discharge in approximately the middle of the inhalation phase; the frequency maximum is reached in them toward the end of inspiration or at the start of expiration; and toward the middle of expiration the volleys stop. The volley of some neurons can begin at the start of inspiration but the maximum frequency is reached toward the end of the inspiration phase, and the volley ends at different moments of the expiration phase, as a rule during the period of expiratory pause. The expiratory-inspiratory neurons begin to work during the expiration phase, in the majority of instances directly after cessation of expiratory movements; the frequency maximum is reached in the middle or at the end of inspiration, and the volley abruptly disappears.

We should especially note the fact that in the medulla oblongata with a great diversity of character of the inspiratory-expiratory volleys all authors stress the stability of the types of volley belonging to the individual neurons. In the pons, on the other hand, one can record changes of the types of volleys of the same neuron over a period of several minutes. Most often it is possible to observe neurons which at first exhibit an expiratory rhythm and then turn into inspiratory-expiratory neurons, and vice versa.

The second class of respiratory neurons, those which work continuously, we can also divide into inspiratory and expiratory, inspiratory-expiratory, and expiratory-inspiratory, depending on the phase of the increase of frequency of their rhythm. These neurons are very unstable in the nature of their activity. We can often see the spontaneous appearance of blank periods which gradually go over to continuous activity.

Unfortunately we still know nothing definite with regard to the functional role of the pons and its respiratory neurons in the origin of respiratory rhythm and in the regulation of the volume of ventilation.

After the work of Marckwald [85, 86] and Marckwald and Kronecker [87] the opinion existed for a long time that for the rise of a rhythmic replacement of inspiration with expiration in vagotomized animals the pons is essential. These authors showed that transsection of the brain stem above the auditory tubercles (above the striae acusticae) and section of the vagi leads to the appearance of inspiratory apnea, or continuous tetanus of the inspiratory muscle. Inspiratory apnea is considered a manifestation of continuous spontaneous activity of the inspiratory center, which in these conditions is freed from the inhibiting action of the vagi and the centers of the pons.

Data gradually began to be accumulated which showed that if the pons is cut exactly along the border with the medulla oblongata, that is, below the striae acusticae, then subsequent vagotomy does not lead to inspiratory apnea, and rhythmic respiration is maintained [57–59, 81, 82, 143].

Moreover, it was established that inspiratory apnea does not develop right after a section above the striae acusticae [138]; having arisen, apneic respiration can disappear and change into rhythmic after brief artificial ventilation [56]; against the background of apneic respiration one can very often see rhythmic respiration of a low amplitude [22–24, 73, 101, 136, 143]; apneic respiration disappears and passes over to rhythmic during deterioration of the preparations [59]; apneic respiration can rise with the pons totally intact, for example, during hemisection of the cervical section of the spinal cord [89]. All these data attest to the fact that in conditions of the removal of the pons and section of the vagi, the mechanisms ensuring the appearance of rhythmic respiration remain unaffected, and they function on the level of the medulla oblongata. These data showed that inspiratory apnea is not a manifestation of the spontaneous continuous activity of the inspiratory center; it originates as a result of the active effect of the lower portions of the pons or the neurons of the respiratory center of the medulla oblongata, and possibly also the spinal cord.

Thus, in animals with the pons removed, rhythmic respiration is realized. However, this respiration has a spasmodic character with a rough transition from inspiration to expiration, and it does not ensure adequate ventilation; the animals perish from acidosis [99]. It should be noted that the spasmodic respiration of such medullary preparations does not change after section of the vagi [58, 79, 143, 156]. Wang and Ngai [142] especially stress that for the realization of normal reflexes in the lungs the pons is essential. It is interesting that during study of the activity of respiratory neurons of the pons, Cohen and Wang [31] repeatedly observed changes of the mode of activity of the neurons (for example, with an increase of frequency, the transition of continuous activity to periodic) during disturbances of the composition of the air inhaled (a change of the CO_2 concentration or O_2 supplementation). Moreover, they often failed to register a parallel change of activity of the diaphragmatic motor neurons.

The instability of the activity of the respiratory neurons of the pons, the existence of so-called "transitional" neurons, the slight change of the phase of activity under the action CO_2, and other factors enables one to assume that these neurons as the chief ones receive information about the need of a respiration adjustment to changed conditions, and accordingly they change the neuronal activity of the respiratory center. Possibly the "transitional" neurons guarantee the smooth replacement of inhalation with expiration. We have already mentioned that in the medullary preparations, inspiration and expiration are spasmodic and change very abruptly.

The pons is undoubtedly involved in the development of the valuable adjusting reactions of respiration, but its function and connection with the neurons of the medulla oblongata and spinal cord still remain unclear.

5. THE LOCALIZATION, NATURE OF ACTIVITY, AND FUNCTIONAL PROPERTIES OF THE RESPIRATORY NEURONS OF THE SPINAL CORD

In the spinal cord respiratory neurons are encountered at the level of the cervical and thoracic regions. In precisely these areas the motor neurons of the diaphragmatic, intercostal, and accessory respiratory muscles are located. Respiratory neurons of all types, situated on the spinal cord level, lose their rhythmic volley activity upon elimination of their connections with neurons of the medulla oblongata [49, 125].

In the cervical region of the spinal cord is located the diaphragmatic nucleus, which has a very complicated histological structure [20, 74, 145]. In it a great number of neurons was discovered functioning with volleys with different durations and frequencies of impulses. Often neurons are encountered the activity of which is continuous but becoming more frequent during the inspiratory phase [16, 51, 113, 114]. During intracellular examination of the diaphragmatic nucleus, very few motor neurons were found; the majority of the neurons proved to be intercalary. In the motor neurons rhythmic volleys arise against a background of slow depolarization the maximum size of which is not more than 7 mV [51]. The dimensions of the diaphragmatic motor neurons are apparently very small; their membrane potential is unstable; and the membrane resistance is in the neighborhood of 2 to 4.2 megohms, that is, approximately 2.5 times greater than the resistance of the motor neurons of the lumbosacral segments. The membrane potential fluctuates from 40 to 70 mV, the action potential from 40 to 93 mV. The diaphragmatic motor neurons probably have a high safety factor for the transmission of an impulse from the initial segment to the somadendritic membrane, since on the upward part of the antidromic action potential one can see a very weakly expressed flexion. The action potential has an afterdepolarization (1.5–7.5 msec) after which hyperpolarization follows, with a duration of about 85 msec. The time constant of the membrane of the diaphragmatic neurons is 1.8 ± 0.5 msec.

In experiments on spinal cats and in conditions of hyperventilation, when the motor neurons of the diaphragmatic nucleus are not active, depolarization of their membrane by an intracellular electrode leads to the appearance of impulses which cease after the current is turned off. In these neurons accommodation is very slight. During normal respiration of the animal, in approximately 40 percent of the diaphragmatic motor neurons examined, a volley of action potentials is absent, but at the same time the intracellular electrode records the periodic appearance of a slow depolarization which reaches the maximum toward the end of inhalation. During the inspiratory phase it is possible to see synaptic noise, which is absent during the expiratory phase. If a depolarizing current of subliminal strength is applied through a microelectrode, then against the background of the natural slow depolarization develop action potentials the quantity and frequency of which depend on the strength of the applied current. Thus, the number of impulses in a volley and the frequency of the discharge depend on the character of the development of the slow depolarization, while the trigger threshold in the diaphragmatic neuron during the volley remains steady; consequently, accommodation does not play a significant role in the limitation of the volley of these neurons [51].

Of great interest are the data concerning the action of CO_2 directly on the diaphragmatic motor neurons. On the spinal cats inspiration of a mixture of 6 percent CO_2 with O_2 leads to a significant reduction of the excitability of the diaphragmatic motor neurons and to the disappearance in them of responses to stimulation of the descending tracts and to the direct action of the depolarizing current (the response disappears on the average in 30 sec). Moreover, it was discovered that in conditions of the action of 6 percent CO_2 the antidromic response increases slightly in amplitude, a phenomenon which indicates an increase of the membrane potential of the motor neurons. In this respect the diaphragmatic motor neurons are similar to motor neurons of the lower segments of the spinal cord. All these data show that the reactions of the diaphragm observed during hypercapnia and hypocapnia are determined solely by the influence of the respiratory center of the medulla oblongata.

During stimulation of the descending reticular tracts and different regions of the reticular formation of the medulla oblongata in the diaphragmatic motor neurons an excitatory postsynaptic potential (EPSP) occurs after which, as a rule, an inhibiting postsynaptic potential (IPSP) follows. The latency of the EPSP fluctuates from 4 to 6 msec and the IPSP from 6 to 14 msec. The duration of the EPSP equals on the average 30 msec, and the IPSP from 50 to 190 msec. An IPSP with a latency of 7 msec and a duration up to 40 msec can be observed also during stimulation of the contralateral diaphragmatic nerve (in the intact posterior rootlets). In certain cases a small EPSP was recorded before the IPSP [50].

At the present time it can be asserted that the behavior of the respiratory neurons of the diaphragmatic nucleus is almost completely subject to the neurons of the medulla oblongata. The diaphragmatic nucleus behaves as if it were functionally isolated in the spinal cord. It is known, for example, that in the same systems where the diaphragmatic nucleus is situated, a mass of motor neurons of the brachial plexus is found. However, for example, stimulation of the median nerve, which evokes a response in the somatic nerve of the same segment after 3 to 4 msec, in the diaphragmatic nerve produces a response after 13 to 22 msec, depending on the phase of the respiratory cycle. It is quite obvious that this reflex response has a supraspinal origin [28]. In expiration, the latency fluctuates from 19 to 22 msec; the response is very small in amplitude and arises sporadically; its duration is 4 to 6 msec. During inspiration the latency of the response upon stimulation of the median nerve fluctuates from 13 to 17 msec; the shortest latency and the response greatest in amplitude and duration (10–15 msec) were observed at the end of inspiration. It is known that in the motor neurons of the diaphragmatic nerve there is always a periodic slow depolarization [51], which apparently renders an effect on the reflex response.

The diaphragm is not involved in any other function whatever; in particular it does not take part in the support of posture. It carries out only the respiratory function. Possibly the complexity of the organization of the diaphragmatic nucleus guarantees its individualization, that is, a certain part of the numerous intercalary neurons perform a protecting function with respect to intraspinal influences. True, ever increasing data have been accumulated recently about the fact that at the level of the diaphragmatic nucleus, reflexes from the lungs and respiratory muscles are realized [52–54, 75, 95]. It should be stressed that in spite of the close connection of diaphragmatic activity with the respiratory center, direct excitatory actions of the motor cortex and extrapyramidal tracts on the diaphragmatic motor neurons are possible. These influences result in conditions of apnea from hyperventilation. The latency of the response (6–9 msec) also attests to the direct links of the higher levels of the central nervous system with the diaphragmatic nucleus [32, 33].

As already mentioned in the first part of this chapter, after destruction of the anterior columns and the anterior parts of the lateral columns at the C_1 level, Pitts [105] observed retrograde degeneration in the ventral part of the middle zone of the respiratory center, that is, at a site where by the stimulation method the inspiratory center was localized, and where there are not any neurons active with respiratory volleys. However, in 1960, Baumgarten and coworkers [12] succeeded in detecting retrograde degeneration in a small number of neurons of the lateral zone of the respiratory center during hemisection of the spinal cord at the C_2 level. Recently, Baumgarten and Nakayama [15, 93] recorded antidromic responses of the inspiratory and expiratory neurons in the lateral zone of the medulla oblongata during stimulation of the spinal cord at the C_1–C_5 level. The majority of inspiratory descending fibers cross in the lower part of the medulla oblongata. Both the crossed and the uncrossed fibers go in the ventral part of the lateral columns and in the dorsolateral part of the anterior columns. Almost all the expiratory fibers pass to the opposite side and descend in the reticular tract of the dorsolateral part of the anterior columns. The average rate of conduction of the descending respiratory fibers is 40 m/sec.

Thus, on the C_1–C_5 level, fibers are found going from inspiratory neurons of the R_α and R_β types, from expiratory neurons, and from the middle zone of the respiratory center. Consequently, both the lateral and middle zones of the respiratory center provide an exit to the spinal cord. What the effect is of these fibers on the nucleus of the diaphragmatic nerve, and what their functional significance is, remain unknown. The question concerning the connections of the neurons of the respiratory center with the diaphragmatic nucleus requires more thorough research.

Besides the diaphragmatic nucleus, respiratory neurons were discovered in the thoracic region of the spinal cord. Here there are inspiratory and expiratory neurons which work either with volleys, or continuously, but with an increase of frequency in inspiration (inspiratory) or expiration (expiratory). Of 106 respiratory neurons recorded in one of these studies, only 10 proved to be motor neurons [130, 131]. They produced an antidromic response to stimulation of the ipsilateral intercostal nerve of the same segment; moreover, this response never arose between volleys. With an increase of the strength of stimulation, following an antidromic response within 18 msec a second orthodromic response arose, and sometimes also subsequent action potentials with a small latent period, that is, involvement of the intercalary neurons, occurred. The intercalary respiratory neurons (inspiratory, 55 and expiratory, 51) responded to stimulation of the ipsilateral nerve within 2 to 20 msec regardless of whether stimulation was applied during a volley or between volleys. The latency of the response and the number of action potentials depended on the strength of the stimulation. With rhythmic stimulation the intercalary respiratory neurons began to function continuously.

With intracellular recording [38] there was success in recording membrane potentials of inspiratory and expiratory motor neurons which were identified by an antidromic response. The membrane potential, which is very unstable, on the average is 55 mV; the duration of the afterhyperpolarization is approximately 90 msec, and its size constitutes 5 percent of the amplitude of the action potential, which is approximately equal to the membrane potential at rest. The reason for the low amplitude of the action potentials remains unknown.

Recently, several works of Sears [125, 127–129] have appeared in which there was studied at length the behavior and functional properties of the motor neurons of inspiratory and expiratory intercostal muscles. During intracellular recording Sears, just as Eccles and others [38], observed slow rhythmic (in the respiratory rhythm) changes of the membrane potential of the inspiratory and expiratory motor neurons. In these motor neurons the volley arises against a background of slow depolarization; moreover, it is shown that an approximately linear relation (on the average, 4 impulses to 1 millivolt) exists between the size of this depolariza- and the frequency of the discharge. In the inspiratory motor neurons the maximum amplitude of the slow depolarization (5–12 mV) is greater than in the expiratory (0.5–6 mV). This difference is connected with the correspondingly small amplitude of the synaptic noise in the expiratory neurons. During deafferentation of its own and the two nearest segments, slow potentials are retained. With section of the spinal cord above the level of recording, they completely disappear.

The entire series of Sears' experiments [129] proved that in all the motor neurons of the intercostal muscles there occurred not only a rhythmic increase and decrease of excitatory synaptic bombardment, but also an increase and decrease of inhibitory synaptic effects. After slow depolarization, against the background of which a discharge arises, there follow repolarization and hyperpolarization of the motor neuron, that is, in every motor neuron during normal respiration there is a phase of active inhibition. Depolarization of the inspiratory motor neurons occurs parallel with hyperpolarization of the expiratory ones, and vice versa. Moreover, the appearance of inspiratory activity is always connected with more powerful inhibition of the expiratory motor neurons.

The stimulation of low-threshold afferent fibers of the ipsilateral nerve evokes a monosynaptic excitatory potential (from 0.5 to 5.2 mV in different motor neurons). It is also possible to obtain a response to stimulation of a nerve of a neighboring segment. Responses to stimulation of contralateral nerves was never observed. During strong stimulation of the ipsilateral nerve an inhibitory PSP was recorded which prevented the discharge from arising during involvement of the low-threshold afferents [38, 128]. In spinal cats during artificial respiration it is possible to see an inhibition of the posterior root reflex at the T_6–T_7 level during stimulation of the muscle nerves. This inhibition, with the choice of appropriate intervals of stimulation, reached 100 msec. [77].

The intercostal muscles by their very properties occupy an intermediate position between the muscles of the type of the m. soleus and m. tibialis [1, 21]. They are rich in muscle spindles [7]; these spindles, just as in other muscles, are innervated by γ motor neurons [34, 35, 39, 125]. The recording of separate fibers of the anterior and posterior rootlets of the inspiratory and expiratory intercostal nerves showed that (1) the γ fibers are activated sooner than the α fibers; (2) a static continuous activity of the γ fibers exists which becomes more frequent in the inspiratory nerves during inspiration, and in the expiratory in expiration, and at the same time a volley activity synchronous either with inspiration or with expiration; (3) this activity is maintained after curarization; (4) it varies accordingly during hyperventilation, hypercapnia, and asphyxia; (5) in spinal animals during passive lengthening of the extrafusal fibers, activity arises in the γ efferents; during artificial respiration in approximately 50 percent of the cases the spindles maintain the same phase dependence between the character of the discharge and the respiratory movements which exists during spontaneous respiration; (6) stimulation of the anterior rootlet of the neighboring segment or mechanical stimulation of the intercostal muscles of the neighboring segments activates the γ motor neurons; (7) stimulation of the cutaneous receptors activates the γ motor neurons; for inspiratory and expiratory intercostal motor neurons the reflexogenic cutaneous fields are approximately 100 times greater than for the primary afferents; convergence at the inspiratory motor neurons is significantly greater than at the expiratory [132], yet the threshold of the expiratory neurons is lower than that of the inspiratory ones.

Thus, unlike the γ motor neurons of the limbs, the γ motor neurons of the intercostal muscles are activated by proprioceptors [34, 35, 39, 91, 148]. This fact indicates undoubtedly that besides the supraspinal mechanisms, mechanisms at the spinal level also participate in the regulation of the volume of ventilation.

We will now consider what happens in respiratory neurons of the spinal level, for example, toward the end of expiration and during inspiration. We know that the γ efferents are activated earlier than the α efferents. (We do not know the direct sources of their activation; it is possible that they are of a central origin.) But by this moment there occurs also the proprioceptive excitation of the γ motor neurons of the respiratory muscles which are still found in a somewhat extended condition.This leads to the increase of the volume of inspiration. The stronger the inspiration, the more pronounced at this point is the inhibition of the expiratory motor neurons. If this did not happen, then the dynamic component of an intense lengthening of the expiratory muscles at the depth of inspiration could lead to the sudden activation of the expiratory motor neurons and a disturbance of rhythm. In the inspiratory muscles active inhibition, as a rule, is expressed more weakly. This is justified since in normal conditions the expiratory muscles contract slightly. It is possible to suppose that active inhibition of the respiratory neurons occurs at the spinal level. However, this question requires more detailed research.

Thus, the spinal level undoubtedly takes an active part in the regulation of ventilation. Some authors think that the reflexes to the expiratory muscles are realized mainly at the level of the spinal cord [115–117]. In experiments with the deglutition reflex it is also possible to be persuaded that the spinal respiratory neurons are not completely subordinate to the neurons of the respiratory center. Sumi [132–135] described in detail the types of reactions of the expiratory and inspiratory neurons of the medulla oblongata during the swallowing reflex. A part of the neurons respond reciprocally, that is, with the appearance at the moment of swallowing of a volley in the inspiratory neurons, the expiratory neurons are inhibited, and vice versa. In other groups there is no such reciprocity; these neurons are simultaneously inhibited or excited; and also, the appearance of subsequent volleys depends on the time of the rise of the reflex. At the spinal level the reaction of the inspiratory and expiratory motor neurons is surprisingly constant and meagre. The inspiratory neurons always produce a short discharge of 1 or 2 impulses, while the following volley arises within the same time interval regardless of the stage of the appearance of the reflex; the expiratory neurons always exhibit a lengthening of the volley, but this phenomenon is significantly less pronounced here than in the respiratory neurons of the medulla oblongata.

In connection with research on the immediate sources of γ and α afferentation, the study of the functional role of the medial and the lateral zones of the respiratory center, each of which has outlets to the spinal cord, seems interesting.

6. CONCLUSION

Recent experimental data make it possible to assert that the control of adequate rhythmic ventilation of the lungs is accomplished by particular interrelations of the three levels of the central nerve system—the medulla oblongata, the pons, and the spinal cord. Study of the nature of the activity, the functional properties, and the behavior of the individual neurons still does not make it possible to construct a concretely defined model of the systems of the respiratory neurons of each level. For the construction of such a model it is essential to search for approaches to the study of the interneural interactions within the systems and among the systems at different levels.

REFERENCES

1.
A. Alderson and J. Maclagan, *J. Physiol.* 1962:162, 23–24.
2.
E. Amoroso, J. Bainbridge, and F. Bell, *Nature* 1951:167, 603.
3.
———, F. Bell, and H. Rosenberg, *Proc. Roy. Soc.* [*Biol.*] 1951:139, 128.
4.
———, F. Bell, and H. Rosenberg, *J. Physiol.* 1954:126, 86.
5.
L. Bach, *Amer. J. Physiol.* 1952:171, 417.
6.
A. Bauer, H. Matzke, and J. Brown, *Arch. Neurol. Psychiat.* 1950:63, 257.
7.
D. Barker, ed., *Symposium on Muscle Receptors*, Hong Kong University Press, 1962.
8.
H. Batsel, *Fed. Proc.* 1963:22, 337.

226   I. A. KEDER-STEPANOVA

9.
————, *Exp. Neurol.* 1964:9, 410.
10.
————, *Exp. Neurol.* 1965:11, 341.
11.
R. Baumgarten, *Pflüg. Arch. Ges. Physiol.* 1956:262, 573.
12.
————, K. Baltasar, and H. Koepchen, *Pflüg. Arch. Ges. Physiol.* 1960: 270, 504.
13.
————, A. Baumgarten, and K. Schaeffer, *Pflüg. Arch. Ges. Physiol.* 1957:264, 217.
14.
———— and E. Kanzow, *Arch. Ital. Biol.* 1958:96, 361.
15.
———— and S. Nakayama, *Pflüg. Arch. Ges. Physiol.* 1964:281, 245.
16.
————, H. Schmiedt, and N. Dodich, *Ann. N. Y. Acad. Sci.* 1963:109, 536.
17.
D. Baxter and J. Olszewski, *J. Neurophysiol.* 1955:18, 276.
18.
L. Beaton and H. Magoun, *Amer. J. Physiol.* 1941:134, 177.
19.
R. Bianconi and F. Raschi, *Arch. Ital. Biol.* 1964:102, 56.
20.
V. Bijlani and N. Keswani, *Ind. J. Med. Res.* 1961:5, 49, 648.
21.
T. Biscoe, *J. Physiol.* 1962:164, 189.
22.
C. Breckenridge and H. Hoff, *Amer. J. Physiol.* 1950:160, 385.
23.
———— and H. Hoff, *Amer. J. Physiol.* 1954:178, 521.
24.
———— and H. Hoff, *Amer. J. Physiol.* 1955:180, 219.
25.
A. Brodal, *Retikulyarnaya formatsiya mozgovogo stvola* (*The Reticular Formation of the Brain Stem*), Moscow: Medgiz, 1960.
26.
J. Brookhart, *Amer. J. Physiol.* 1940:129, 709.
27.
D. Burns and G. Salmoiraghi, *J. Neurophysiol.* 1960:23, 27.

28.
I. Calma, *J. Physiol.* 1952:117, 9.
29.
P. Chatfield and D. Purpura, *Amer. J. Physiol.* 1953:173, 632.
30.
M. N. Cohen, *Fed. Proc.* 1958:21 (2), 349.
31.
M. Cohen and S. Wang, *J. Neurophysiol.* 1959:22, 33.
32.
J. Colle and J. Massion, *Arch. Int. Physiol.* 1958:66, 496.
33.
———— and J. Massion, *Acta Neurophysiol.* 1959:67, 223.
34.
V. Critchlow and C. Euler, *Experientia* 1962:18, 426.
35.
———— and C. Euler, *J. Physiol.* 1963:168, 820.
36.
M. Dirken and S. Woldring, *J. Neurophysiol.* 1951:14, 211.
37.
A. Dontas and R. Gesell, *Arch. Int. Physiol.* 1955:63, 305.
38.
R. Eccles, T. Sears, and C. Shealy, *Nature* 1962:193, 818, 42.
39.
G. Eklund, S. Euler, and S. Rutkowski, *Acta Physiol. Scand.* 1963:57, 481.
40.
S. Euler and U. Soderberg, *J. Physiol.* 1952a:118, 545.
41.
———— and U. Soderberg, *J. Physiol.* 1952b:118, 554.
42.
C. Evzaguirre and J. Taylor, *J. Neurophysiol.* 1963:26, 61.
43.
R. Gesell, *Ergebn. Physiol.* 1940:43, 475.
44.
————, A. Atkinson, and R. Brown, *Amer. J. Physiol.* 1940:128, 629.
45.
————, J. Bricker, and S. Magee, *Amer. J. Physiol.* 1936:117, 423.
46.
———— and A. Dontas, *Amer. J. Physiol.* 1952b:170, 690.
47.
———— and A. Dontas, *Amer. J. Physiol.* 1952a:170, 702.

48.
———, S. Magee, and J. Bricker, *Amer. J. Physiol.* 1940:128, 615.
49.
P. Gill, *J. Physiol.* 1963:168, 239.
50.
——— and M. Kuno, *J. Physiol.* 1963a:168, 274.
51.
——— and M. Kuno, *J. Physiol.* 1963b:168, 358.
52.
V. D. Glebovskiy, *Fiziol. Zh. SSSR* 1962:48, no. 6, 545.
53.
———, *Fiziol. Zh. SSSR* 1964:50, no. 9, 1158.
54.
——— and N. A. Pavlova, *Fiziol. Zh. SSSR* 1962:48, no. 12, 1444.
55.
E. Haber, K. Kohn, S. Ngai, D. Holaday, and S. Wang, *Amer. J. Physiol.* 1957:190, 350.
56.
V. Henderson and T. Sweet, *Amer. J. Physiol.* 1929:91, 94.
57.
H. Hoff and C. Breckenridge, *Amer. J. Physiol.* 1949:158, 157.
58.
——— and C. Breckenridge, *J. Neurophysiol.* 1952:15, 47.
59.
——— and C. Breckenridge, *Arch. Neurol. Psychiat.* 1954:72, 11.
60.
T. Hukuchara, *Brain Nerve* (Tokyo) 1952:4, 269.
61.
———, S. Nakayama, and H. Okada, *Jap. J. Physiol.* 1954:4, 145.
62.
———, H. Okada, and S. Nakayama, *Jap. J. Physiol.* 1956:6, 87.
63.
———, T. Sumi, and H. Okada, *Jap. J. Physiol.* 1953:3, 138.
64.
K. P. Ivanov, *Fiziol. Zh. SSSR* 1955:41, no. 6, 775.
65.
Z. N. Ivanova, in collection *Issledovaniya po farmakologii retikulyarnoy formatsii i sinapticheskoy peredachi* (*Research on the Pharmacology of the Reticular Formation and Synaptic Transmission*) edited by A. V. Valdmana 1961, pp. 176–192.

66.
F. Johnson and G. Russel, *Anat. Rec.* 1952:112, 348.
67.
I. A. Keder-Stepanova, in collection *Elektrofiziologiya nervnoy sistemy* Materialy IV vses. elektrofiziol. Konf. Tezisy dokladov Ph/D (*The Electrophysiology of the Nervous System*, material of the IV All-Union Electrophysiological Conference, Ph.D. Theses reports) 1963.
68.
—— and G. A. Kurella, *Fiziol. Zh. SSSR* 1957:43, no. 8, 721.
69.
—— and V. A. Ponomarev, *Biofizika* 1965:10, issue (vyp) 2, 324.
70.
——, V. A. Ponomarev, and A. N. Chetaev, *Biofizika* 1966:11, issue (vyp) 1.
71. ·
A. Keller, *Amer. J. Physiol.* 1929:89, 289.
72.
D. Kerr, *Aust. J. Exp. Biol. Med. Sci.* 1950:28, 421.
73.
——, C. Dunlop, E. Bert, and J. Mullner, *Amer. J. Physiol.* 1954:176, 508.
74.
N. Keswani, R. Groat, and W. Hollinshead, *J. Anat. Soc.* (India), 1954:3, 82.
75.
D. A. Kocherga, *Byull. Eksperim. Biol. i Med.* 1955:39, no. 12, 7.
76.
K. Koizumi, J. Ushiyama, and C. Brooks, *Amer. J. Physiol.* 1961:200, 679.
77.
——, J. Ushiyama, C. Brooks, *Amer. J. Physiol.* 1961:200, 694.
78.
A. Liljestrand, *Acta Physiol. Scand.* 1953:29, 106.
79.
——, *Physiol. Rev.* 1958:38, 691.
80.
Yu. P. Limanskiy, *Fiziol. Zh. SSSR* 1961:47, no. 6, 671.
81.
T. Lumsden, *J. Physiol.* 1923:57, 153.
82.
——, *J. Physiol.* 1923–1924:58, 81, Parts 1, 2.

83.
F. Magni and W. Willis, *Arch. Ital. Biol.* 1963:101, 681.
84.
H. Magoun and L. Beaton, *Amer. J. Physiol.* 1941:134, 186.
85.
M. Marckwald, *Z. Biol.* 1886:23, 149.
86.
————, *The Movements of Respiration and their Innervation in the Rabbit*, London: Blackie, 1888.
87.
———— and N. Kronecker, *Arch. Anat. Physiol.* (Physiol. Abt.) 1880: 1–2, 441.
88.
M. Ye. Marshak, *Regulyatsiya dykhaniya u cheloveka* (*The Regulation of Respiration in Man*), Moscow: Medgiz, 1961.
89.
N. A. Merkulova, *Byull. Eksperim. Biol. i Med.* 1960:50, 16, 9, 41.
90.
N. A. Mislavskiy, *O dykhatel'nom tsentre—izbr. provizv.* (*The Respiratory Center—selected works*), Moscow: Medgiz, 1952.
91.
E. Mortimer and K. Akert, *Amer. J. Physiol.* 1961:40, 228.
92.
J. Morton, *Amer. J. Physiol.* 1958:195, 23.
93.
T. Nakayama and R. Baumgarten, *Pflüg. Arch. Ges. Physiol.* 1964:281, 231.
94.
———— and T. Hori, *Jap. J. Physiol.* 1964:14, 147.
95.
P. Nathan and T. Sears, *J. Neurol. Nerosurg. Psychiat.* 1960:23, 590.
96.
J. Nelson, *J. Neurophysiol.* 1959:22, 590.
97.
R. Nesland and J. Nelson, *Fed. Proc.* 1963:22, 337.
98.
———— and F. Plum, *Exp. Neurol.* 1965:12, 337.
99.
S. Ngai, *Amer. J. Physiol.* 1957:190, 356.
100.
————, M. Frumin, and S. Wang, *Fed. Proc.* 1952:11, 112.

101.
H. Nicholson and J. Hong, *Fed. Proc.* 1939:1, 63.
102.
R. Oberholzer and W. Tofani, *Handbook of Physiol. SI.*, *Neurophysiology*, 1960, vol. 2, pp. 1111–1129.
103.
D. Ondina, W. Yamamoto, and W. Masland, *Amer. J. Physiol.* 1960:198, 389.
104.
B. Ya. Peskov, *Fiziol Zh. SSSR* 1962:48, no. 11, 1368.
105.
T. Pitts, *J. Comp. Neurol.* 1940:72, 605.
106.
R. Pitts, *Amer. J. Physiol.* 1941:134, 192.
107.
——, *J. Neurophysiol.* 1942:5, 75.
108.
——, *J. Neurophysiol.* 1942:5, 403.
109.
——, *Physiol. Rev.* 1946:26, 609.
110.
——, H. Magoun, and S. Beaton, *Amer. J. Physiol.* 1939:126, 673.
111.
——, H. Magoun, and S. Ranson, *Amer. J. Physiol.* 1939:126, 689.
112.
——, H. Magoun, and S. Ranson, *Amer. J. Physiol.* 1939:129, 654.
113.
D. Purpura and P. Chatfield, *J. Neurophysiol.* 1952:15, 281.
114.
—— and P. Chatfield, *J. Neurophysiol.* 1953:16, 85.
115.
G. Ramos, *Acta Physiol. Lat. Amer.* 1959:9, 246.
116.
——, *Acta Physiol. Lat. Amer.* 1960:10, 104.
117.
—— and E. Mendosa, *Acta Physiol. Lat. Amer.* 1959:9, 257.
118.
G. Salmoiraghi, *Ann. N.Y. Acad. Sci.* 1963:109, 571.
119.
—— and R. Baumgarten, *J. Neurophysiol.* 1961:24, 203.

232 I. A. KEDER-STEPANOVA

120.
—— and D. Burns, *J. Neurophysiol.* 1960:23, 2.
121.
—— and D. Burns, *J. Neurophysiol.* 1960:23, 14.
122.
—— and F. Steiner, *J. Neurophysiol.* 1963:26, 581.
123.
M. Scheibel, A. Scheibel, A. Mollica, and G. Moruzzi, *J. Neurophysiol.* 1955:18, 309.
124.
T. Sears, *J. Physiol.* 1958:142, 35P.
125.
——, *Nature* 1963:197, 4871, 1013.
126.
——, *J. Physiol.* 1964a:173, 150.
127.
——, *J. Physiol.* 1964b:174, 295.
128.
——, *J. Physiol.* 1964c:175, 386.
129.
——, *J. Physiol.* 1964d:175, 404.
130.
T. Sumi, *Fed. Proc.* 1962:21, 349.
131.
——, *J. Neurophysiol.* 1963:26, 467.
132.
——, *J. Neurophysiol.* 1963b:26, 478.
133.
——, *Pflüg. Arch. Ges. Physiol.* 1963:278, 172.
134.
——, *Ann. N.Y. Acad. Sci.* 1963:109, 551.
135.
——, *Pflüg. Arch. Ges. Physiol.* 1964:278, 467.
136.
P. Tang, *Amer. J. Physiol.* 1953:172, 645.
137.
—— and T. Ruch, *Amer. J. Physiol.* 1951:167, 830.
138.
A. G. Teregulov, *Russ. Fiziol. Zh. im. I. M. Sechenova* 1928:11, 259.
139.
A. Torvik, *Comp. Neurol.* 1956:106, 51.

140.
———— and A. Brodal, *Anat. Rec.* 1957:128, 113.
141.
A. V. Valdman and Ma Ts'uan-geng, *Fiziol Zh. SSSR* 1964:48, no. 7, 793.
142.
S. Wang and S. Ngai, *Ann. N. Y. Acad. Sci.* 1963:109, 550.
143.
————, S. Nagai, and M. Frumin, *Amer. J. Physiol.* 1957:190, 333.
144.
H. Wang and S. Wang, *Ann. Rev. Physiol.* 1959:21, 151.
145.
R. Warwick and G. Mitchell, *J. Comp. Neurol.* 1956:105, 553.
146.
S. Woldring, *Acta Physiol. Pharmacol. Neerl.* 1951:2, 140.
147.
———— and M. Dirken, *J. Neurophysiol.* 1951:14, 227.
148.
R. Wuerker and F. Henneman, *J. Neurophysiol.* 1963:26, 539.
149.
O. Wyss, *Pflüg. Arch. Ges. Physiol.* 1939:242, 215.
150.
————, *Pflüg. Arch. Ges. Physiol.* 1940:243, 456.
151.
————, *Pflüg. Arch. Ges. Physiol.* 1941:244, 712.
152.
————, *Helv. Physiol. Pharm. Acta* 1954:S10, 5.
153.
————, *Schweiz. Med. Wochenschr.* 1957:87, 814.
154.
————, *Acta Neurophysiol.* (Paris), 1959:135.
155.
————, *Ann. Rev. Physiol.* 1963:25, 180.
156.
Yu. N. Zubilov, *Fiziol. Zh. SSSR* 1962:48, no. 5, 554.

7

A Model of a System of Neurons   I. A. Keder-Stepanova
with Periodic Volley Activity   N. N. Rikko
Resistant to Random Afferent
Influences

In the lateral regions of the respiratory center of the medulla oblongata neurons are located the volley activity of which determine the respiration rhythm. A regular rhythmic volley activity of the neurons of the medulla oblongata, synchronous with some phase of the respiratory cycle, is maintained after elimination of all respiratory afferentation and connections with neurons of other levels. It is possible to single out the neurons the volleys of which are synchronous with the inspiration phase (the inspiratory ones) or with the expiration phase (expiratory ones).

The question naturally arises concerning the generation mechanism of periodic volleys in the system of the respiratory neurons. The existing points of view cannot explain the whole complex of known physiological facts. Some authors assume that the appearance of periodic volleys is determined by the different properties and reciprocal interrelations of the inspiratory and expiratory neurons. Nowadays the most widespread viewpoint is that of Burns and Salmoiraghi [1] and Salmoiraghi and Baumgarten [3], who think that two types of respiratory neurons form two self-exciting chains; within each of them exist abundant excitatory connections, and between the two chains there are inhibitory connections. Any afferent impulse of sufficient strength can start the activity of one of the self-exciting chains which contain neurons with a somewhat different excitability and refractoriness. The presence of self-limiting mechanisms, in particular neurons of the Renshaw cell type, and the increase of the threshold of neurons in the course of the development of a volley leads to a reduction of the overall activity of the chain. This promotes the activity of the second reciprocal chain, which once and for all inhibits the work of the first.

We will note that such a model neither explains a whole series of experimental data (see Chapter 6) nor takes into consideration, moreover, the role of the continuously working cells among which the respiratory neurons are located. The fact is that under certain conditions the number of neurons producing rhythmic volleys coinciding with the phases of respiration increases, such that there is basis to assume that initially these cells did not function in a respiratory mode.

The reticular formation of the medulla oblongata within which the respiratory neurons are found possesses a high degree of activity; its neurons are able to respond with volleys to a single afferent stimulus. The majority of the cells work continuously or with irregular volleys. One can assume that all the neurons of the reticular formation are somehow connected with one another, but, besides that, they form several functionally different groups, for example, a group of respiratory neurons within which there are specific and nonspecific interneural interrelations. It is entirely probable that in each such group there is a convergence of specific and nonspecific afferent influences. The specificity of the afferent and efferent connections determines the functional importance of the group.

In the present article we will try to describe some model concepts about the origin of volley activity in the systems of the two types of neurons. The neurons of the first type form a medium in which cells of the second type are immersed. Between the neurons of the medium there are only nonspecific connections. The neurons of the second type can also have nonspecific connections with the cells of the medium, but besides that, specific links do exist among them. We should at once note that these neurons have different properties. As a result of the interaction of the two types of neurons, there arises in such a system rhythmic volley activity. In constructing a model, we have used certain properties of continuous-control systems and ideas suggested by I. M. Gelfand and M. L. Tsetlin [2].

The model concepts were assumed on the basis of a program for the electronic digital computer which made it possible to carry out a series of computer experiments.

1. THE ELEMENTS OF THE MODEL

The model consists of two types of cells. Cells of type A are located in the nodes of a plane rectangular network and are connected only with their own four immediate neighbors, thus forming a model of a continuous medium. Cells of type B are immersed in this continuous medium. Each of them is connected with a large number of type A cells located in its immediate vicinity. There are many times fewer type B cells than type A cells.

1.1. THE TYPE A CELL

The type A cell is capable of being excited, and the excitation lasts one unit of time. After excitation the cell is refractory R units of time. At moment $t + 1$, for excitation of a type A nonrefractory cell it is sufficient if at moment t only one of the cells of the medium connected with it is excited. In this way, excitation spreads over the medium at a unitary rate. Excitation can be caused by an external impulse. In the absence of spontaneous activity an excitation which arises in one cell under the influence of an exterior signal spreads from it over the medium. Within time K the front of excitation (the aggregate of simultaneously excited points) will be found within K cells. If on the path of the excitation wave refractory cells are met, then the wave will bypass them. The cells immediately behind the excitation front become refractory. With the movement of excitation over the medium, the refractory zone moves behind the excitation front. If excitation fronts move toward one another, then the excitation disappears where these fronts meet. Type A cells can have spontaneous activity, that is, the capacity to become excited spontaneously with a period T: within T units of time after excitation the cell again is excited, if, of course, during time T it was not previously excited under the influence of the activity of the neighboring cells or exterior impulses.

1.2. TYPE B CELLS

Just as do type A cells, type B cells possess the capacity to be excited and to be refractory. However, for excitation of a type B cell at moment $t + 1$ it is necessary that at moment t a specific number P_t of type A cells, connected with it by excitatory connections, are excited. The size of P_t—we will call it the *excitation threshold*—depends on how the type B cell was working prior to moment t. We reckon that the size of P_t for a refractory cell is infinite. It is assumed that if at moment t_0 the cell was excited, then $P_{t_0+1} = \cdots = P_{t_0+R} = \infty$; $P_{t_0+R+1} = P_{t_0} + h$. Thus, excitation raises the threshold h units. In the event of the absence of excitation the size of the threshold drops at the rate r to the value P_0. If at moment t the cell was not excited and was not refractory, then

$$P_{t+1} = \begin{cases} P_t - r, & \text{if } P_t - r > P_0, \\ P_0, & \text{if } P_\sigma - r \leqslant P_0. \end{cases}$$

The value P_0 is called the *initial threshold*.

2. THE PROPERTIES OF THE MODEL

The initial threshold determines the initial level of excitation of the medium of type A cells sufficient to activate a type B cell. An important characteristic of the B cell is the structure of its surroundings, that is, the geometry of its connections with type A cells. We will describe the geometry of the connections in the following manner: we will imagine that excitation spreads out from some A cell. The front of excitation, passing over the medium, will also pass over type A cells connected with a B cell. For the B type cell it is possible to construct a diagram of the connections with regard to any fixed type A cell which proves to be a source of excitation. Along the abscissa is reckoned the time which has passed from the moment of the excitation of the A cell, and along the axes of the ordinates, the number of A cells connected with the B cell which at the given moment were found to be excited (Fig. 3).

We will cite an example: in Fig. 1, a case is depicted when external excitation is applied to the center of the B cell's neighborhood (small circle). The position of the excitation front at subsequent moments of time is given by the dotted line. The type of A cells able to transmit excitation to the type

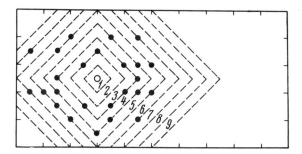

Fig. 1. Position of the front of excitation at subsequent moments of time (the dotted line). The source of excitation is in the center of the neighborhood. By the 7th moment there are no excited A cells.

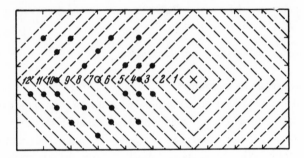

Fig. 2. Position of the excitation front at subsequent moments of time (dotted line). The source of excitation is the cell marked by a small cross. By moment 14 there are no excited A cells.

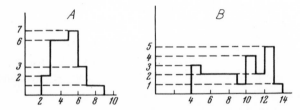

Fig. 3. Distribution of the excitatory connections of a type B cell relative to the source of excitation found (A) in the center of the neighborhood (see Fig. 1), and (B) outside the neighborhood (see Fig. 2).

B cells are marked by dots. The remaining cells are found in other nodes of the rectangular network. A diagram characterizing the connections of the B cell was constructed for this case in Fig. 3A. In Fig. 2 the excitation source is the cell marked by the ×, the B cell is situated at the point marked by the circle. A corresponding diagram is presented in Fig. 3B.

Thus, the behavior of the B cell depends on the following parameters: the number and distribution of connections with type A cells which excite the B cell, and the sizes of h, r, and R, that is, the relationship of the threshold change and the refractoriness. These parameters determine the chances of appearance of a volley in the type B cell, and its character, that is, the length of the volley, the number and distribution of the impulses in the volley. We will examine the example of the activity of the type B cell with the geometry of connections represented in Figs. 1 and 2, and with the parameters $P_0 = 3, h = 2, r = 1,$ and $R = 1.$ Along the ordinate we will plot the values of P_t, assuming that when $t = 0$ the threshold value is equal to P_0. From Fig. 4A it is apparent why in the case when at moment $t + 0$ the center of the region of the B cell is excited, the spreading excitation forces the B cell to trigger within 3 and 5 time units after external excitation (in this case the excitation of the B cell is not transmitted to the A cells). In 10 time units the threshold P_t falls to the value $P_0 = 3.$

In the diagrams depicting the changes of threshold P_t, a diagram is drawn by a dotted line of the distribution of the connections of the B cell relative to the center of excitation. If we excite the A cell marked by the $\times$ in Fig. 2, and assume that at the moment of excitation $t = 0, P_{t+0} = P_0$ then the B cell in this instance will trigger in 4, 10, and 12 time units after excitation, and the value P_0 will be attained by the moment $t = 17.$ The maximum threshold value in both cases will be 7.

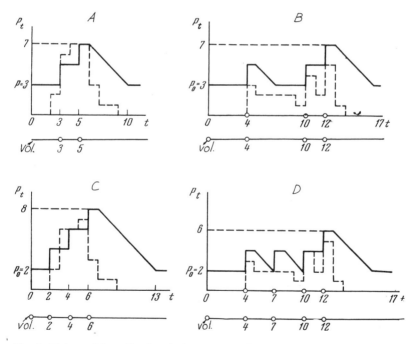

Fig. 4. Nature of the volley in relation to the mode of behavior.

For the case $P_0 = 2$, corresponding diagrams are represented in Figs. 4C and 4D. It is not difficult to see that the decrease of P_0 leads to the lengthening of the volley. In a certain sense the decrease of P_0 is equivalent to the enlargement of the neighborhood. An analytical description of the proposed model is extremely difficult. Therefore, for its analysis a program was worked out for the digital computer and a series of computer experiments were done. These experiments made it possible to clarify the influence of the parameters of the B type cell on its activity and the interrelationship of these parameters. In the programming, a field was used 30 × 48 type A cells and 1 to 3 cells of type B. For one and the same neighborhood, the influence on the characteristics of the volley of different sizes of the parameters h and r under the same initial conditions for type A cells was examined. Here it was made clear that during a change of these parameters the volley remains constant if the ratio h/r is maintained. With the increase of h the volley becomes more infrequent and shortens. An increase of r makes the volley more frequent and lengthens it.

So far, we have examined cases when excitation from the B cell is not transmitted to the type A cells adjacent with it. In these conditions it is obvious that the discontinuance of the volley is related to the exit of the excitation front from the range of the cell's neighborhood. The case is interesting when the B cell is included in the medium; that is, from its triggering the four type A cells adjacent with it become excited. As the experiments showed, the inclusion of the B cell in the medium causes an abrupt change of the nature of the volley. The volley lengthens by several times; during the volley threshold $P\sigma$ manages to grow to a considerably larger value. In the instance when the B cell is included in the medium the discontinuance of the volley is found to be related not only with the exit of the excitation front from the neighborhood of the cell but also with the growth of the threshold P_t.

The volley can cease if P_t becomes so high that the number of excited A cells is found insufficient for the excitation of the type B cell. As an example we will cite the volley depicted in Fig. 6, for an instance when the initial wave goes from the center marked by a small cross as in Fig. 2. The parameters were selected as follows: $h = 2, r = 1, R = 1, P_0 = 3$. Let us compare Fig. 6 with 4B. In the Fig. 6 volley are 9 impulses instead of 3; the

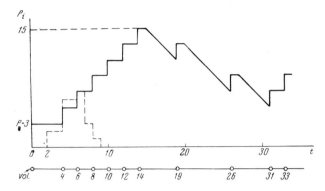

Fig. 5. Type B cell included in the medium. The source of excitation is in the center of the neighborhood (see Fig. 1). $P_0 = 3$; $h = 2$; $r = 1$; $R = 1$.

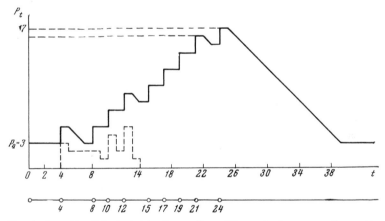

Fig. 6. Type B cell included in the medium. The excitation source is outside the neighborhood (see Fig. 2). $P_0 = 3$; $h = 2$; $r = 1$; $R = 1$.

spacing between the extreme impulses is 22, in comparison with 8; P_t max $= 17$ instead of 7. The increase of P_t leads to an increase of the spacing between impulses at the end of the volley, that is, to a decrease in frequency of the volley toward the end (Figs. 5 and 6).

So far, we have reported the results of experiments for the case of only

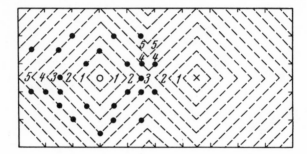

Fig. 7. Mutual extinguishing of excitations. By moment 7 there are no excited A cells.

one front of excitation. In the event of the passage of several fronts the picture becomes more complicated. For instance, two fronts moving from opposite directions extinguish one another. If we visualize that at moment 0 the two points shown in Fig. 7 ($\bigcirc$, $\times$) are simultaneously excited, then at moment 4 the fronts will meet and begin their mutual extinguishing. The intervals between each two impulses of this volley will lengthen, the volley will become shorter, since the extinguishing of the front in a certain sense is equivalent to a reduction of the number of the connections of the B cell.

We will now pass to the question concerning the appearance in the model of a rhythmic volley's working behavior and the stability of this mode. Spontaneous activity of the medium can cause repetition of the volleys, or external afferentation entering into the medium from without and synchronized with its volley activity. In the presence of spontaneous activity the volley of B cells both included in the medium and not included in the medium can periodically recur. In case the B cell is not included in the medium, any chance excitation changes the shape of the volley since the shape of the volley depends on which A cell the excitation comes from. If the B cell is included in the medium, then although the shape of the volley is disturbed by a chance excitation, the reestablishment of the shape of the volley occurs in the next volley since spontaneous activity of the medium always develops in the same place—in the immediate vicinity of the B cell included in the medium. Indeed, suppose at moment t the last impulse of the volley occurred, that is, for the last time in this volley the

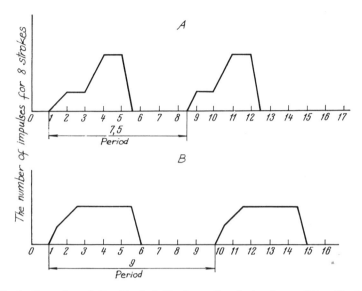

Fig. 8. Operation of a B cell included in the medium in the absence (A) and in the presence (B) of afferentation (i.e., during feeding of the volley of a B cell once again into the medium with some delay). Each division along the axis of the abscissae equals 8 units.

B cell was excited. Then the A cells adjacent with the B cell will become excited at the moment $t + 1$, and the rest will become excited later. Consequently the spontaneous excitation of the A cell adjacent with the B cell will occur at the moment $t + 1 + T$, that is, earlier than the excitation of the other type A cells. If the volley rising in the B cell moves again into the medium with some delay (afferentation), then a fixed pattern is established and the volley behavior is found resistant with respect to random excitations.

An analysis of the working of behavior of B cells on the computer is done as follows: for the B cell there are assigned the parameters P_0, h, r, and R and the geometry of the connections, and for the type A cells the sizes of R and T are assigned. Excitation proceeded from the center of the neighborhood. For each set of parameters the stable form of the volley and the repetition period of the volley are found. The calculation was repeated twice: for cases when the B cells are not included in the medium, and for cases where they are so included. Then the parameters were changed and the experiment was repeated. Thus, the examples were obtained which we cite below; these examples illustrate the role of the parameters in the determination of the characteristics of the volley and of the working behavior of the volley.

An example of the operation of a single B cell included in the medium, during the absence and presence of afferentation (i.e., with the feeding of the B cell volley with some delay once again into the medium), is presented in Fig. 8. With the inclusion of afferentation the volley is shorter and the frequency of the impulses increases toward the end. With the exclusion of afferentation the volley and the total period (volley plus pause) are lengthened. The volley of the B cell becomes less frequent, which is typical for transition from a mode of behavior with a period which is determined by the size of the delay in the afferent paths to a mode which is characterized by spontaneous activity.

Now we will pass to cases when two or several type B cells were included in a medium of type A cells, and we will examine the summary volley generated by these cells. Already in the case of two cells we observe a new phenomenon—synchronization of their behavior. By synchronization we comprehend the sort of behavior where the volleys of the individual B type cells are not independent.

In an experiment with two type B cells with nonintersecting neighborhoods we observed that during simultaneous initial excitation the cells worked independently of one another since the mutual extinguishing of the waves occurred outside the neighborhood of the cells. In this instance the summary volley coincided with the volley of one cell (their characteristics were identical). However, such a mode is unstable. Random excitation will change it to another possible mode which is found even to be stable. This stable mode can be described as follows: if during initial excitation one of two identical type B cells become excited, then during behavior with a steady periodic volley we will always see the same summary volley. Either of the two cells having begun to work first will force the second cell to trigger within a fixed time. The volley of the cell which began to work first will differ from the volley of the second. If the neighborhoods of the cells intersect, then mutual extinguishing of the waves

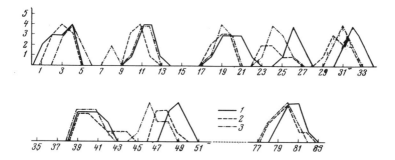

Fig. 9. Example of synchronization of three type B cells. One division along the axis of the abscissae is 8 units. Along the axis of the ordinates are the number of impulses in a given length of time. (1) = B_1; (2) = B_2; (3) = B_3.

occurs in the neighborhood of the cells. This causes a decrease in frequency and lengthening of the volley, but in this instance the summary volley is stable.

We will present the example of three cells included in the medium. In Fig. 9, three B cells (we will call them B_1, B_2, and B_3) are located in a row, their neighborhoods are identical, and they intersect; $r_1 = r_2 = r_3 = 2$, $R_1 = R_2 = R_3 = 1$, $h_1 = h_3 = 8$, $h_2 = 9$. The B cells begin to function with $P_t = P_0$. The fixed volley has a length of approximately 50 units, the volley of each of the B cells ≈ 40 units. For these cells not included in the medium, the length of the summary volley strongly depends on which point of the field of A cells the initial excitation comes into. It is not difficult to see that the length of the volley of each of these cells does not exceed the radius of the circle circumscribing the neighborhood. This length is in the neighborhood of 20 units. The summary volley for cells not included in the medium is in the vicinity of 30 units and its shape depends on which cell of the medium the excitation moves from.

The degree of synchronization depends on the degree of "intersectedness" of the neighborhoods. We will submit that the cells included in the medium differ only in their h/r ratio. Then the length and most frequent volley will belong to that cell which has the smallest h/r ratio. This means that the last waves over the medium pass from it. That cell proves to be the leader in the system of cells. The volleys of the remaining cells will begin within a fixed time after the start of work of the "leading" cell. The summary volley is resistant to random excitations both in the presence of a feedback and in its absence. We will cite an example when three type B cells with identical neighborhoods but different h/r are in the medium.

In Fig. 10 an example of the behavior of three such cells—B_1, B_2, and B_3—is presented. The cell with the parameters in which h/r is the least (other things being equal) will always lead the volley. The volley is synchronized according to the start and length of the volley of the leading cell. When $h_1 = h_3 = 8$, $h_2 = 9$, and $r_1 = r_2 = r_3 = 2$, either cell B_1 or cell B_2 will conduct, depending on the site of application of the external excitation. The summary volley has a length of 40 units, the volley of the leading cell is 32 units, the volley of the guided one is 24 units. With the parameters $h_1 = h_3 = 8$, $h_2 = 5$, the summary volley continues to be 40 units, the volley of the leading cell is 40 units, the volley of the guided ones 32 units. With the further increase $h_1 = h_3 = 11$, $h_2 = 8$ the summary volley is 24 units, the volley of the leading cell is 24 units, and the volley of the guided cell is 20 units. For $h_1 = h_3 = 15$, $h_2 = 11$, the summary volley is 20 units, the length of the volley of the guided cells is 16 units.

In Fig. 11, an example is presented illustrating the influence of the degree of "intersectedness" of the neighborhoods of different cells on their synchronization in the summary volley. At the beginning 3 cells were working differing only in the parameter h ($h_1 = h_3 = 8$, $h_2 = 9$). In the summary volley the last cell, B_3, was the leader. After the change of h ($h_1 = h_3 = 15$, $h_2 = 9$), the middle one, B_2, became the leader. After exclusion of the middle cell from the medium (moment 68) the degree of the synchronization of the cells changed. The spacing between initial volleys in the summary volley increased and the neighborhoods of the extreme cells intersected slightly (moment 93). After the middle cell, B_2, was again included, the original working behavior of the three cells (see above) was established. The beginnings of the volleys converged and the volleys lengthened a little, aligning themselves to the volley of the middle cell, B_2.

The experiments were carried out on a Gamma-drum machine. The length of the program constituted one thousand one-address commands. Since in the machine the program and data are registered on the drum, and the directions are received from the buffer memory in an arithmetical

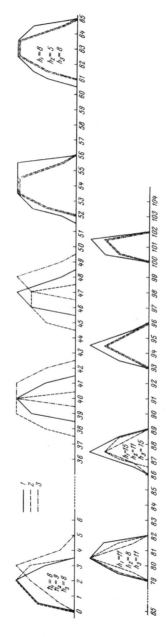

Fig. 10. Example of the operation of three type B cells with identical neighborhoods but different h/r; one division along the axis of the abscissae equals 8 units. (1) = B_2; (2) = B_1; (3) = B_3.

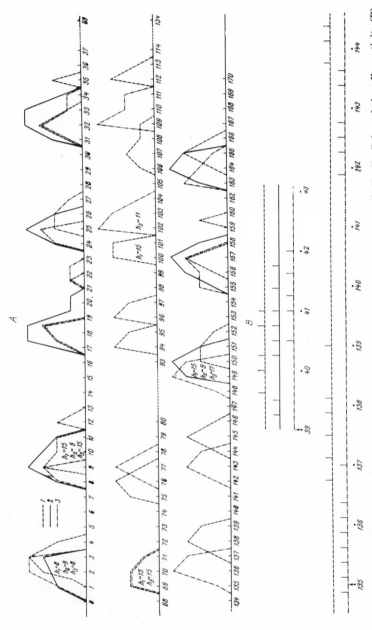

Fig. 11. Influence of the degree of "intersectedness" of the neighborhoods of type B cells (B_1, B_2, B_3) on their volley activity (B) and synchronization in the summary volley (A) (see text). (1) — B_1; (2) — B_2; (3) — B_3.

arrangement, on each unit or stroke of the operation of the system approximately 12 seconds are spent for the entire computation of all the cells. For print-out of the whole field, 5 seconds are needed.

3. CONCLUSIONS

1. The model described of a system of heterogeneous neurons represents one of the possible mechanisms of generation of rhythmic volley activity. This activity is resistant to random external effects. The computer experiments carried out on the model made it possible to evaluate the influence of different parameters on the stability, the frequency of the appearance of volleys, and their duration and configuration.

2. The volley activity arises as a result of the interaction of two types of neurons. The neurons of the first type (A) make up the medium in which neurons of the second type (B) are included. It is assumed that the type B neurons have a varying triggering threshold, and all the connections are excitatory.

3. The appearance of periods of volley activity can be explained either by feedback, which operates with some delay, or by spontaneous activity of the cells of the medium. In the latter case, especially with the inclusion of a B cell in the medium, the system is found most resistant to random external effects.

4. If random afferentation enters the medium, then the cells of the medium located near the B cell work in a volley behavior the same as that of the B cell. The outlying type A cells will function continuously with volleys of a different period or even keep silent. In this sense the behavior of the A cells is analogous to the behavior of the cells of the reticular formation surrounding the respiratory neurons and constituting with them a single functional group.

5. Phenomena of synchronization were uncovered in the model, and it was established that the degree of synchronization depends on the degree of "intersectedness" of the neighborhood of the type B cells, while the cell having the longest volley initially always turns out as the leader in such a system.

6. The system described here cannot of course serve as a complete model of the respiratory center. In it there are not taken into consideration

such important features of its structures as the existence of inspiratory and expiratory neurons. Also not taken into consideration is the presence of inhibitory connections, which very likely are in all systems of neurons.

REFERENCES

1.
D. Burns and G. Salmoiraghi, *J. Neurophysiol.* 1960:23, 270.
2.
I. M. Gelfand and M. L. Tsetlin, *Dokl. AN, SSSR* 1960:131.
3.
G. Salmoiraghi and R. Baumgarten, *J. Neurophysiol.* 1961:24, 203.

8

Some Special Features of
Organization of the Cerebellar
Cortex

V. V. Smolyaninov

This chapter reflects some attempts to approach an investigation of the organization of the cerebellar cortex from geometric standpoints. There will be described here the results of a quantitative investigation of cortical folding (see section 1) and the nature of the three-dimensional organization chiefly of the system of the basic neurons: the granule cells—the parallel fibers—and the Purkinje cells (sections 2–5); in the sixth section, simple electrophysiological models are discussed. The structure and characteristics of distribution of other neurons—the stellate and basket cells, and the Golgi cells—are examined very briefly, to the extent essential for an analysis conducted on the organization of the cortex.

In Cajal's classical work [6] we can find almost everything which the contemporary neurohistologist is able to see in the light microscope, but there still of course are not enough similar descriptions for an understanding of the intrinsic mechanisms of the organization of the system for the construction of even the most simple structural models. Recent quantitative research substantially supplements and develops the older classical concepts. As we know, many of the ideas and trends of the quantitative study of the morphology of the nervous system originates from the works of Bok [3] and Sholl [37], who were concerned with the examination of the cortex of the cerebral hemispheres.

At present the opportunity for a discussion of the organization of the cerebellar cortex in "quantitative language" is in many respects related to the recent research of Fox and Barnard [15], Braitenberg and Atwood [5], Friede [18], and Szentágothai [40]. A quantitative approach to the questions of the organization of the system makes it possible to come nearer to a geometric description of its structure, a formulation of the laws of organization, and partly perhaps an understanding of the inherent causes determining the given structure.

The question, really, about the geometry of the cerebellar cortex was first raised by Braitenberg and Atwood [5], who formulated five basic invariants (constant characteristics) of this geometry. The research of these authors was carried out on the human cerebellar cortex. In order to evaluate the significance of one or another of the characteristics of the geometry of the cortex, it seemed essential to us to extend the idea of the analysis of the organization of the system from the standpoint of detection of quantitative and spatial invariants in a phylogenetic series. The idea of similar comparisons is, of course, not new, and this idea, as we know, is only one of the possible paths of study of the variability of the organization. It is also interesting, of course, to trace the change of the structure in ontogenesis, but the range of phylogenetic changes are both broader and richer, and it is necessary to add that this means of procedure is simpler; therefore, we limited for the time being our research to phylogenesis.

Initially our interest in the cerebellar cortex was evoked by the granular layer in connection with concepts about continuous media examined in the work of I. M. Gelfand and M. L. Tsetlin [19]. The structure and function of the cerebellum, on the basis of similar concepts, were discussed in physiological seminars conducted by I. M. Gelfand.

We should note here that an examination of the functional structure of the cerebellar cortex as a continuous medium does not contradict the obvious assumption that from a morphological viewpoint it, like the other parts of the nervous system, consists of individual neurons. The existence of electrotonic interaction among neurons, discovered recently, indicates one of the possible ways of guaranteeing functional continuity in the system of neurons. But even for structures not having similar connections it is possibly expedient in some cases to carry out a functional description proceeding from the hypothesis of "continuous media."

In the illustrations, and sometimes also in the text, the following abbreviations are used:

ML   molecular layer
GL   granular layer
PL   layer of Purkinje cell bodies
PC   Purkinje cell
GC   granule cell
SC   stellate cell
BC   basket cell
GoC   Golgi cell

⊥-section   a section perpendicular to the axis of the lobe and parallel to the fibers

‖ -section   a section parallel to the axis of the lobe and perpendicular to the layer of Purkinje cell bodies

All the tables with the results of the measurements and tabulations are given at the end of the article (appendix) and, in addition, a separate list of letter symbols of basic sizes. An explanation of certain methods of measurements is also given at the end of the article.

1. THE MACROSCOPIC ORGANIZATION OF THE CORTEX

The homogeneity of the cortex, the stability of its internal structure in phylogenesis, and the harmony of the overall pattern of cortical folding make it possible to assume that the causes and laws of the formation of the convolutions are common for all animals, and that the diversity of the shape of the cerebellum is the consequence of the different degree of development both of the whole cortex and its individual parts.

In this section we will summarize the basic concepts concerning the principles of folding of the cerebellar cortex introduced in a series of papers [5, 15, 17, etc.], we will describe certain geometric characteristics of phylogenetic diversity of the forms of the cerebellum, we will cite the results of our measurements, and will discuss the possible causes of cortical folding. Bok [3] studied similar problems of the geometry of the cortical folding of the cortex of the cerebral hemispheres.

1.1. THE STRAIGHTENED CORTEX

One of the most popularized explanations of cortical folding (of both the cerebellum and the cerebral hemispheres) reduces to the following: the growth of the neural plate in ontogenesis is checked by the cranium, and the cortex is forced to form folds. Although such an interpretation of the causes of distortion by itself, possibly, is not even correct (see below), the prime concepts contained in it about the cortex as an elastic plate able to be folded by external effects are very convenient for the description of the geometry of the convolutions. Later on we will use these concepts without stipulations. Thus, as follows from the indicated hypothesis of folding, if the cortex developed in an open space, then the plate of gray matter formed by it would not have folds and such a plate (a straightened cortex) in the simplest case could be flat and have the appearance of a hemisphere, a saddle, and so forth. It is in fact possible to carry out this conceivable experiment by means of special geometric transformations of portions of the real cortex.

The diagrams of the cerebellum which are usually used are an unfolding into a plane of only the exposed part of the cortex, that is, that which is visible with examination of the cerebellum from all sides [26–28, 32]. Of course such charts give only a "superficial" idea of the distribution of convolutions on the cortex because a large part of the cortex of mammals is hidden in fissures.

The shape ("model") of the straightened cortex is a convenient object for clarification of its two-dimensional (in the plane of the plate) macro-organization and micro-organization. The principal possibility of such a transformation results from the fact that folds arising in the layers in the folded sections are clearly determined by the sign and the amount of bending. Moreover, the fact that the volume of the layers does not change during folding is significant.

1.2. VOLUME AND AREA

For a long time we have known that in the ascending order of vertebrates the volume of the cerebellum progressively increases, though not steadily. A sufficiently large amount of work exists in which the volume (or weight) of the cerebellum has been carefully measured and comparisons conducted with the volume (weight) of other parts of the central nervous system. It seems to us, however, that we should compare such formations as the cortices in area and thickness if as a matter of fact it makes sense to compare them with respect to macroindices.

It is relatively simple to measure the thickness of the cortex and the volume of the cerebellum, while the well-known methods of measuring the cortical area (of mammals) leads to a vast amount of measurements and recalculations [2]. We can suggest a quite simple method of finding the area of the cortex, based on the fact that its volume is the invariant of the folding. If the volume of the cortex (of the cerebellum) V_{ccb} is known, then its area is

$$S = V_{ccb}/(H + h), \tag{1}$$

where H and h are the thicknesses of the molecular and granular layers for an unfolded cortex. Consequently, the area determined by this expression is the area of an unfolded (or "straightened") cortex.

A basic difficulty is in the measurement of V_{CCb} because in practice usually (by immersion in liquid) the volume is measured of the whole cerebellum (V_{Cb}) cut off at the level of the peduncle from the medulla oblongata, that is, in the V_{Cb}, besides the V_{CCb}, there enters the volume of the central nuclei, the white matter, and part of the peduncles.

As is obvious from sagittal sections in the region of the vermis, the cortex, folding repeatedly, makes up for almost the whole volume (Fig. 1). From our measurements on sections it follows that the portion of the volume of white matter in the region of the vermis does not exceed 1 percent and all the white matter with the nuclei and peduncles constitute approximately 10 percent, that is, $V_{CCb} \approx 0.9 V_{Cb}$. In an equivalent concept such a relationship of volumes indicates the following: if the whole cerebellum is a sphere with a single radius, then all the white matter with nuclei will occupy the central part with the mean radius approximately 0.45.

The results of the measurements of the volumes of the thickness of the layers and the values of the cortical area estimated according to [1] are presented in Table 1 of the Appendix.

1.3. THE LAWS OF CORTICAL FOLDING

First let us note [see also 5, 15, 17] the basic characteristics of the layers, the first three of which it is easy to observe even during visual study of the slices (see Fig. 1): (1) the thickness of the molecular layer H is everywhere constant; (2) the thickness of the granular layer is determined by the sign and degree of cortical folding. We will label it $h_{\pm}$ with concave ($+$) or convex ($-$) folds and h for nonfolded sections of the cortex. Then $h_{+} < h < h_{-}$; (3) the spacing between the bodies of the Purkinje cells in a $\perp$-section increases in concave sections and lessens in convex ones: $a_{-} < a < a_{+}$; (4) the volume of the cortex and the volumes of each of the layers are invariants of the folding. Owing to the cylindricality of cortical folding there hence follows the invariance of the areas of the layers, and vice versa.

Further, from the invariance of the areas S_{ML} and S_{GL} there results

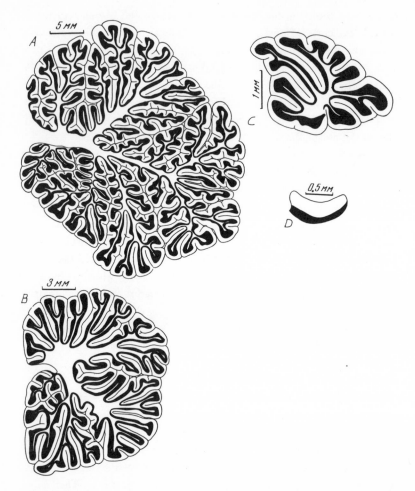

Fig. 1. Sagittal section of the cerebellum ("arbor vitae"). (A) man; (B) cat; (C) mouse; (D) frog.

$(S_{ML}/S_{GL}) = \text{const} = (H/h)$. The constancy of the last proportions within the cerebellum alone is a trivial consequence of (4). However, it is remarkable that these proportions are constants of phylogenesis. This interesting characteristic of the cerebellar cortex of mammals and birds was recently discovered by Friede [18]. From our material we could also be convinced of the accuracy [of the statement] that for the cerebellum of birds and mammals

$$H/h \simeq 1.5. \tag{2}$$

From the constancy of the thickness of the molecular layer and the invariance of its volume or area (below we will examine only $\perp$-sections of the cortex) there follows a very simple, but important for the understanding of the geometry of distortion, circumstance: (5) the molecular layer and hence also all the cortex are folded around the middle level of the molecular layer.

Therefore, the spacing in the layer of Purkinje cell bodies will increase in concave sections and decrease in convex ones (Fig. 2). In the first approximation we can describe the change of the average spacing between the Purkinje cells as $a_\pm = 2a \cdot r_\pm / (2r_\pm \mp H)$ where $r_\pm$ is the radius of the curvature of the layer of Purkinje cell bodies.

As a rule, concave sections of the cortex have a sharp breaking off of the outside boundaries—the bend of the molecular layer by $180°$ (Figs. 1 and 2), therefore $r_+ \approx H$ and $a_+ \approx 2a$.

In accordance with the increase and decrease of spacings at the boundary between the molecular and granular layers and because of the constancy of the ratio of these layers and also the constancy of the thickness of the molecular layer, changes of the thickness of the granular layer should occur. Suppose that in unfolded cortical sections the isolated areas of the layers are defined thus: $S_{\text{ML}} = aH$, $S_{\text{GL}} = ah$. Then for the folded cortex (Fig. 2), $S_{\text{ML}} = aH(1 \pm H/2r_\mp)$, $S_{\text{GL}} = ah_\mp(1 \pm h_\mp/2r_\mp)$. Since during folding the proportion $S_{\text{ML}}/S_{\text{GL}} = H/h$ is constant, for the determination of $h_\mp$ we have the equation

$$h_\mp{}^2 \mp 2r_\mp h_\mp + (H \pm 2r_\mp)h = 0. \tag{3}$$

As already noted, the concave sections of the cortex are arranged in a rather standard way, and the radius of the concave folding r_+ assumes for them the minimum of possible values: $r_+ \approx H$. Therefore, the minimum thickness of the granular layer is

$$\min h_+ = (\sqrt{(1 + h/H)} - 1) \cdot H.$$

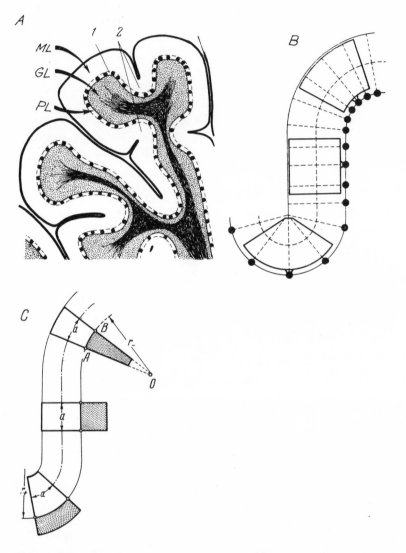

Fig. 2. (A) Cortical layers of the cerebellum; (1) convex section, (2) concave section O (B, C) diagram of molecular layer (ML) deformations and rectangular areas of the branching of the Purkinje cell dendrites [GL: granular layer; PL: Purkinje cell layer].

The behavior of the granular layer on convex sections is more interesting. The fact is that the thickness of it cannot increase ad infinitum and for r_- there is a lower limit: it is (theoretically) possible to decrease the radius of the curvature until the granular layer does not fill the triangle OAB (Fig. 2C) which satisfies the condition $r_- = h_-$. Hence, from (3),

$$\max h_- = (\sqrt{(1 + H/h)} + 1)h.$$

If we take into consideration (2), then

$$\min h_+ \simeq 0.42h, \quad \max h_- \simeq 2.58h. \tag{4}$$

The first relation (4) is well realized in the actual cortex and is easily checked experimentally. The second relation holds true "on the average" and, as will be obvious subsequently, makes it possible with good approximation to evaluate the average pattern of cortical folding, to establish the connection between the length of the cortex and the number of lobes.

Usually the simplest single cortical fold between two adjacent fissures is called a *folium*. Therefore, with the calculation of the number of folia in sagittal sections of the cerebellum (see below), we assume for the boundaries of the folium only concave sections the outer cortical surface of which forms a fissure (Figs. 3A, B, C). We call the *perimeter of the folium* the length of the cortex in a $\perp$-section measured along the layer of Purkinje cell bodies (l), or along the middle level of the molecular layer (l^*). We will observe that l^* is precisely the actual length of the cortex and the actual perimeter of the folium because the middle level of the molecular layer remains fixed during distortion of the cortex. Yet in an actual cortex the lengths l^* and l are usually equal, that is, compressions into the layer of Purkinje cell bodies in convex sections are compensated for by expansions in the concave ones.

Let us assume that there is a two-layer strip, imitating the cortex, and let us form from it a circular ring. Obviously the length of the strip l^* is related to the radius of the curvature r_- thus: $l^* = 2\pi(r_- + H/2)$. The minimum length of the strip from which it is possible to form a complete ring is determined by the condition

$$\min l^* = 2\pi\left(\max h_- + \frac{H}{2}\right),$$

or, using (4), we obtain

$$\frac{(\min l^*)}{(H + h)} \simeq 8.4. \tag{5}$$

The shape of actual folia and their perimeters rather sharply vary within each cerebellum. In Fig. 3, examples are given of typical folia: a, incomplete; b, normal complete; and c, complete extended; and a histogram is presented of the distribution of the perimeters of the folia for a sagittal section of a human cerebellum.

Condition (5) determines the perimeter of the smallest complete folium, we call such a folium *optimum*. For the cortex of a human being $H + h = 0.58$ mm, and from (5) the optimum perimeter $l \simeq 4.8$ mm, which conforms with the average perimeter for an actual cortex (see the histogram, Fig. 3).

1.4. ARBOR VITAE

The cerebellum appears for the first time in the phylogenetic series in lampreys in the form of a small undistorted plate. According to the degree of increase of the length of the cortex of higher vertebrates, there appears in the cerebellum one, two, etc., convolutions of the folia. If we compare the arbor vitae of different animals [besides our own data we used abundant data of 26–28 and 32] then with the increase of the number of folia the following changes occur: at first basic convolutions form—the folia of the first order; then on the folia of the first order appear *invaginations*, that is, a second order of distortion emerges and the original folia change into

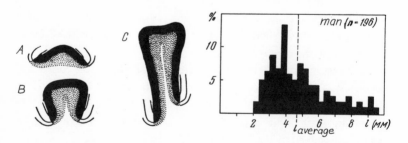

Fig. 3. Shape of folia of different expansions in ⊥-section (A, B, C); a histogram of the perimeters of the lobes.

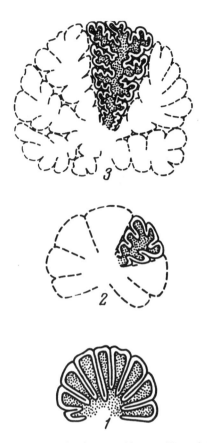

Fig. 4. Diagrams of cerebella having one (1), two (2), and three (3) orders of distortion.

lobules; finally, in higher mammals the third order of folding appears. Diagrams of cerebella with one, two, and three orders of folding are presented in Fig. 4. On the basis of all the material which we have been able to see, the connection between the number of folia n in a medial sagittal section and the appearance of orders of folding is probably thus: the second order appears when $n > 10$, and the third, when $n > 60$.

Moreover, a clear connection exists between n and the *fixed length of the cortex*, that is, the ratio $L/(H + h)$. In our own experimental material we were repeatedly convinced (some measurements are cited in Table 1) that

$$\frac{L}{(H + h)} \simeq 9n. \tag{6}$$

Thus, the formation of each new folium occurs, in the first approximation, when the longitudinal cortical expansion increases to a length 10 times greater than its overall thickness. Of course this rule is based on the feasibility of the existence of optimum folia [29].

In addition to this it turns out that almost regardless of the size of L, $H + h$, and n the cortex of the cerebellum is always folded in some "optimum" way.

The cases cited in Table 1 constitute with regard to all three parameters, length, thickness, and number of folia, a monotonic series. The impression can arise that this joint monotonicity is an essential property of phylogenesis. The fact that this is not so is obvious, for example, from comparison of the cerebella of the gray duck ($L = 56$ mm; $H + h = 0.22$ mm $+ 0.14$ mm; $n = 17$) and the hobby ($L = 90$ mm; $H + h = 0.3$ mm $+ 0.2$ mm; $n = 17$). The cerebella of these two birds coincide in the number of folia, but they differ noticeably in the thickness of the cortex, which, according to [36], also should lead to differences in L.

1.5. EXPOSED AND CONCEALED SURFACES

In this section we will make a simplified evaluation of what portion of the area of the cortex is hidden in the depth of the cerebellar fissures.

Since "cylindricality" is characteristic of cortical folding, the ratio of the area of the exposed surface S_{ex} to the area of the entire cortex S is equal in the first approximation to the ratio of the perimeter of the cerebellum L_{ex} in a sagittal section to the full length L of the cortex in the same section: $(S_{ex}/S) \simeq (L_{ex}/L)$. From the measurements it follows (the values of L are presented in Table 1 in the Appendix), that for a mouse, cat, and human being this ratio, respectively, equals 0.43, 0.16, and 0.1, that is, in a human being approximately 90 percent of the cortical area is concealed.

Based on the invariance of the cortical volume (or the area for a $\perp$-section) we can make an evaluation of the ratio S_{ex}/S (or L_{ex}/L for a $\perp$-section) as follows. The area of the slice in the sagittal section is equal to approximately $L(H + h)$, since the volume of the white matter in the vermis is very small (see section 1.2) or, if we assume that the shape of the sagittal section is near to circular, $L_{ex}^2/4\pi$, therefore

$$L_{ex} \approx 2\sqrt{\pi L(H + h)},$$

and taking into consideration [36],

$$L_{ex}/L \simeq 2\sqrt{\pi(H + h)}/L = 1.2/\sqrt{n}. \tag{7}$$

Supposing that $n = 200$ for man, this relation yields $L_{ex}/L \approx 0.085$.

1.6. THE CAUSES OF CORTICAL FOLDING

The basis of a *mechanical hypothesis* of cortical folding described in section 1.1 is, of course, the idea of the necessity of the compact arrangement of the brain in the cranium [6, 25]. Some authors [36] do not agree that in a given case only the "packing determines the shape of the contents," and this viewpoint is understandable. We also think that the *principle of compactness* as a requirement for convenience of transportation of the brain is not the only or even the basic cause of cortical folding. But what then "pulls" the cortex inward, forcing it during development to invaginate and to form creases? It is not excluded that the connections, and primarily the afferent and efferent fibers, play a basic organizing role in the folding of the cortex.

As follows from the foregoing, in a straightened cortex the layers should be balanced in thickness, the Purkinje cells are distributed more regularly, the deformations of their dendritic regions (Fig. 2B) disappear, and so forth; that is, possibly in the first approximation for the inner organization of the cortex it is not essential whether folding occurs or not. But for afferent and efferent fibers this process is extremely important, since their total length is related to the character and degree of folding. In the example of a cerebellum that has folia of the first order, it is not difficult to figure out the optimum number of folia in the presence of which the total length of the fiber will be smallest. This problem resembles the problem concerning the optimization of the transport routes inside a city or large plant where all the houses or shops would make up a single continuous building —a model of the cortex.

In this way we can think that one of the causes of cortical distortion is the *principle of economy* expressing itself in the minimization of the total length of afferent and efferent fibers, that is, in the minimization of the volume of white matter. The results of rough estimates, on the whole, confirm such an assumption.

Possibly, too, one of the essential moments of organization of cortical activity is the simultaneous excitation of several far-dispersed regions; it is possible to assume this considering the nature of the distribution of mossy afferents (see section 2.4). Inasmuch as in a distorted cortex such a problem is resolved more easily, it is not ruled out that the *principle of synchronism* participates to some extent in the organization of its folds.

Both of the last principles should also play a fairly large role in the organization of the long associative (intracerebellar) connections. A drawing together of cortical regions is "utilized" even in the short associative connections within one folium. For example, the passing of an axon of a Golgi cell of the first type from one side of the folium over the white matter to the other (see Fig. 8) is described by Cajal [6]. We observed in the dendrites of the Golgi cells a similar passing over the white matter within the folium. All this indicates that the distortion of the cortex in evolution could prove to be helpful for a series of various problems of organization of the connections.

It is not ruled out that all the principles noted above are only superficial reflections of those mechanisms of intercellular interactions which actually regulate the cortical formation in ontogenesis. Most of the deformations of the cortical structure bear a mechanical nature, as if the cortex had initially developed flat, and then "somebody" folded it and put it into the cranium.

Thus, to understand the causes of cortical distortion partly means to find the ontogenetic way of attaining such a result.

2. THE ORGANIZATION OF THE GRANULAR LAYER

Homogeneity of organization is not so rare a phenomenon in the central nervous system, but the granular layer of the cerebellum very likely represents in this regard the most surprising formation.

Possibly, the whole specific character of organization of the cerebellar cortex which distinguishes it from the many other central formations is determined by the existence of the granule cells, or more accurately by the existence of the granular layer formed by them. In any case, such characteristics of the cortex as homogeneity, anisotropy, and "two-layeredness" are connected, undoubtedly, with the peculiarities of formation of the granule cells. The latter form not only with their own bodies and dendrites a homogeneous and isotropic medium—the granular layer—but also, outside the granular layer, out of the branches of their own axons, a homogeneous system of parallel fibers which also constitutes the basis of the molecular layer (section 4).

The contents of this section is made up of a brief description of the formation of the granule cells and glomeruli, of the formulation of the *principle of the continuity of the layer*; of the results of our measurements of the density of the granule cell and attempts to estimate the density of the glomeruli; of the examination of ways of the branching of the mossy afferents in the layer; of a brief description of the other neurons of the layer—the fusiform cells and the Golgi cells.

2.1. THE GRANULE CELLS AND THE GLOMERULI

The body of the granule cell generally has a round or oval shape. The nucleus occupies almost the entire space of the body; according to some data, a nucleolus is absent [7]; the thickness of the layer of protoplasm is 0.5–1.00 μ. There branch out from the body 4 to 5 thin (1 μ), and in the majority of cases short (20–40 μ), dendritic shoots. At the distal end the dendrites form a characteristic cluster of fingered expansions by means of which the granule cell comes in contact with the endings of the mossy fibers. The diameter of a granule cell body in man, and in the majority of mammals and birds, is 5 μ. Friede [18] linked certain deviations from this "standard" with the sizes of the animals; in the horse, for example, the diameter of the granule cell is 7 μ, but in the canary it is 4 μ. But it is doubtful whether the size of the animal is the significant cause of the change of the granule cell size, since in the frog the size of these cells is 7 to 8 μ.

In the light microscope (Nissl preparations) the granular layer appears as a dense, uniform mass of round granules. With high magnification it is apparent that these granules closely gather around spaces—the *cerebellar islets*. In nondistorted cortical regions the bodies of the granule cells form

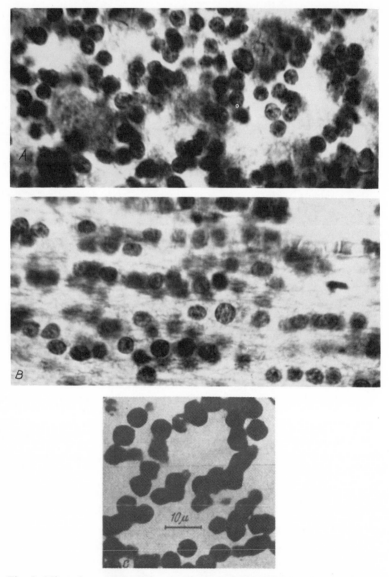

Fig. 5. Microphotographs of the granular layer (Nissl stain) (A) rings formed by the granule cell bodies (macaque cerebellum, a modification of [15]; (B) and (C) granular layer in undistorted and concave sections of the human cortex, respectively.

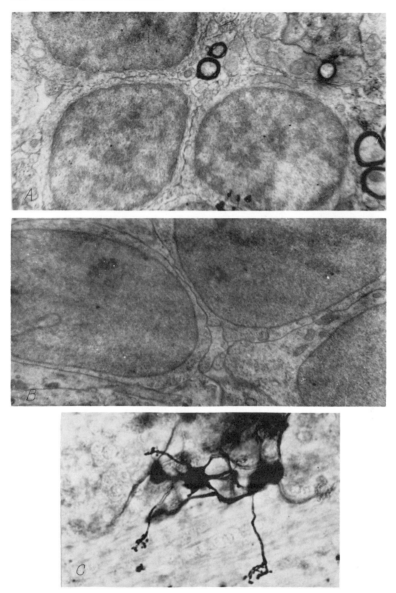

Fig. 6. Soma-somatic contacting of granule cells (A) mouse and (B) frog (electron microscopy, stained with lead and uranyl acetate); (C): "bound together" granule cells in Golgi preparations (light microscopy).

unique rings (Fig. 5, A and B), but in concave ones the rings extend into *chains* (Fig. 5C), a process which involves the extension of the layer into those regions described above. The cerebellar islets correspond to loci of arrangement of glomeruli—complex synaptic formations. In electron microscopic specimens it is still more clearly apparent that the granular layer consists mainly of complexes of granule cell bodies, closely adjoining one another (Fig. 6, A and B) and regions rich with synaptic formations, that is, glomeruli. The basis of a glomerule is an expansion of the mossy fiber into the concave sections of which the fingered swellings of the granule cell dendrites take root, and also the axon and possibly dendritic processes of certain types of Golgi cells.

Such a make-up of the *parenchymatous glomeruli* was described by Cajal [6] and recently confirmed by the electron microscopic research of Gray [21] and Szentágothai [39]. Moreover, Gray indicated the possibility of dendro-dendritic contacts of the granule cells with the aid of the fingered protuberances. But synapses are absent on the granule cell bodies [7, 21]; our search for somatic synapses on electron microscopic specimens were also unsuccessful.

2.2. SOMA-SOMATIC CONTACT

The soma-somatic contact of the granule cells is of particular interest for the question of the organization of the granular layer. The membranes of the cells closely adjoining one another, according to electron-microscopic observations of a series of authors [7, 10, 20] and our own observations, are separated by a narrow gap of 100–200 A. Sometimes for a short distance the membranes are separated by glial processes but in the majority of cases we fail to detect any building up of formations in the intermembrane fissure. The total area of the contacting surface of the individual granule cell is generally large—one-half and more of the whole cellular surface. In view of this circumstance we can assume that the soma-somatic contact of the granule cell is of functional significance and is intended, possibly, for the transmission of excitation from one cell to another. Estable [13] proposes that in a given case a synaptic contact is realized between the granule cell bodies. But the absence in the place of contact of typical synaptic formations—an expansion of membranes, vesicles, etc.— suggests that if transmission of excitation does occur, then the mechanism of it is not chemical. In a given case we could suspect realization of an electrotonic means of interaction of the granule cells.

For the organization of the granular layer the assumption of a functional stipulation of soma-somatic granule cell contacting leads to the following hypothesis: that bodies of the granule cells form a continuous network as regards the spreading of excitation and the whole granular layer is organized according to the *principle of continuity*.

It would be helpful to understand what concrete morphological consequences result from this principle, that is, in what way should the distribution patterns of granule cell bodies differ depending on whether a somatic network is realized? Another viewpoint on these contacts can be based on the *principle of close packings*. We can point to the simple theoretical consequences of the principle of continuity: (1) there is no granule cell not making contact soma-somatically with cells similar to it; (2) each cell makes contact with not less than two neighbors, and cells exist which make contact with more than two neighbors. But as yet there has been no success in finding simple, practical criteria distinguishing these two principles. Possibly, however, these criteria do not exist. Thus, if, for example, the presence of an intercellular contact automatically leads to the establishment of intercellular interaction, then the contacts established in the early stages of ontogenesis according to the principle of close packings can later lead to functional continuity.

In Golgi preparations, we can also see granule cells contacting somasomatically; generally there are complexes of two or three cells bound together by their bodies (Fig. 6C), but it is possible to encounter complexes consisting of 5, 6, and even more cells.

As is known, contacting cells can be stained in Golgi preparations both separately and together. In this connection separate staining still does not prove the absence of contact; and combined staining, conversely. On the strength of this characteristic of staining in Golgi preparations, it is possible to see separately, for example, single granule cells, the branches of the mossy fibers, and also the branches on the expansions of which the granule cell dendrites make contact. In exactly the same manner, side by side with individual granule cells, it is possible to see complexes of several granule cells. However, the frequency of occurrence of separate granule cells and complexes is not constant from preparation to preparation and depends on the variations in processing, or, possibly, some other reasons. (We are not especially concerned with this question, but, on the basis of the available comprehensive material, we failed to observe a correlation of changes of processing of the Golgi material with the degree of the stainability of one or the other of the cortical elements.) At any rate, in individual blocks or series of sections, almost all the stained granule cells can be isolated; and in the others, on the other hand, a large percent of the complexes.

It is possible to make similar observations also with regard to the selectivity of the stainability in the Golgi method of all other cortical neurons. For example, there are series of our specimens in which only the stellate cells were stained, or mainly the Purkinje cells, etc. Probably it is possible to learn to control by the selectivity of staining of one or the other types of neurons, and this knowledge would be very useful. But for the present we must bear in mind the basic characteristic of the Golgi method: the absence of some element even in a large number of specimens, treated in similar conditions, is not proof of the absence of this element in the cortex. So, for example, conducting thorough research of the distribution and connections of liana-shaped afferents in the cortex. M. Scheibel and A. Scheibel [35] did not discover in their Golgi specimens contact of the granule cell dendrites with the expansions of the mossy fibers; in connection with this, unjustifiable doubts arose [see also 7, 25] with regard to the plausibility of the original ideas about the presence and role of these connections.

In the Golgi preparations we can see the granular cells, linked by the "feet" of their dendrites, which correspond to the structure of the glomeruli described. In the plan of the concepts developed here the proposition of V. M. Bekhterev is extremely interesting. Bekhterev wrote about having observed in Golgi specimens dendro-dendritic linkages of granule cells: "... the nerve cells, contrary to the generally accepted opinion, can enter among themselves into contact by means of their own protoplasmic processes. In all probability, the linkage with the protoplasmic processes of the adjoining cells determines a union of activity of many cells of the granular layer, owing to the transmission of excitation from one cell to another." Furthermore, the author notes that dendro-dendritic contacts are observed also in neurons of the spinal cord and that generally similar contacts, possibly, "... serve for the union of activity of functionally homogeneous cells" [1, p. 16].

2.3. THE DENSITY OF THE GRANULE CELLS

The granule cells and the glomeruli are the basic elements of the granular layer; the density of the other structures (the Golgi cells, glia, and so forth) is comparatively low. The granule cells are evenly distributed over the whole layer and their density, as appears from our testings, does not depend on the distortion of the cortex: calculations of the number of granule cells in different regions of the convolutions give values that are in very good agreement. For the best orientation in factual data we will try first to evaluate theoretically the maximum density of granule cells. We will consider the cell body a sphere with a diameter 5 μ and suppose the cell has four dendritic branches 25 μ in length and 1 μ in diameter. The total volume of such a cell equals 163 μ, which constitutes $6.1 \cdot 10^{-6}$ parts of 1 mm^3. Thus, in 1 mm^3 it is possible to locate (theoretically) $6.1 \cdot 10^6$ cells. Later on we will label the density of the granule cells p_{GC}, but we will understand by this quantity the number of cells not in 1 mm^3, but the number in a volume 10^9 times smaller. The quantity introduced, p_{GC}, is the "number" of cells in 1 μ^3. Thus, max $p_{GC} \simeq 6.1 \cdot 10^{-3}$.

In the human cerebellum $p_{GC} = 2.1 \cdot 10^{-3}$, and in the mouse cerebellum, $3.5 \cdot 10^{-3}$. It is possible to imagine that in the ascending order of mammals (see Table 2) the changes of the granule cell density obeys the general tendency of the nerve cells—the density decreases are in phylogenesis, however, in comparison with the Purkinje cells (see the same table) the granule cell density changes little.

In bony fish (carp) $p_{GC} = 3.4 \cdot 10^{-3}$, that is, its granule cell density is the same as in the mouse. However, in an intermediate representative of the phylogenetic series—in the frog—the granule cell density is clearly undersized; it is less than half the p_{GC} of a human being (Table 2). This is partly determined by the increased size of the granule cell bodies of the frog (see section 2.1) and by the fact that the fibers which generally form the white matter here extend inside the layer stratifying it. Therefore, if the soma-somatic contacting of the granule cell is not important functionally, then in the more thinned-out granular layer of a frog's cerebellum this contact should not be so regular and general as in the cerebella of mammals. However, our electron-microscopic observations, carried out together with V. L. Boroviagin, showed that the soma-somatic granule cell contacting in the frog is as common and is relatively no less extensive than in the mouse (Fig. 6, A and B). Apparently, the granule cell bodies of a frog simply form a "wider-meshed network."

2.4. THE DENSITY OF GLOMERULI

We can suppose that the glomeruli, like the granule cells, are distributed evenly in the granular layer, and that the number of granule cell dendrites converging from one glomerule is on the average constant. We will note that the presence in the layer of glomeruli automatically brings into the distribution of the granule cell bodies a definite organization. The latter, snugly adjoining the glomerule and one another, form in the space a certain continuous net. Therefore, depending on the nature of the distribution of the glomeruli and their density, the structures of the nets can be different.

In practice it is impossible to determine the density of the glomeruli by the parameters because of the indistinctness of the outlines of their contours. We will try to estimate this size by an indirect method, using an assumption concerning their uniform distribution. If we take into consideration the peculiarity of the organization of the layer of granule cells, then three different approaches are possible here.

THE FIRST APPROACH. The maximum density of glomeruli can be estimated in the following manner. As follows from the foregoing, max $p_{GC} \approx 6.1 \cdot 10^{-3}$, and in the actual cortex, for example human, the granule cell density is three times less. Therefore, in each isolated volume of the layer, granule cells occupy one-third, and the remaining two-thirds can be given up to the glomeruli. According to approximate estimates, we can consider the glomerule a sphere with a diameter of 15 μ. Having divided the remaining volume by the volume of the glomerule, we obtain max $p_{Glo} \approx 0.4 \cdot 10^{-3}$. In this instance for each glomerule there will be on the average a number of granule cells equal to $\gamma = p_{GC}/\text{max } p_{Glo} \approx 5$.

THE SECOND APPROACH. In accordance with the pictures which we can observe in Nissl preparations (Fig. 5), we will consider a glomerule together with the granule cells adjoining it as a certain unit complex. If for simplicity we consider the glomerule a sphere with a diameter 15 μ, then we could enclose each glomerule together with the adjoining granule cell bodies into a sphere with a diameter 25 μ. If in a space we tightly pack these spheres then $p_{Glo} \approx (0.085 \div 0.13) \cdot 10^{-4}$. Hence, the number of granule cells per one glomerule for the actual human cortex is $\gamma = p_{GC}/p_{Glo} \approx 2.1 \cdot 10^{-3}/p_{Glo} \approx 15 \div 25$. However, with the given dimensions of a glomerule and the granule cell bodies within an isolated complex it would be possible to arrange around a glomerule with the tightest packing (disregarding the dendrites) about fifty cells, that is, in the model described of the layer with $\gamma = 15 \div 25$, within one complex there is possible a multitude of ways of interdistribution of the granule cells. If we take into consideration the principle of continuity then the picture of the granule cell distribution appears as the following.

Around the glomerule—a sphere with a diameter of 15 μ—we can construct a continuous ring (conforming to a large circle) using 12 granule cell bodies with a diameter of 5 μ. For the formation of a continuous three-dimensional net around each glomerule it is sufficient to have only two continuous rings in two mutually perpendicular planes (Fig. 7A). Then in each plane the rings will form a continuous net represented in Fig. 7B. Obviously, both rings will include altogether $\gamma = 22$ granule cells, which corresponds to the minimum number of granule cells around one glomerule in a given model of the continuous granular layer.

THE THIRD APPROACH. Using the principle of continuity, we can try to

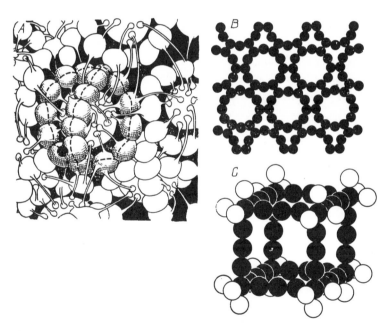

Fig. 7. Models of the granular layer. (A) three-dimensional diagram of the layer. The granule cell complex around one glomerule is isolated by a dotted line; (B) A plane net of granule cell bodies; (C) cubic network of granule cell bodies (see text). The solid black cells make up a single unit.

solve the reverse problem: what continuous (for simplicity, rectilinear) nets of granule cell bodies are possible if the density of the granule cells and their diameter are given?

The edge in such a net represents a column of granule cell bodies, and the length of the edge determines the dimension of the cell and hence the number of these cells, for example, in 1 mm³. If we can locate a glomerule in each cell, then we will establish thereby their density. For example, if we construct a cubic net of such cells as shown in Fig. 7C, each edge comprising four granule cells, then we will obtain $p_{GC} \approx 2.5 \cdot 10^{-3}$, $p_{Glo} \simeq 0.125 \cdot 10^{-3}$, and $\gamma = 20$.

2.5. THE MOSSY FIBERS

The afferents of this type terminate exclusively in the granular layer; as already noted, their terminal expansions, forming the glomeruli base, establish synaptic contacts with the dendrites of the granule cells. The individual mossy fiber can ramify in the white substance, sending secondary branches into the different folia (Fig. 8A); further, in the white matter within the same folium, secondary branches form several tertiary offshoots which enter the granular layer in different places finally and form terminal branchings with their characteristic expansions. From this picture it is apparent that the *secondary branches* of the mossy afferents are related to the presence of the folia, and we can assume that in the straightened cortex they would be absent: in such a model of the cortex the mossy fiber, coursing along the lower boundary of the granular layer, would form, within certain intervals, only branches producing terminal branchings; that is to say, probably the only important circumstance is that one mossy fiber is able to excite in the cortex a certain number of the local zones dispersed sometimes from one another by a rather great distance, if we measure along the cortex.

The terminal *tertiary branches* often ramify within the granular layer as a compact cluster (Fig. 8, A and B). Such branches are able to excite at once a rather large number of granule cells, but within small (local) zones. The tertiary branches of the mossy fiber entering the granular layer and ending here as a compact cluster, we will call, for brevity, *local branches*. The branching laws of the mossy fibers have been little studied, a fact which is related to a considerable degree to the difficulties of their delinea-

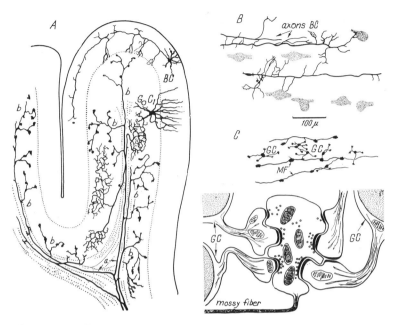

Fig. 8. Mossy fibers (MF). (A) Branching of mossy fibers within the folium and the white matter; (a) branches going into different folia; (b) terminal (local) branches branching into the granular layer [6] (Golgi method). BC = basket cell; GoCI = Golgi type I cell. (B) Tangential cortical section on the Purkinje cell layer level (Golgi method, cat). (C) Diagram of the contacting of a terminal enlargement of the mossy fiber granule cell dendrites [21]. GC = granule cell; BC = basket cell; GoCI = Golgi type I cell.

tion by the Golgi method. For an understanding of the organization of the cortex it would be very helpful to answer, for example, such questions as how many local branches are able to form an individual mossy fiber; whether these local *sources of excitation* are distributed at some determined spacings from one another; how many glomeruli form an individual local branch, etc. Unfortunately, at present it does not seem possible to answer the greater part of such questions.

According to the nature of the branching and the three-dimensional orientation we can divide the local branches into the following three types:

(1) terminal expansions of a cluster distributed through the whole thickness of granule cells and located mainly on the plane of the $\perp$-section; (2) the same, only the plane of primary branching coincides with the $\parallel$-section; (3) the fibers forming the terminal expansions spread horizontally, parallel to the granule cell boundaries, and can go at different levels of the layer (as a rule, they are stretched out along $\perp$-sections).

It seems to us that the branches of types 1 and 3 are most common in the cortex; that is, it is more characteristic for the mossy fibers to branch in sections parallel to the region of the branching of the dendrites of the Purkinje cells; this is apparent not only from comparison of Golgi preparations orientated along and across the folia, but also in tangential preparations. Braitenberg and Atwood [5] for the most part record sagittal orientation of the terminal branches of the mossy fibers.

The cluster-shaped branches of the first and second indicated types are, possibly, used for the simultaneous excitation of some aggregates of granule cells distributed through the entire layer. Because of the compressed nature of the branching of these branches and the small length of the granule cell dendrites, the region of simultaneously excitable granule cells also is compressed. The fibers of the third type possibly also serve the objectives of the directed "firing" of the granular layer, not through the whole thickness, but incompletely, for example, only of the upper or lower half. In all these cases it is a significant circumstance that every time in the granular layer a compact region is fired—all the cells are of some closed region.

If we assume, according to the estimates of the previous sections, that in 1 cubic millimeter there are 10^5 glomeruli, and that one granule cell has on the average 4 dendritic processes, then the number of granule cells making contact with one glomerule is approximately 80. We tried to calculate by the Golgi preparations how many glomeruli form one local branch. As a result of these attempts the following figures seem to us probable: in a rat the branches of the first and second type form 4 to 5 expansions, and in a cat, 5 to 8 expansions. The horizontal branches (third type) have 2 to 3 expansions in a rat and 3 to 5 expansions in a cat (fibers of this type frequently differ in thickness from one another; they are both as thick as 2–3 μ, and as thin as 0.6–1 μ). Thus, one cluster-shaped branch, possibly, excites simultaneously approximately 300 to 400 granule cells in the cerebellum of a rat, and 400 to 600 granule cells in that of a cat.

Branches of the first type distribute their expansions through the entire thickness of the layer, and the horizontal dimension (in a ⊥-section) of the region occupied by them coincides with the horizontal dimension of the dendrites of the Purkinje cells. The latter makes possible the following assumptions: local branches of the first and second type excite simultaneously as many granule cells as there are of them in one Purkinje cell. This number is designated $m_{GC/PC}$ (see Appendix), and its value for different cerebella is presented in Table 2. Based on the assumption made and using the known sizes $m_{GC/PC}$, the number of glomeruli formed by one cluster-shaped branch of a mossy fiber should be: man, 16; macaque, 10; cat, 6; rat, 4; mouse, 2.

Proceeding from these assumptions, in the simplest model of a cerebellar cortex (see section 6.2), in which mossy afferents form only local branches of the first type, it is possible to examine the granule cell aggregate per Purkinje cell as a certain functional unit all the cells of which, possibly become excited simultaneously by an afferent impulse.

2.6. OTHER NEURONS OF THE GRANULAR LAYER

In the middle of a homogeneous mass of granule cells now and then other larger neurons are encountered. They are diffusely distributed in the layer at different levels and as yet there has been no success in establishing any regularity in their mutual arrangement. According to observations on Golgi preparations [1, 16, 24], all these neurons, it is taken for granted, divide into two groups: the fusiform cells and the multipolar or Golgi cells.

The fusiform cells have an elongated body with a diameter of 20 μ, from the poles of which branch out two dendritic processes. At a distance of from 100 to 150 μ from the body each process forks, and later one branch ends in the Purkinje layer, and the other branch, in the granular layer. Possibly, the upper branches make contact with axons of the basket cells [14], and the lower terminate in the glomeruli. The dendrites are usually strictly oriented on the ⊥-section plane, their total spread extending up to 1,000 μ. The nature of the branching axon of the fusiform cells and its connections are not known for certain; Fox [14] assumes that the axon branches out from the body and then rises into the molecular layer. The initial cone of the axon, according to our observations, is located in the primary dendritic trunk or generally there is no success in detecting it.

It is assumed that the Golgi cells divide into two types [1, 16, 24]: to the first (GoC I) belong the cells the axon of which forms within the granular layer and, as a rule, not far from the body a dense net of multiple branchings; the second type (GoC II) are the associative cells, their axon at once emerges into the white matter sometimes forming recurrent collaterals, and heads for other regions of the cortex. Each of these types has a multitude of subtypes, if we carry out a supplementary classification based on the ways of distribution of their dendrites. In certain of these cells the dendrites are distributed mainly in the granular layer; in others they gather and terminate in the Purkinje layer; and in a third they are almost completely found in the molecular layer. There are differences also in the nature of the three-dimensional distribution of the dendritic branches: together with the plane regions of branching of dendrites which are also characteristic to other cortical neurons, there are Golgi cells the dendrites of which branch within cylindrical regions or diffusely.

We will dwell briefly on the ways of axon distribution of the type I Golgi cells. Cajal [6] already had divided these cells into four subtypes: (1) the axon branching over the whole depth of the layer spreading in a horizontal direction to distances two times greater than the thickness of the layer; (2) the axon distributed only in the upper half of the granular layer; (3) the axon distributed in the lower half of the granular layer; (4) the axon forming several scattered zones of branching that occasionally can be found in neighboring lobes.

On the basis of our own Golgi preparations we could be convinced of the existence of all these subtypes, except the fourth. A diagrammatic representation of the subtypes of type I (GoC I) is given in Fig. 9. A cell of the fourth subtype drawn by Cajal is presented in Fig. 8A.

The existence of variations of displaced arrangement of the cell body and dendrites with regard to the axon region (see variations of subtype I in Fig. 9) is justified, if it is necessary to realize *lateral inhibition* or *contrast* at the granular layer level (see section 6.2). It is interesting that the presence of all subtypes of the type I Golgi cell conform with the presence of different types of distribution of the local branches of the mossy afferents.

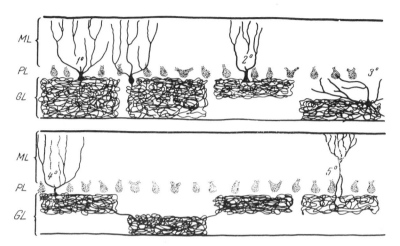

Fig. 9. Different subtypes of type I Golgi cells (GoC I) with a short axon. [ML = mossy fiber layer; PL = Purkinje fiber layer; GL = granule cell layer.]

The cell marked in Fig. 9 as subtype 5 is called by Cajal a "displaced" Golgi cell. He assumed that such an arrangement of the body and dendrites with respect to the branching region of the axon is probably an "artifact of histogenesis" rather than an individual type of cell, since he succeeded in observing similar cells only in preparations of the rabbit cerebellum. But it is not excluded that these cells are a subtype of type I Golgi cells or even stellate cells of the molecular layer, because we observed them also in a rather large number in the Golgi specimens of a macaque cerebellum.

We will again note that the presence of the mutually complementary cells of subtypes 2 and 3 indicate, possibly, the functional stratification of the granular layer, at least at two levels—the upper and lower substrata (see also section 6.3).

3. THE PURKINJE CELLS
The homogeneity of the system of parallel fibers, determined by the homogeneity of the granular layer producing it, extends further into the whole aggregate of neurons connected with it and, primarily, into the Purkinje

cells which determine with the parallel fibers the greatest number of contacts. Owing to the homogeneity of structure of the Purkinje cells they are located (in the molecular layer) at approximately fixed distances from one another, forming thereby a kind of network.

In this section a description is given of our attempts to comprehend the structure of the network formed by the Purkinje cells, and to estimate its parameters. Moreover, a description is given of the geometry of the branching of the dendrites of these cells and the results of the calculation of the total number of spinules. Finally, the question about the effect of the number of granule cells per Purkinje cell, on the density of the distribution of the latter, in the molecular layer will be discussed. But first, based on data in the literature, a brief description of the axon and collaterals of the Purkinje cells will be given.

3.1. THE SOMA, THE AXON, AND COLLATERALS

In the majority of cases the bodies of the Purkinje cells have a characteristic drop-shaped form. The diameter of the body of the cell is fixed in each animal, and in phylogenesis the diameter changes proportionally to the vertical expansion of the dendrites, that is, proportionally to the thickness of the molecular layer [18].

The axon branches out from the lower pole of the soma or from the side, penetrating the granular layer, emerges into the white matter and goes further into the cerebellar nuclei. At a distance of 20 to 30 μ from the body the axon is sheathed with myelin.

According to Cajal's description [6], within the granular layer of mammals there branch out from the axon approximately 3 or 4 collaterals, which, staying in the $\perp$-section, ascend to the bodies of the Purkinje cells. The branches of the collaterals form two, and sometimes three [24], layers of fibers: the lower layer is directly under the bodies of the Purkinje cells, where the branches extend horizontally in the $\perp$-section; the upper layer is above the bodies of these cells (Fig. 10). Having reached the upper layer, the branches of the collaterals again split off and then go horizontally in the $\parallel$-section, that is, in exactly the same manner as the parallel fibers of the granular cells. The fibers of the upper layer form synapses on the main dendritic trunks of the Purkinje cells. The other connections of the collaterals are unknown. For its full length to the synaptic twig the collaterals

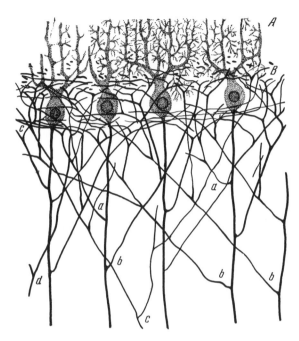

Fig. 10. Recurrent collaterals (a, b, c, d) of the axons of the Purkinje cells [6] (Ehrlich method, cat).

do not lose their myelin sheaths. The lower the collateral develops the further it branches out after this from the given Purkinje cell, and the greater number of times it branches (Fig. 10). The collaterals can also branch out from the axon in the white matter, in this instance they obey the same principle—they go to the more distant cells of the layer. The number of collaterals formed by one axon drops to two in birds (see Fig. 10), in reptiles there is one collateral, and in amphibia and fish there is only one or not even one.

Some authors [16, 39] think that part of the associative cortical connections, realized strictly on the sagittal plane between distant folia, is formed by the Purkinje cell collaterals. The collateral branching out from the axon of the lower boundary of the granular layer can have terminal branchings at a distance four or five cells from the given Purkinje cell.

However, reliable data concerning the maximum distance of the collaterals, and for the length of the longitudinal fibers formed by them in the second layer, are lacking.

3.2. THE DENSITY OF THE CELLS

The bodies of the Purkinje cells in mammals and birds, in each located in one layer, is different in various regions of the cortex; in comparison with undistorted sections the density in convex regions increases, and it diminishes in concave regions. However, the volume of the molecular layer per Purkinje cell for a given cerebellum is always constant (see section 2.2).

We can determine the average Purkinje cell density for an undistorted cortex (p_{PC}) by tangential sections across the layer of the bodies of these cells. The results of our measurements p_{PC} for certain representatives of the phylogenetic order are presented in Table 2 ($p_{PC} \cdot 10^6$—the number of cells on an area 1 mm²). As appears from these figures, the Purkinje cell density rather strongly changes in the order of vertebrates: in man the p_{PC} is almost an order of magnitude smaller than in the mouse (and differs by more than an order of magnitude, in comparison with the frog).

If we determine the Purkinje cell density not from tangential preparations alone, but from transverse and longitudinal (to the axis of the folium) preparations, then a weak anisotropy develops in the distribution of the cells. Our measurements from celloidin sections of a human cerebellum of 20 μ thickness give the following result: on a cortical slice 1 mm long, in a $\perp$-section there are approximately 5 cells; and in a $\parallel$-section, 7 cells; that is, if we determine the Purkinje cell density only by sagittal or only by frontal preparations, then instead of an actual value of the density of 300 cells per 1 mm², we will obtain 250 or 350. This difference from the average value is not so really great, approximately 17 percent, but we take them into consideration during analysis of the structure of the network formed by the Purkinje cells.

Proceeding from these results we can assume that the Purkinje cells form a rectangular network with the coefficient of anisotropy, $a = 1.185$, where $\alpha = a/b$ and a, b are the spacing between cells in $\perp$- and $\parallel$-sections, respectively:

$$a = \frac{10^3}{\sqrt{250}} \simeq 63 \ \mu, \qquad b = \frac{10^3}{\sqrt{350}} \simeq 53 \ \mu. \tag{7}$$

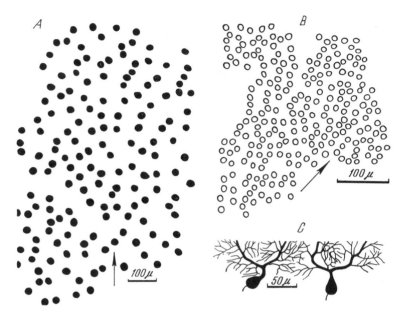

Fig. 11. Actual network of the Purkinje cells of man (A) and mouse (B) (the arrow is oriented along the axis of the folium; sketch from Nissl preparations). (C) Example of the displacement of the body of the cell with regard to the dendrites (Golgi method, mouse).

However, we will examine the question about the structure of the network in much greater detail.

3.3. THE STRUCTURE OF THE NETWORK

If we analyze tangential preparations across a layer of Purkinje cell bodies, then we fail to detect rectilinear rows of cells, that is, any rectilinear network (Fig. 11). We can treat this circumstance, of course, in two ways: (1) a rectilinear network does not exist, the Purkinje cells are distributed randomly, but with a constant average density, and with some anisotropy for ∥- and ⊥-sections: (2) a rectilinear network exists, but actual pictures do not indicate its probability because of the random displacement of both the individual cells and the rows (one can, for example, even think that the network as a whole is more exactly realized on the Purkinje cell dendrites, and its distortions are caused, moreover, by the random displacements of the bodies with regard to the dendrites [Fig. 11], but it is practically impossible to verify this statement).

The question about the rectangularity of the network is closely related to the question concerning the uniformity of the Purkinje cells (see section 3.5); however, we can try to solve it independently on the basis of Nissl preparations. If we proceeded from assumption (2) then we must indicate, showing statistical regularity in the distributions of the intervals between two Purkinje cell bodies (for ⊥- and ‖-sections), of what type is the network most adequate to the actual case. For detection of the characteristic spacings between Purkinje cells in ⊥- and ‖-sections, we constructed histograms of the distribution of the intervals (according to continuous series of measurements of cortical sections, containing a rather large number of Purkinje cells, each of the histograms in Fig. 12 indicate no less than five hundred similar measurements).

Although the average interval between two Purkinje cells is constant in

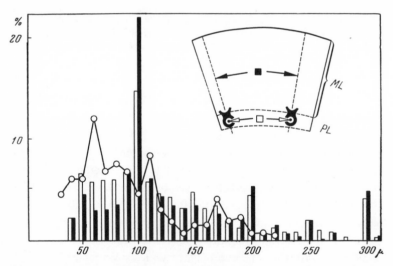

Fig. 12. Histograms of the distribution of intervals between the Purkinje cells (Nissl preparations, man, celloidin slices 20 μ); for ⊥-section, the white and black columns (different in the measuring methods of the intervals, see diagram); for ‖-section, the small circles.

sections of a given thickness [see also 5 and 15], a network of type 1 cannot occur. Histograms of the distribution of the intervals between two Purkinje cells show (Fig. 12) that the cells have the tendency to be located in ⊥-sections at a distance 100 μ (human cerebellum). We will construct a network considering this characteristic spacing for Purkinje cells as the actual spacing between the cells in the direction of the ⊥-section (with straightening of the cortex, that is, in measurement of the intervals between Purkinje cells not in a layer of the bodies of these cells but along the median line, this characteristic spacing of 100 μ is still more emphasized) (see section 1.2). We will call a *row* an aggregate of cells located in the direction of a ⊥-section, the average spacing between which is 100 μ (a *step down the row*). Now let us determine the width of the row. The histogram presented in Fig. 12 was constructed for sections of the same thickness (20 μ), along which anisotropy was established (see section 3.2). But in the sections of this thickness on a piece of cortex 1 mm long there were, on the average, located 5 cells; that is, the average interval between each two cells equal 200 μ. We can expect that the average interval will equal 100 μ for sections the thickness of which are approximately two times greater. We will consider as the width of the row the thickness of the section in which the average interval between the cells equals the step down the row. Carrying out a more exact estimate of the width of the row (it is convenient to label this size in terms of $b/2$) by the known size p_{PC}, we will obtain $b/2 = 33.5 \mu$.

If now we consider the row as a "rigid" formation and construct a network of the rows, then on the whole one of two following types of rectilinear networks are possible: the cells of the adjacent rows line up behind one another (a rectangular network); the adjacent rows are out of alignment with regard to one another by a half-step (the network is of a hexagonal or "checkerboard" type). Both visual observations and certain histograms indicate the great probability of a network of the hexagonal type.

Many of the histograms constructed by us for ∥-sections yield a rather diffuse maximum for intervals of 40 to 90 μ (we should stipulate that it is possible to select a ∥-section of folia on frontal or tangential cerebellar specimens, but because of the distortion of the folia in the laterolateral direction, the length of suitable sections obtained is comparatively small, whereas cortical sections of an adequate expansion and "exactly" per- pendicular to folia are easier to obtain), in successful cases this maximum narrows down to intervals of 60 to 70 μ, but emphasis of an interval of 30 to 40 μ is never obtained (we will note again that the size 30 to 35 μ corresponds to the horizontal dimension of the Purkinje cell body). One of the "successful" histograms for ‖-sections is presented in Fig. 12.

Thus we can assume that Purkinje cells form in the cortex a hexagonal type of network. We will call a *rank* (Fig. 13) an aggregate of cells aligned in the direction of ‖-sections and distant from one another by a spacing equal to double the thickness of the row. For the human cerebellum the step down row a equals 100 μ, and the step down rank b equals 67 μ.

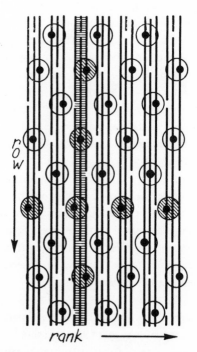

Fig. 13. Diagram of the network of Purkinje cells. A single row and single rank are singled out by shading.

These parameters of the Purkinje cell network for some other mammals estimated according to histograms and density are presented in Table 2 in the Appendix to this chapter.

Returning to the question concerning the anisotropy of the network, we note the following phenomena. We can examine a network of a hexagonal type as an aggregate of two rectangular networks enclosed in one another. The directionals of the rectangular networks correspond to the directions of the rows and ranks, the coefficient of anisotropy is $\alpha = a/b \approx 1.5$. Obviously, the sizes of a and b are related to the density of the Purkinje cells as follows:

$$p_{PC} = \frac{2}{ab}. \tag{8}$$

Hence, for a theoretical determination of the steps down the row and the rank we derive the formula

$$a = \sqrt{2\alpha/p_{PC}}, \qquad b = \sqrt{2/\alpha p_{PC}}. \tag{9}$$

For the cerebellum of a cat, $p_{PC} = 6.3 \cdot 10^{-4}$ (see Table 2), assuming that the coefficient α is constant in phylogenesis, we estimate the quantity a:

$$a = \sqrt{2(1.5)10^4/6.3} \simeq 69 \ \mu. \tag{10}$$

Approximately this interval is indeed emphasized in histograms for the $\perp$-sections of a cat's cerebellum.

As we know, the problem about the tight arrangement of the bodies on the surface leads to a hexagonal type network. From the solution of this problem it follows that the spacing in two mutually perpendicular directions should satisfy the equation $a = \sqrt{3}b$, that is, the *coefficient of anisotropy* $\alpha = \sqrt{3} \approx 1.73$. Therefore, if the Purkinje cells were located according to a principle of maximum density, then for the cerebellum of a human being instead of $a = 100 \ \mu$ and $b = 67 \ \mu$, we would obtain according to (9) 108 and 62.4 μ. And, hence, the actual arrangement of Purkinje cells is rather close to the last case.

Pezard [31] determined the density and the total quantity of Purkinje cells for many varieties of birds and some mammals. It is interesting that, basing his method of determination of the density of Purkinje cells from sagittal sections, he proceeded from a hexagonal type network, but assumed that $a = b$.

3.4. THE DENDRITES

In the Purkinje cells, two types of dendritic branches sharply differ: the *smooth dendrites* represented by the main trunks and thick branches (6–3 μ), and the *spiny dendrites* or thin branches (1 μ), bearing little spinules and uniformly filling the whole region of branching. The dimensions and shape of the dendritic branching region are different in various mammals (Fig. 14) and can differ within the cortex of a single animal. These differences are determined mainly by the branching nature of the smooth dendrites.

The dendritic branches branch out from the upper part of the body either at once in the form of two trunks, or initially as one shoot. Then these basic branches go aslant, forming secondary and tertiary branchings. Comparing a drawing of the branching of the dendrites of many Purkinje cells (according to Golgi preparations and specimens of reduced silver, the latter frequently clearly showing the smooth Purkinje cell dendrites), we can see a striking tendency of the smooth dendrites of these cells to maintain definite angles of inclination. In the cerebellum of the cat and macaque, the basic trunks usually go at an angle 30° to the lower boundary of the molecular layer, and the secondary and tertiary branches maintain one of the directions parallel to the primary branches (Fig. 15).

However, together with such "characteristic" Purkinje cells (for the cerebella of the mouse, rat, cat, dog, and macaque), there are encountered cells the dendrites of which have a substantially greater angle of inclination and cells with almost vertical dendritic branches. In the human cerebellum, Purkinje cells are encountered with inclined dendrites; but more often than in other mammals, cells are encountered with vertically orientated branches (the well-known Purkinje cells of a human cerebellum drawn by Cajal [6], illustrated in Fig. 14); we know for fish only a single type of Purkinje cell according to Schaper [33] (Fig. 14). The question about whether there are in fact the indicated varieties of Purkinje cells with these different types of cells and in what ratios they are encountered in the cortex requires special research; unfortunately, it does not seem for the time being possible to answer this question. Why such a question about the

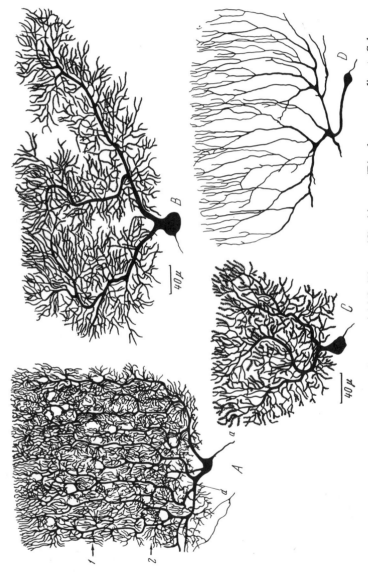

Fig. 14. Purkinje cells. (A) man according to Cajal [6]; (B) cat; (C) white rat; (D) salmon, according to Schaper [33]. Golgi method; (1) and (2) clusters of spiny dendrites.

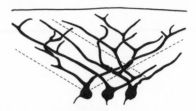

Fig. 15. Inclination of smooth dendrites of the Purkinje cells (sketch from Golgi preparations, cat; the dotted straight lines are drawn at 30° to the lower boundary of the layer).

types of Purkinje cell arises will be obvious subsequently (see section 3.5). Let us note that in speaking about different inclinations of the smooth Purkinje cell dendrites we are not taking into consideration those characteristic changes which these cells undergo in connection with deformations of the molecular layer in different distorted cortical sections (see section 1.2).

The spiny branches are formed along the whole length of the smooth dendrites. The length of the individual ramule is usually small (20–50 μ), and they branch in the $\perp$-section plane, diffusely and uniformly filling up the whole region of branching of the dendrites.

For a comparison of Purkinje cell dendrites of different animals, taking into consideration the planar nature of their branching, we can introduce a sort of quantitative parameter as the density of the branching of the spiny branches, σ_B: the ratio of the total length of the spiny branches to the area on which these branches are distributed in their projection on the plane (this determination indicates the method of measurement according to the Golgi preparations; see section 4, p. 298). Although for each of the mammals examined by us (mouse, white rat, cat, macaque, man) the size of the σ_B of dimension $1/\mu$ varies from 0.15 to 0.4, and in individual cases even greater deviations are obtained, nevertheless the most characteristic interval of the values of this size is 0.2 to 0.3. Moreover, the value $\sigma = 0.3$ is more characteristic for man, and $\sigma = 0.2$ for the mouse. In visual observations such small changes of σ_B in a series of mammals are almost unnoticeable, but it corresponds to the general tendency of a decrease of this quantity in a series of vertebrates. Judging from the descriptions and sketches by Cajal and others [see 6, Chapter IV], the branching density of spiny branches noticeably diminishes in birds, reptiles, amphibia, and fish (see Fig. 14).

In order to understand the real significance of the values 0.2 to 0.3, we will do a simple estimate [15]. The diameter of a spiny branch equals 1 μ; the spinules are distributed along the shoot on two sides, and their length is also approximately 1 μ. Therefore, the total width of the branch together with the spinules equals 3 μ. If on some plane region we closely pack "tubules" of a diameter of 3 μ in one layer, then the ratio of their total length to the area of the region will be 0.3. Thus, the spiny branches of the Purkinje cells fill a region of branching with a density close to the maximum density of filling in one layer.

The spinules are distributed along the shoots rather evenly, by our calculations the number of spinules on a branch 10 μ is 10 to 15, that is, on a length 1 μ there are located approximately $n_s = 1 \div 1.5$ spinules. The tendency is also observed for the value of n_s to increase in the ascending phylogenetic order. Thus, the value $n_s = 1.0$ is more characteristic for the mouse, and $n_s = 1.5$ for the macaque and man.

For the distribution characteristics of spinules on the branching plane of the Purkinje cell dendrites we can substitute the two quantities σ_B and n_s with one: $\sigma = \sigma_B' \cdot n_s$. The ratio is of the number of spinules to the length of the branch, the quantity n_s characterizes the *linear density* of the distribution of spinules, but the quantity σ is the ratio of the total number of spinules along the branches filling some region to the area of this region, and hence establishes the number of spinules per unit of area. It also makes sense to define precisely the density of the spinules for the Purkinje cells by means of the quantity σ. Granting the phylogenetic tendencies of σ_B and n_s as approximate average values of σ for the mammals examined, it is possible to accept, it seems to us from observations and calculations, the values cited in Table 3 in the Appendix to this chapter.

If now we measure by a separate method the area S_{PC} of the branching region of the dendrites of a single Purkinje cell, then the total number of spinules N_0 of a single cell is simply

$$N_0 = \sigma S_{PC}. \tag{11}$$

Typical values of S_{PC} and the values of N_0 calculated according to [29] are presented in Table 3.

Our estimate for the Purkinje cell of the macaque differs from the estimate made by Fox and Barnard [15]; they obtained $N = 61,000$. The reason for the divergence of the results is explained as follows. Fox and Barnard in the determination of N started from a Purkinje cell the branching density of which equals approximately 0.5. In our measurements a similar branching density occurred only for certain portions of certain cells. Therefore, it is impossible to consider, all the more so it seems to us, a value of the branching density of 0.5 as typical for the whole branching region of the single cell.

If we compare the Purkinje cell dendrites of different mammals, then it is impossible not to note that in the nature of the formation of the spiny branches of higher mammals an additional organization appears. In a human being the tendency of spiny branches to group into individual clusters is the most clearly expressed. These clusters are located along the smooth dendrites at some distance from one another (Fig. 14A) and along both sides. Along the height of the molecular layer we can count altogether on one side of a smooth branch up to 10 clusters. For the rat and cat the clusters are not so characteristic, and frequently there branch out from the smooth dendrites simply single spiny branches. Possibly a role of individual structural units (see section 5.3) is assigned to the clusters of spiny branches in phylogenesis.

3.5 OVERLAPPING OF DENDRITES IN A ROW

In all mammals the horizontal dimension a_0 of the branching region of the Purkinje cell dendrites is greater than a step down row a. For example, in man, $a_0 \approx 300\ \mu$ while $a \approx 100\ \mu$. This means that the Purkinje cells forming one row are overlapped by the branching regions of the dendrites (Fig. 13). In this instance the cells with a rectangular branching region are overlapped uniformly and at any place of the row dendrites of the three adjoining cells are overlapping (Fig. 16A).

For cells with a triangular shape of branching region (Fig. 16B) we have a completely different situation. If all the Purkinje cells forming a given row have the same triangular shape, then at different heights of the molecular layer the number of overlapping regions will be different. And hence the density of the contacts made by the cells of the row with the system of

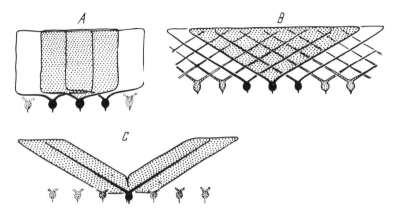

Fig. 16. Diagrams of the overlapping of the dendrites of Purkinje cells having a different shape of the branching region.

parallel fibers increases according to the height of the layer (in proportion to the overlapping, because σ for the cell is a constant).

However, we can make the overlapping homogeneous even for cells with slanting dendrites, if the shape of the branching region has the sort of form shown in Fig. 16C, that is, each slanting branch of smooth dendrites forms a zone of spiny branches. If the width of the zone is equal to the step down the row, then the overlapping will be homogeneous everywhere and at each place of the row the three cells will overlap equally.

If all the actual Purkinje cells, for example, of a cat, were to have the rectilinear, symmetrical shape of our model (Fig. 16C), then we would be able to talk about the homogeneity of the row, and moreover, about the fact that the smooth dendrites form a certain rectilinear net, as about the simple corollary of the constancy of the intervals among the cells in the row. In this instance the idea itself, a row of Purkinje cells, would be in all only a conditional idea defining an aggregate of Purkinje cells which line up in the direction of ⊥-sections (as a row of Purkinje cells was initially defined in section 3.3). However, proceeding from actual situations which we observed in the Golgi sagittal sections in places of the appearance of groups of Purkinje cells the view point for a row of Purkinje cells was completely different. In spite of the fact that the individual Purkinje cells differ from one another in the shape of the branching region of the dendrites, the smooth dendrites of a row of adjacent cells form a network.

Of course this network of smooth dendrites is not strict, but changes the spacing between the nearest parallel branches of the dendrites forming the network; we obtained for the most part the values 30 to 35 μ. With a 30° angle of inclination of the dendrites (see Fig. 15) these values correspond to the spacing between cells in a row 60 to 70 μ long, which is close to the estimate of the step down the row conforming to the density of the Purkinje cell [4].

Stöhr's hypothesis [38], proposed by him even in 1923, explains the naturalness of the formation of a dendritic network. Studying reduced silver preparations, Stöhr turned his attention to the fact that the smooth dendrites of adjacent Purkinje cells make contact with one another and he stated the assumption that the dendrites of a row form a continuous net. (We will call for brevity this assumption the *hypothesis of rows*; below in section 5 we will formulate the *hypothesis of ranks*). In Golgi preparations and in preparations of reduced silver one can actually observe a "conglomeration" of smooth Purkinje cell dendrites. However, the necessity still remains to show the existence of a contact by electron-microscopic methods and to prove the electrical continuity of the dendritic net by electrophysiological methods. Theoretically, a dendro-dendritic connection of an electrotonic nature between neurons is possible; this was recently proved in other preparations [10].

The hypothesis of rows in the cortex changes not only our concept about the functioning of the Purkinje cell but also our concept about the organization of these cells. From the viewpoint of this hypothesis the conditional idea emerges not of rows but of the Purkinje cell as an individual element, and therefore it makes sense to examine not an aggregate of Purkinje cells but a system of these cells. Forming in an aggregate a continuous net—a network, the "individual" cells can differ from one another by the pattern of their smooth dendrites, and then on the whole the network can be built according to the *principle of mutual complementation*. Moreover, uniformity of the cells can appear only on the average, for the preservation of the homogeneity of the row, and with respect to such integral parameters as the area of the branching region, the total number of spinules, etc. However, with respect to these parameters, they must differ little from our model (Fig. 16C) for which the area of the branching region is easily determined on the whole as $S_{PC} = 4aH - a^2tg\phi$. For a cat's cerebellum $a \approx 60 - 70 \mu$, $H = 220 \mu$, and $\phi = 30°$; therefore, $S_{PC} \approx 60,000 \mu^2$, which agrees with the branching area of the actual cells.

If a row actually is a rigid and "homogeneous" formation, then to a certain degree the idea of a network of Purkinje cells (see section 3.3) loses meaning, because the rows can be shifted with regard to one another at random and then the difficulty of obtaining "nondiffuse" histograms of the distribution of the intervals between the Purkinje cells in ‖ sections is explained by this.

We can explain by means of the hypothesis of rows, moreover, the experimental differences in the sizes of the branching density of the spiny branches. Just as in Golgi preparations the granule cells, thanks to soma-somatic contact, frequently appear "bound together" (Fig. 6C), dendrites of adjacent Purkinje cells can appear fragmentarily, for example, of one entirely, and from adjacent ones only pieces are above the site of contact.

In the structural plan, we can advance only one "objection" to the hypothesis of rows: how a continuous net can be built with Purkinje cells with slanting dendrites is understandable, but how such a net is realized of cells with dendrites going vertically is not very clear. Of course, we can devise a continuous net even for cells with such dendrites as in Fig. 14A, but such a net will have its own laws of construction, and it is difficult to "join" it with the slanting one. It is possible in the same way that the rows in the cortex are "sectionally continuous" and cells with vertical dendrites are located between continuous sections of the rows.

3.6 THE NUMBER OF GRANULE CELLS PER PURKINJE CELL

This number is one of the chief parameters of the cortical organization of the cerebellum. We will examine its significance for the Purkinje cell network. For brevity we will label this number $m_{GC/PC}$.

If p_{PC} is the density of the Purkinje cells, then the reciprocal value $1/p_{PC}$ is the area in a tangential cortical section for one Purkinje cell, the *eigen* or *characteristic* area of one cell. Then H/p_{PC} is the eigenvolume of one Purkinje cell in the molecular layer, and $v = h/p_{PC}$ is the eigenvolume of the granular layer for one Purkinje cell. In the ascending order of mammals both eigenvolumes increase, while these changes proceed (or can proceed) independently according to two variables: (1) because of the increase of the thickness of the layer (we will recall that $H/h \approx 1.5$, see section 1.2) or (2) because of the decrease of p_{PC}. Theoretically, there is not excluded the existence of such groups of animals within which the changes of the eigenvolumes of the Purkinje cells occur due to changes of only one size—the thickness of the layer or the density of the Purkinje cells.

The eigenvolume of the granular layer of one Purkinje cell, obviously, contains a number of granule cells equal $m_{GC/PC}$, that is,

$$m_{GC/PC} = V \cdot p_{GC} = hp_{GC}/p_{PC}. \tag{12}$$

The numerical values of this size for certain vertebrates are presented in Table 2. The quantity p_{PC} gives the chief contribution to the phylogenetic change of $m_{GC/PC}$ for animals from this list, because, in comparison with it, the quantities h and p_{GC} not only change slightly, but their changes are partly compensated for in equation (12), in mammals a decrease of p_{GC} usually occurs with an increase of h.

In the model of the cortex in which p_{GC} is constant, we can change the value of $m_{GC/PC}$ by the same two independent ways we did the eigenvolume: by (1) "trimming" or "building up" the granular layer, that is, varying h; and by (2) moving apart or bringing together the Purkinje cells, that is, changing p_{PC}. But it is doubtful whether for the phylogenesis of the cortex of the cerebellum the change of the values of h and p_{PC} has independent significance. Rather the size $m_{GC/PC}$ is the "independent variable," the change of which also causes a change of h or p_{PC} or both h and p_{PC}. It is significant that with the change of $m_{GC/PC}$ each of the quantities h and p_{PC} can change independently of the other; therefore we can examine the change of one of them, while holding the other fixed.[1]

If the Purkinje cell network is held fixed, that is, the p_{PC}, then the change of the thickness of the granular layer should lead, on the strength of the

[1] Speaking of the change of any cortical parameters, we imply the existence of that set of cerebella on the great number of which corresponding values of these parameters are realized.

constancy of H/h, to the change of the vertical expansion of the Purkinje cell dendrites. It would be helpful to understand what the simplest model of the cortex it is possible to arrive at, if $m_{GC/PC}$ becomes smaller and smaller. In the given case p_{PC} is fixed—the limitations on the minimum $m_{GC/PC}$ should be dictated by the organization singularities both of the Purkinje cell dendrites and the granular layer (see section 2.2), then we can reduce its thickness, generally speaking, to the size of glomeruli, realizing from the granule cell bodies a plane network of, for example, the sort of type as shown in Fig. 7B.

When the thickness of the layers are held fixed, the decrease of $m_{GC/PC}$ should lead to the drawing together of the Purkinje cells—an increase of p_{PC}. Let us assume that, decreasing $m_{GC/PC}$, we reached a critical situation: the Purkinje cell bodies tightly abut one another. (A similar "crisis" for the Purkinje cell dendrites should set in earlier than for the bodies.) How can we now lessen $m_{GC/PC}$ with h constant? Obviously, only by means of lessening the granule cell density. But then we will enter into contradiction with the principle of the continuity of the granular layer. In order that this does not occur, lessening $m_{GC/PC}$, should increase the diameter of the granule cell bodies.

Possibly the frog cerebellum is an example of a situation when the $m_{GC/PC}$ size is smaller than the critical. With respect to the density of the granule cells and the density of the Purkinje cells, the frog clearly stands out among the other representatives of the phylogenetic order (see Table 2). Together with this in the frog the bodies of the Purkinje cells violate the usual single-layer arrangement, and the diameter of the granule cells is increased up to 7 to 8 μ. The diameter of the bodies of the Purkinje cells is rather small (10 to 13 μ), and they could be laid out in a single layer, almost adjoining one another; but this does not happen, possibly because of the dendrites.

4. THE SYSTEM OF PARALLEL FIBERS

Climbing into the molecular layer, the granule cell axon at some level branches, forming two horizontal branches (the parallel fibers, in Fig. 17), subsequently running parallel to the axis of the folium. According to the degree of movement in the molecular layer, each ∥-fiber forms synaptic

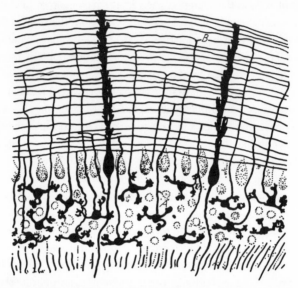

Fig. 17. Axons of the granule cells, from [6]. (A) ascending part; (B) the point of bifurcation (‖-section).

contacts with spiny dendrites of the Purkinje cells, stellate cells, and with dendrites of the Golgi cells entering here.

Until recently the question of the length of the parallel fibers remained open and on that account different hypotheses were formed [1, 6, 34]. The most prevalent hypothesis was the following, proceeding from Cajal's point of view: parallel fibers within a folium have unlimited length. The first estimate of this length was the electrophysiological result of Dow [8], who showed that during local stimulation of the folium a wave of excitation spreads along its axis no more than for 5 mm.

Braitenberg and Atwood [5], estimating the volume occupied by the parallel fibers, came to the following conclusion: the most probable length of these fibers is in the range of 1 to 10 mm (human cerebellum). Tracing the parallel fibers in Golgi preparations, and conducting quantitative comparisons in the granular cells—the Purkinje cells system—Fox and Barnard [15] give a more exact estimate of the length: 1–3 mm (macaque cerebellum). Finally Szentágothai's observations [39] of traces of degeneration of these fibers in isolated strips of a folium also indicate their length is 2 mm (cat cerebellum).

Taking into account the phylogenetic tendencies of the granular cells and of the Purkinje cells described in the preceding paragraphs, it seems to us useful to analyze the characteristics of the system of the parallel fibers in greater detail. The simple and natural hypotheses from which it seems to us one can proceed are the following: (1) *the Purkinje cells, within the cortex of a given animal, are identical with respect to the dimensions of the branching region and the total number of spinules*; (2) *the number of spinules correspond to the number of parallel fibers with which the Purkinje cell makes contacts.*

Having established these hypotheses, we can solve questions concerning the average number of contacts made by one parallel fiber, that is, the number of Purkinje cells with which this fiber interacts, and concerning the average period of contacting of the parallel fibers and certain other characteristics of this system.

We will for the time being assume (until section 4.3), that all parallel fibers have some fixed length λ_0 equal to the average length of these fibers for the cerebellum.

4.1. THE AVERAGE NUMBER OF CONTACTS

According to hypothesis (2) above, each Purkinje cell forms with a system of parallel fibers N_0 contacts. But the branching region of the dendrites of one Purkinje cell penetrates a very large number (N) of parallel fibers. Suppose that each fiber (length λ_0) penetrates all together $2q$ Purkinje cell dendrites (at the rate of q Purkinje cells on each side from the point of bifurcation), producing $2q_0$ contacts.

If each fiber produces as many contacts as cells it penetrates, that is, if $q_0 = q$, then each Purkinje cell should have all together N spinules; if $q_0 < q$, then it should have q/q_0 times less spinules. Hence, the total number of spinules of a single Purkinje cell is

$$N_0 = (N/q)q_0. \tag{12}$$

Each eigenvolume of the granular layer of one Purkinje cell forms in the molecular layer a number of parallel fibers equal to $m_{GC/PC}$. If each eigenvolume contained one granule cell, then the number of fibers piercing the dendrites of one Purkinje cell would be $m_{GC/PC}$ less and would be equal to the number of cells penetrated by one fiber, that is,

$$\frac{N}{m_{GC/PC}} = 2q, \quad \text{or} \quad \frac{N}{q} = 2m_{GC/PC}.$$

Substituting this into (12), we obtain

$$N_0 = 2q_0 \cdot m_{GC/PC}. \tag{13}$$

Both sizes N_0 and $m_{GC/PC}$ are determined experimentally and independently (see section 3, and Tables 2 and 3 in the Appendix to this chapter). Therefore, we can find another unknown of the geometry of the cerebellar cortex—the average number of contacts q_0 produced by one parallel fiber with the Purkinje cell dendrites. The values of q_0 calculated by this formula for certain mammals are presented in Table 4 in the Appendix. The figures obtained give the following interesting result: (3) *the average number of contacts produced by one parallel fiber with the Purkinje cells is a constant in phylogenesis.*

4.2.   THE AVERAGE PERIOD OF CONTACTING

If a fiber λ_0 long produces $2q_0$ contacts, then the ratio $\lambda_0/2q_0$ is "the average period of contacting" of the fiber. In order to determine this size for a given cerebellum, it is necessary, obviously, to know the average length of the parallel fibers. But as is obvious from the history presented above of the experimental determination of the length of parallel fibers, on the whole, we have at our disposal only approximate estimates. A more or less exact value of λ_0 is known only for the cerebellum of the macaque [15]: $\lambda_0 \approx 2$ mm.

Therefore, the average contacting period of parallel fibers in the cerebellum of the macaque has the following value:

$$T = \frac{\lambda_0}{2q_0} = \frac{2,000}{33} \approx 61 \ \mu.$$

If we now return to the results for the network of Purkinje cells (section 3.3, Table 2), then it is apparent that the value T obtained is greater than the spacing between the rows and corresponds rather well with the step along the rank:

$$T \approx b. \tag{14}$$

The contacting pattern of parallel fibers with dendrites of Purkinje cells can be in fact very complex and not periodic: individual parallel fibers can make contact at once with two spinules of one cell, as Gray observed [21], or establish contacts with several cells of one row, but the result obtained makes it possible to determine the simplest circuit of interaction—the interaction of parallel fibers with one rank of Purkinje cells.

A simple cortical model of a cerebellum which we can assume, proceeding from this, is the following: (4) *the granule cells of one eigenvolume of a Purkinje cell are connected via parallel fibers only with one given rank.*

Thus, we assume here that the contacting period of the parallel fibers is always constant and equal to the step down the rank. The simplest structural unit, moreover, which we can isolate is an aggregate of Purkinje cells connected with the parallel fibers of one eigenvolume—the *elementary rank*.

Condition (14) obtained for the cerebellum of the macaque, possibly occurs also for other mammals. On the strength of the constancy of the average number of contacts C this is equivalent to the hypothesis that the phylogenetic change of the average length of the parallel fibers λ_0 is caused only by a change of density of the Purkinje cells. The values of λ_0, calculated according to the formula $\lambda_0 = 2q_0 T$, where $T = b$, are presented in Table 4.

4.3. THE LENGTH AND DIAMETER OF THE PARALLEL FIBERS

The length and diameter of parallel fibers is not constant in the cortex of a given cerebellum, but in the distribution of these fibers a tendency to a definite orderliness is observed [14, 15]: the lower the fiber is located in the molecular layer, the greater length and diameter it has. A definite orderliness is also observed in the arrangement of the granule cells: the closer the granule cell is situated to the molecular layer, the higher the point of bifurcation of its axon in the molecular layer.

A vertical correspondence in the arrangement of granule cells and parallel fibers arises as a result of a particular sequence of formation of the granular layer during the period of the embryonic development of the cortex [6]. The cells of the external granular layer (the layer of nondifferentiated cells located during development of the cortex above the still unformed molecular layer), gradually slipping deep into the cortex, forms at a certain level horizontal shoots. These shoots subsequently change into parallel fibers, and the cells themselves sink lower and gradually assume the characteristic shape of granule cells. The next cells and their parallel fibers are located above the previous ones.

Idealizing the described characteristics of the parallel fibers we can assume (5) *the points of bifurcation repeat in the molecular layer the arrangement of granule cells in their own layer*: (6) *the diameter and length of the parallel fibers steadily decrease with elevation*. Thus, the parallel fibers produced by the eigenvolume of one Purkinje cell possibly in a ∥-section of the molecular layer fill a certain *trapezoid* (Fig. 18).

In the macaque cerebellum [15] the diameter (d) and the length (λ) of the parallel fibers at the boundaries of the molecular layer have the following values:

$$d = 1.0 - 0.2 \, \mu; \qquad \lambda = 3,000 - 1,000 \, \mu.$$

We obtained the same interval ($1.0 - 0.2 \, \mu$) of change of the diameters of parallel fibers in a cat cerebellum; in a rat it is less: $d = 0.6 - 0.25 \, \mu$. The circumstance that the diameter of the ascending part of the granule cell axon is constant and coincides with the diameter of the horizontal branches attracts attention. The initial part of the thick axons constitute an exception: directly at the cell body (in the course of 10–$30 \, \mu$) the diameter of the axon is smaller and equal to approximately $0.2 \, \mu$ (an analogous

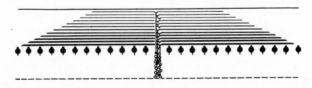

Fig. 18. Diagram of the changes of the length and diameter of parallel fibers according to the height in the molecular layer; represented is the part of the fibers formed by the eigenvolume (for one Purkinje cell, see section 3.6) of the granular layer.

tapering of the initial part, but more sharply expressed, is observed in the axons of the basket and stellate cells (see section 5.1).

The difficulties of obtaining reliable results with regard to the length of the parallel fibers by means of direct changes obliges us to look for indirect ways. Thus, Braitenberg and Atwood [5] for an estimate of the average length of the parallel fibers use the calculation of the volume occupied by these fibers. Our estimate of the average length is based on the calculation of the average number of contacts and the average period of contacting (see above). It is interesting to make the next step—to find the criteria of the estimate of the gradual changes in the system of the parallel fibers. One of the simplest interpretations, like the search for the relation between the change of the diameter and the change of the length of the fibers, proceeds from the examination of the electrophysiological picture of the spreading of excitation along the parallel fibers.

Suppose all the granule cells of the eigenvolume of one Purkinje cell (Fig. 18) at some moment are excited (from a mossy afferent, see section 2.5). Within a certain time excitation reaches the points of bifurcations of the axons of these granule cells. Let us assume that all these points are excited simultaneously. Then the excitation will spread along the parallel fibers in opposite directions in the shape of two symmetrical *fronts*. The inclination of the front (in the simplest cases fronts can have the shape of segments of straight line, see Fig. 24), will change in proportion to the distance from the points of bifurcation (because of the gradient of the diameters of the parallel fibers, the spreading rates of excitation at different heights satisfy the condition that the excitation front reach the ends of all these fibers simultaneously. This condition provides the following relation between the diameter of the fiber and its length:

$$\lambda = \sqrt{d} = \text{const}$$

(the parallel fibers are unmyelinated, therefore the rate of spreading along them is proportional to the square root of the diameter, $v = \text{const} \sqrt{d}$). Thus, the ratio of the lengths should be equal to the square root of the ratio of the diameters

$$\max \lambda/\min \lambda = \sqrt{\max d/\min d}. \tag{15}$$

For the macaque cerebellum we have max λ/min λ = 3,000/1,000 = 3, $\sqrt{\text{max } d}$/min d = $\sqrt{1.0}$/0.2 $\approx$ 2.24. However, if we take into account that the ascending part of the granule cell axon coincides with the diameter of the horizontal branches, that is, during simultaneous excitation of the granule cell of one eigenvolume, the corresponding points of bifurcation do not become excited simultaneously, but with a certain gradual lag according to the height, then instead of (4.4) we must write:

$$(\text{max } \lambda + 2h)/(\text{min } \lambda + 2H) = \sqrt{\text{max } d}/\text{min } d \qquad (16)$$

It is not difficult to be persuaded that for the cerebellum of the macaque the ratio in the left part of this expression equals approximately 2.2. Such a good coincidence of the results still does not of course prove the accuracy of the basic hypothesis, but to a certain degree justifies the development of model representations in this direction.

5. THE ORGANIZATION OF THE MOLECULAR LAYER

The interaction of neurons of the molecular layer with the system of parallel fibers introduces a definite peculiarity in the nature of their distribution and organization of connections. The connection of the dendrites of the Purkinje cells with the parallel fibers is predominant in the organization of the layer, and all the remaining neurons, the basket and stellate cells, only supplement and complicate it. In section 5.3 we will show what further conclusions about the organization of the molecular layer we can arrive at, analyzing the connection of the Purkinje cell dendrites with the parallel fibers, but at first we will dwell on other elements of the layer. The greater part of our investigations of the basket and stellate cells—of the peculiarities of their distribution in the layer, the different kind of quantitative characteristics, the phylogenetic tendencies, and others—is not completed at the present moment; therefore we will confine ourselves to brief descriptions based on compiled data and certain corresponding observations.

5.1. THE BASKET AND STELLATE CELLS

Neurons of this type appear in the cerebella of almost all vertebrates [6]. The basket cells are absent in the frog cerebellum [20]. The dimension of the basket cell body can vary from 8 to 12 μ. The initial section of the axon is very thin (0.1–0.2 μ); then, before formation of the collaterals, it sharply thickens (to 3 μ) and then, almost without changing the diameter, runs horizontally along the row of Purkinje cells, that is, in the ‖-section plane. The length of the axon of different basket cells can be varied [6]; however, Szentágothai [39], using a method of degeneration, shows that long-axon cells are significantly smaller than the ordinary ones, that is, the sort the axon which covers a distance of 10 to 12 Purkinje cells of a row.

Besides the thick descending collaterals from which the *basket plexuses* form around the Purkinje cell bodies (as a result of convergence from several basket cells), the axon forms along the whole length a small number of thin collaterals both ascending and descending (Fig. 19). The thin collaterals terminate on the Purkinje cell dendrites, and on the bodies and dendrites of the stellate cells [6, 25, 34]. The thick collaterals descend to the axon cone of the Purkinje cells and form here synaptic contacts [6] which according to electron microscopic observations [39] are synapses of the second type, according to Gray.

The larger part of the collaterals is concentrated nearer to the soma (Fig. 19) and their density decreases toward the end. According to the form of the ending of the axons, Cajal distinguishes two types of basket cells: (1) the density of the descending collaterals gradually diminishes toward the periphery and the axon itself ends in a basket; and (2) the axon

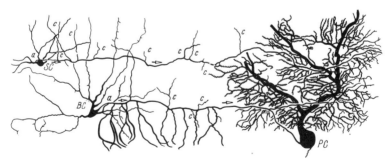

Fig. 19. Sketch from a Golgi preparation of basket (BC), stellate (SC), and Purkinje cells (PC) (the actual mutual arrangement of the cells was maintained; macaque).

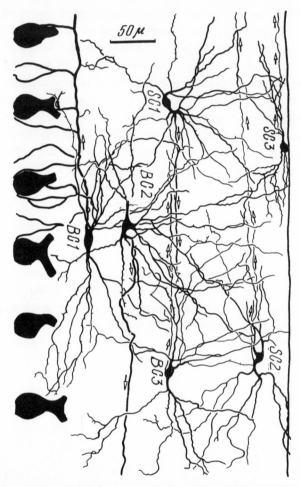

Fig. 20. Dendrites of the basket (BC) and stellate (SC) cells (a combined drawing Golgi method, macaque). The arrows indicate the distal direction of the axons.

forms collaterals only in the first third of its run, and at the end of the course forms a tassel of thin collaterals distributed in the molecular layer. With respect to direction, the axons of the basket cells are, naturally, only of two types: the *right* and the *left* (Fig. 20, BC 1 and 2). Occasionally the axon, branching out from a cell in one direction, at a certain distance turns in the opposite direction (Fig. 22, BC 3).

In tangent Golgi preparations we can see that the collaterals of basket cell axons are distributed in several Purkinje cell rows (Figs. 21 and 8B). According to Szentágothai's calculations, one basket cell affects approximately 70 Purkinje cells—10 cells along the row and 7 in the cross direction.

According to the nature of the axon branching there are two types of stellate cells: (1) the cell with a long—up to 1.5 mm (macaque cerebellum) —horizontal axon which differs mainly from the basket cell axon by the absence of the "basket" collaterals (Fig. 19); all the collaterals formed are thin, with a characteristic varicosity located mainly near the body. With the help of the collaterals the stellate cell axons form synaptic contacts with their own dendrites, with Purkinje cell dendrites and with those of neighboring stellate cells, moreover, they produce axo-axonal contacts with the parallel fibers [34]; and (2) the cells with a short axon which near the neuron form a net of thin collaterals.

The basket cell and stellate cell dendrites branch in the ⊥-section plane, forming long thin (1–2 μ) shoots. Fig. 20 gives some idea of the branching nature of the dendrites of the basket cells and stellate cells with a horizontal axon. According to the number of primary dendritic shoots branching from the body, the basket cells and stellate cells are two and three-polar (compare BC 1 and BC 2; SC 1–SC 2 in Fig. 20).

There are rather few spinules on the basket cell and stellate cell dendrites, they are located sparsely on the branches at intervals of 10 to 30 μ.

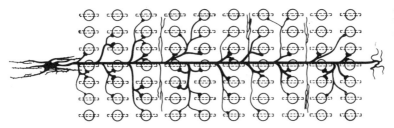

Fig. 21. Diagram of the distribution of collaterals of a basket cell axon relative to the network of Purkinje cells [39].

The spinules of these cells, as for the Purkinje cells, are intended for contacting with the parallel fibers.

Our calculations of the average number of these cells on a Purkinje cell give the following results: mouse, 7–12; cat, 25–30; man, 60. In the horizontal direction of a ⊥-section, both the basket cells and stellate cells are distributed evenly. However, with respect to the height of the molecular layer L these neurons are distributed irregularly: if we divide the molecular layer into 10 identical levels, then the number of cells in each level gradually increases toward the periphery and near the upper boundary of the layer the number of cells is usually 2 to 2.5 times greater than that near the lower boundary. However, the density of the stellate and basket neurons does not increase steadily: in the distribution of the densities of these cells, a *trough* is, as a rule, observed near the middle of the layer, while in the remaining portions the density steadily increases; the results of these calculations will be described separately in greater detail.

From numerous observations on Golgi preparations, mainly of such

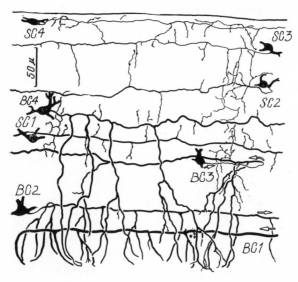

Fig. 22. Axons of the basket and stellate cells—"a system of transverse fibers" (combined drawing, Golgi method, macaque).

regions of these preparations in which only basket cells and stellate cells were abundantly apparent through the whole thickness of the molecular layer, we can conclude the following: the basket cells are located at different levels of the molecular layer, but not above the midline; the stellate cells are located in the upper two-thirds of the layer (the width of the zone of the overlapping of the regions of distribution of basket cells and stellate cells is equal to approximately 1/6 the height of the molecular layer).

The long stellate cell axons together with basket cell axons form a *system of fibers* which is orientated at right angles to the axis of the folium, that is, along the rows of Purkinje cells. The diameter of these fibers decreases according to the height in the molecular layer (Fig. 22).

Measurements at different levels of the layer indicate a more or less gradual change of the diameters of the basket cell axons from 3.5 to 1.5 μ in the cat and macaque and from 1.8 to 0.8 μ in the mouse. Similar changes take place also in the stellate cell axons: from 1.5 to 0.5 μ for the cat and macaque, and from 1 to 0.3 μ for the mouse.

5.2. THE CLIMBING FIBERS

As it is customary to assume, the afferent climbing fibers begin to branch only at the level of the Purkinje cell bodies, subsequently winding around all the smooth dendrites of these cells. In other words, it is customary to consider as one of the characteristic peculiarities of the climbing afferents (mainly on the basis of Gajal's investigations) their one-to-one correspondence (with respect to number) to the Purkinje cells—one fiber terminates on one Purkinje cell. Aside from contacts with the Purkinje cell dendrites, the climbing fiber with the help of numerous collaterals forms synapses on the bodies and dendrites of the basket cell and the stellate cell [34]. The length of these collaterals are small, so that one fiber affects the neighboring neurons only within the branching region of the dendrites of the given Purkinje cell.

At the level of the Purkinje cell bodies the climbing fiber forms a supplementary descending collateral which branches in the granular layer [34]. Collaterals of that sort go mostly along the Purkinje cell row (for a distance approximately 6 cells of a row) and form synapses on the bodies of the Golgi cells [39, 40].

According to recent research [40] the larger part of the climbing fibers arrive at the cortex of the cerebellum from the inferior olives and the smaller part from certain other nuclei of the medulla oblongata. Taking into account these circumstances and the fact that the inferior olives are with respect to their cellular composition a rather homogeneous formation [35] we can try to verify the indicated "one-to-oneness" comparing the number of neurons of the lower olives with the total number of Purkinje cells of the cortex of the cerebellum.

Our calculation, together with S. M. Blinkov's (for the macaque, cat, and mouse) yield a rather unexpected result: there are 10 to 15 times more Purkinje cells in the whole cortex than there are cells in the inferior olives. Hence, we can conclude that either there are sources of climbing fibers in addition to the inferior olives, or the principle of "one-to-oneness" does not occur. The latter seems to us more probable.

5.3. THE SUBLAYERS OF THE MOLECULAR LAYER

As long as the length of all the parallel fibers is assumed to be the same, we can consider the interaction of the system of parallel fibers with the dendrites of the Purkinje cells as homogeneous. Parallel fibers of one eigenvolume form with each Purkinje cell of an elementary rank (see section 4.2) an identical number $m_{GC/PC}$ contacts, and the total number of spinules of one Purkinje cell is the result of the interaction of the given Purkinje cell with $2q_0 \equiv N_0/m_{GC/PC}$ eigenvolumes (each Purkinje cell is involved in the formation $2q_0$ of the elementary ranks). Using the idea of the elementary rank, we can talk about the simplest cortical model constructed by means of *propagation* of these elementary *units*.

With an inconstancy of length of the parallel fibers with respect to the height of the molecular layer, the indicated homogeneity of interaction with the Purkinje cell dendrites is maintained only for that sort of group of parallel fibers within which all these fibers have the same length. For the preservation of homogeneity of the whole cortex we must accept that each eigenvolume produces a certain identical number M of the indicated groups of parallel fibers and in the simplest instance (maintaining a certain homogeneity within the granular layer eigenvolume) all the groups also have the same number of fibers. (And in the event of a nonconstant length of the parallel fibers, it is helpful at first to make clear what rectilinear homogeneous cortical models can occur.)

The whole aggregate of parallel fibers, within which all the fibers have the same length, forms in the molecular layer a distinctive *sublayer*, and the whole molecular layer is divided into M such homogeneous sublayers. If the length of the parallel fibers changes steadily according to the height of the molecular layer, then the sublayers do not overlap, each of them is located on a layer at a specific height and has a specific thickness (Fig. 23). The underlying sublayers by virtue of the great length of the fibers entering them produce a large number of contacts with the Purkinje cells, and since the density of the spinules of these cells is constant, the thickness of the sublayer decreases with the height.

Possibly the tendency of the spiny Purkinje cell dendrites to gather in clusters (see section 3.4) is a reflection of the presence of the sublayers. In the human cerebellum, according to the height of the molecular layer there are located approximately 10 to 20 such clusters, which possibly also correspond to the number of existing sublayers.

On the other hand we can also explain the tendency of the transverse fibers, formed by the axons of the basket and stellate cells, to be parallel to the molecular layer boundary by their relation to specific sublayers of parallel fibers. Moreover, in the transverse fibers, as in the parallel fibers, a gradual decrease of the diameters according to the height in the layer is observed. If the number of stellate cells and basket cells is constant for each sublayer, then the increase of the density of these neurons according

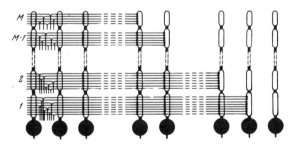

Fig. 23. Sublayers of the molecular layer formed by parallel fibers of different lengths.

to the height in the molecular layer (see section 5.1) is simply the result of the decrease of the thickness of the sublayer toward the periphery of the layer.

The stellate and basket cells can, for example, be arranged in a layer in the following manner: for one Purkinje cell in each sublayer there are two basket cells (up to the middle of the molecular layer) and two stellate cells (in the upper two-thirds of the layer), one of these cells is the *right* one according to the direction of the movement of its axon, and the other is the *left*. The necessity of observing symmetry, that is, the presence in each elementary rank of right and left basket and stellate cells, follows from electrophysiological reasons of contrast (see section 6.2) but it is incomprehensible in this connection that short-axon stellate cells can be distributed along the sublayers.

One of the basic questions of the organization of the molecular layer is this question about the number of sublayers. Unfortunately existing factual material does not make it possible to solve it at all clearly. Even if the maximum and minimum lengths of the parallel fibers at the boundary of the layers are known, as, for example, for the cerebellum of the macaque [15]: max $\lambda = 3$ mm, min $\lambda = 1$ mm, then we can indicate only the maximum number of possible sublayers, when the parallel fibers of the neighboring sublayers differ by the size of the step along the rank (Fig. 24). For the macaque, $b \approx 60\ \mu$, therefore, max $M \approx 15$. Actually, parallel fibers of adjoining sublayers can differ by a greater value and then M will be smaller. In the extreme case, an alternate version of the existence of only 2 types of fibers, max λ and min λ, is possible and then the molecular layer is divided into 2 sublayers. It seems to us that two sublayers or even

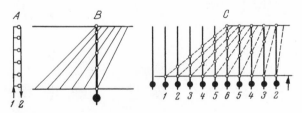

Fig. 24. Diagrams of the interaction of the dendrites of the Purkinje cells with the climbing fibers (A) and the parallel (B, C) fibers.

one is possible in the lower vertebrates. Judging by the quantity of stellate and basket cells (see above), in the mouse we can expect about 5 sublayers; in the cat, 10; in man, 20. Therefore, it is not excluded that sublayers are formed with a minimum step equal to the step along the rank. The constancy of the average number of contacts (section 4.1) imposes a very interesting restriction on the dynamics of the sublayers in phylogenesis. But all these questions require further research and new experimental data about the distribution of the lengths of the parallel fibers according to the height in the layer. In raising the question about the sublayers, we wanted to point to the possibility of a particular specific character of organization of the molecular layer produced by the system of parallel fibers.

We will show how, by the length of its parallel fibers, the thickness of a given sublayer can be determined. The fibers of kth sublayer of λ_k length make contact with $2q_k = \lambda_k/b$ Purkinje cells (in the event if the period of contacting [see section 4.2] is equal to the step along the rank), where $\lambda_1 = \max \lambda$; $\lambda_M = \min \lambda_n$ and $k = 1, 2, \ldots, M$. Thus, $q_k = q_1 - (k - l)m$, if $m = (\lambda_{k+1} - \lambda_k)/2b$, the number of sublayers is determined by the expression

$$M = 1 + (q_1 - qM)/m.$$

Reasoning further, in exactly the same manner as in section 4.1, we will find that the number of spinules formed by the fibers of the kth sublayer on the dendrites of one Purkinje cell equals $N_k = 2(q_k m_{GC/PC})M$. But such a number of spinules, by force of the constancy of their density, should take up a horizontal strip with a width $z_k = HN_k/N_0$, that is the thickness of kth sublayer $z_1 + z_2 + \ldots + z_M = H$. Therefore

$$z_k = \frac{H}{M} \cdot \frac{q_k}{q_0},$$

where $q_0 = (q_1 + q_m)/2$. With the increase of k the size z_k decreases, and hence the overlying sublayers have a smaller thickness than the underlying ones.

5.4. THE HYPOTHESIS OF THE RANKS

Analyzing the connection of the Purkinje cells with the system of parallel fibers, it is convenient to isolate aggregates of Purkinje cells of the sort which are located along the axis of the folium, that is, in one of the ‖-sections. In section 3.3, we conditionally called these Purkinje cell aggregates *ranks*, in order to distinguish them from the Purkinje cell *rows* orientated in the planes of $\perp$-sections. If the smooth Purkinje cell dendrites do not form a continuous network along the rows, as it is possible to assume on the basis of certain observations (see section 3.4 and [38]), that is, if we assume that the hypothesis of rows is not valid, then we can consider the ranks (the hypothesis of ranks) as the basic structural units of the cortex. The simplest cortical model which we can construct in this instance is a cortex obtained by means of propagation of the elementary rank, that is, we can consider one eigenvolume of the granular layer, the aggregate produced by it of parallel fibers, and the Purkinje cells "strung" on these fibers as a structural unit. Introducing into the elementary rank other cortical neurons, for example, since this is done in the preceding paragraph concerning stellate cells and basket cells, we can obtain as a result of propagation of the elementary units that sort of homogeneous and, we may say "crystallized" rectilinear cortex which will in the first approximation satisfy all the known experimental facts. Unfortunately, it is not known how the Golgi cells will be distributed in such a cortex.

Thus, if the hypothesis of ranks is valid, then we can expect that cortical phylogenesis proceeds along the path of the increase of the number of ranks and complication of their inner organization (the latter will be partly connected with the increase of the number of sublayers of the molecular layer). Taking into account cortical homogeneity, we may suppose that simple functional problems are solved right in the individual ranks. The contrast mechanism (see section 6.2) realized by the basket and, perhaps, the stellate cells, constitutes some confirmation of this hypothesis.

6. ELECTROPHYSIOLOGICAL MODELS

Many facts of cortical organization of a quantitative and qualitative nature, as is evident from everything written above, have been little examined. In addition, the scarcity of our knowledge about its functions, about the nature of the transformation of an afferent impulse into an efferent, about the electrophysiological properties of the individual neurons, etc., is a considerable obstacle in the attempts to interpret the cortical organization of the cerebellum.

An attempt to show ways of transformation of an afferent impulse into an efferent one was recently made by Braitenberg [4], who examines the Purkinje cell as a unit of comparison of two types of afferent signals arriving along the climbing and mossy fibers. This model uses the simplest concepts concerning cortical organization. Beginning with the properties of the neurons of the cortex and the system of parallel fibers described above, and on the basis of the recent research of Eccles and coworkers [11, 12], we can come to more differentiated but, unfortunately for the present, ambiguous ideas about the nature of their functioning. In section 6.1 we will briefly examine simple (hypothetical) electrophysiological models of the activity of the cortical neurons of the cerebellum.

6.1. THE ACTIVITY OF THE PURKINJE CELLS

The dendrites of the Purkinje cells have at least two polysynaptic mechanisms capable of having an effect on the activity of this cell: the climbing fiber and the system of parallel fibers. In spite of the difference in constructional appearance, both these mechanisms can render on the Purkinje cell the same type of effect.

The climbing fiber and the smooth dendrite represent two fibers synaptically connected to one another across certain intervals (Fig. 24A). The impulse arriving along the first fiber will sequentially excite from below up portions of the second fiber. As a result of multiple actions the second fiber, from a single stimulus in the first fiber, can respond with a volley.

Such a transformation of a single impulse into a volley is possible provided only that the rate of the spread of excitation along the first fiber is less than along the second (a more precise condition must take into account the duration of the refractory phase τ of the second fiber and the spacing between two synapses Δl: $v_1 < (1 + v_2\tau/\Delta l v_2)$. Taking into consideration the great difference in diameters of the climbing fiber (0.5–1 μ) and smooth Purkinje cell dendrites (3–5 μ), we can expect in response to a single stimulation of the climbing afferents, for example, in the inferior olives, a Purkinje cell volley response (with recording from the cell body). It seems to us that the research of Eccles and coworkers [12], judging by the recordings cited, confirm this hypothesis.

With a single excitation of a mossy afferent a situation is possible that along an aggregate of parallel fibers a single tilting front of excitation (see section 4.3) will spread and hence, in turn, also from below up, different sections of the Purkinje cell dendrites will become excited (Fig. 24B). In this way, and by the system of the parallel fibers, a Purkinje cell is able to respond to a single stimulation of a mossy afferent with a volley.

Possibly, these models demonstrate the prevalent opinion [9] that volley activity is one of the characteristic modes of function of the Purkinje cells.

The mechanism of operation of the Purkinje cell via the climbing fiber is fixed or *soldered*. The nature of the response of the cell can be changed only by means of an alteration of frequency of the stimulation of this afferent: when the stimulation period is greater than the duration of the volley, the cells will respond with individual groups of impulses, and all the groups according to the number of impulses and duration will be standard: with reduction of the stimulation period the groups will draw together and then change into continuous activity.

During functioning of the Purkinje cell the nature of its activity is distinguished from the system of parallel fibers not only by the frequency of the afferent impulse (of the mossy fibers), but also by the remoteness of the cell from a source of excitation. Within one rank, Purkinje cells situated at a different distance from a given active region of the granular layer are able to respond with volleys of different durations (the figures in Fig. 2C are an example of how, during the passage of a single front along the Purkinje cell rank, the number of impulses in a volley can be distributed).

6.2. CONTRAST

On the basis of the research of Eccles and co-workers [11], we can assume that the basket cells exert an inhibitory influence on the body of the Purkinje cells. Taking this circumstance into account, as well as the nature of the distribution of collaterals of the axon of the basket cells, Szentágothai [39] suggested that these cells, inhibiting the neighboring regions of Purkinje cells, contrast in our terminology the activity of the individual ranks at the level of the Purkinje cell bodies.

If we start from the similarity of the stellate cells with a horizontal axon and of the basket cells, then we can assume that these stellate cells render the same influence on the Purkinje cell dendrites at different levels of the molecular layer, that is, they contrast the activity of the individual rank at different heights of the layer. We will note that in these assumptions we can grant the realization of the hypothesis of rows and assume that the system of transverse fibers ensures the separation of individual ranks from continuous Purkinje cell rows. The absence of basket cells in the frog cerebellum [20] indicates the simplest kind of problems the cortex can solve without the contrast mechanism, at any rate at the level of the Purkinje cell bodies.

On the other hand, the system of transverse fibers, taking into account the presence in it of the gradient of the diameters according to the height in the layer, can play the same role as does the system of parallel fibers. It is helpful to have in mind also another possible action mechanism of the system of transverse fibers, namely, an excitatory one. In this case during the volley response to a single excitation of the mossy fiber there will be involved not a linear aggregate of Purkinje cells (a rank), but a two-dimensional one, that is, some matrix of the Purkinje cells composed of several ranks and rows. The presence of gradients of spreading rates of excitation in both systems of fibers—parallel and transverse—ensures a different volley activity of the various Purkinje cells of this matrix. Moreover, if the hypothesis of continuity is realized, then the net of Purkinje cell dendrites play for the system of transverse fibers the same role which the granular layer does for the system of parallel ones.

For the time being we can only hope that very shortly a more definite choice between these models will be made. During attempts of experimental demonstration of the various electrophysiological ideas of the cortical functioning of the cerebellum it is essential to have in mind the feasibility of such contradictory models.

6.3. THE GRANULAR LAYER

At present the opinion that the granular layer is a multitude of individual granular cells for the conducting of excitation is the most prevalent. According to these ideas, the dense adjacency of granule cells to one another is the result of *dense packings*, and a unique problem of these cells is to transmit excitation from the mossy fibers to the molecular layer [6, 39]. However, admitting the interaction of the granular cells with one another, the point of view on the granular layer can differ from the preceding opinion. We have already formulated the differing viewpoint under the title of the principle of continuity, and used it for the solution of several structural problems. We will briefly consider here this principle from the electrophysiological viewpoint. A demonstration of the functional soma-somatic interaction of granule cells by electrophysiological methods is a very difficult methodological problem, and we must not hope for the solution of this question in the very near future. Therefore, although it may be too premature to develop this hypothesis, yet it seems to us helpful to indicate its naturalness.

Thus, we will consider that soma-somatic contacting of the granule cell is intended for the transmission of excitation. Then we can examine the chains of granule cell bodies in the structural models described above of the layer (see Fig. 7), as pieces of nerve fibers (of 5 μ diameter) and the whole layer as a continuous net—a syncytial medium.

For a straightened cortex (see section 1.3) the granular layer is a plate of a continuous medium of a fixed thickness. With the single excitation of some internal granule cell the aggregate of cells that are successively excited will have at first the shape of an expanding "sphere," which, after this, having reached the edges, "breaks apart," and a cylindrical wave of excitation will spread over the plate. The cylindrical edge perpendicular to the boundaries of the layer is the most stable configuration of the excitation wave—in the presence of small obstacles it will flow around them and quickly recover its shape. It is essential that in the horizontal section of the layer the edge can have an unrestricted shape determined, for example, by the configuration of the sources, but in $\perp$-sections and $\parallel$-sections it is always perpendicular to the boundaries of the layer.

It is clear from what has been said that to render any significant influence on a similar wave is possible only by a system of fibers distributed over the whole thickness of the layer. Possibly there is even caused by this circumstance the presence in the cortex of Golgi cells of the first type, which form a dense and rather extensive axon net (see section 2.6). These cells, the axon of which branches over the whole thickness of the layer, can completely block the spreading of excitation along the granular layer, or, inhibiting individual regions of the layer, prevent the spreading of excitation in certain directions. If all the Golgi type I cells distributed their "axon cloud" over the whole granular layer thickness, then we could suppose that the granular layer too becomes excited every time over the whole thickness. In this case we would examine the granular layer in a functional respect as an excitable plane and assume that the thickness of the granular layer is determined only by structural causes: by the necessity for the creation of a system of parallel fibers. Certainly in fact we should represent the granular layer at least as two such planes which can be fired both separately and together.

The presence of different methods of distribution in the layer of the local branches of the mossy fibers and the existence of subtypes of Golgi cells the axons of which branch only in the upper or lower half of the layer (see Fig. 9) indicates the possibility of such ideas of the granular layer. Those sublayers of the molecular layer, about which the question was raised in section 5, are related to the peculiarities of excitation spreading along the system of parallel fibers, etc. The stratification of the granular layer possibly finds reflection in the stratification of the molecular layer into two layers formed by the basket cells.

We will note that this additional cortical stratification complicates the idea about the volley activity of the Purkinje cells from the parallel fibers (see section 6.1) and must hamper the experimental observation of these modes of operation of the cells in a clear way. Therefore, we can assume that volley activity has to be more easily recorded in the Purkinje cells of the frog cerebellum, in which there are no basket cells [20] than in the cat.

In conclusion we will note the following remarkable characteristic of the system of parallel fibers. Any picture of activity taking place at a given moment in the granular layer without distortion is transmitted—is *carried in parallel* into the molecular layer. Descriptively speaking, this system is all the time making instantaneous "photographs" of the activity of the layer and distributing them in a somewhat indistinct form to the nearest "neighbours." Every time in an analysis of each "dot" of the image approximately fifty Purkinje cells, or even more, participate. An analysis of the simplest interactions of afferentation even in simplified cortical models leads to complex constructions which we will not examine here. However, the development of ideas of models in this direction that are able to bring us nearer to an understanding of the need of the propagation of afferentation in the cortex, the role of the individual rows and ranks, the possibilities of the existence of restrictions on the length of parallel fibers, etc., seems to us extremely useful.

1. The preparations which we used for measurements and calculations were prepared according to standard histological methods of staining— these are methods of Nissl and Braché, the quick Golgi method in the modification of Kopsch, Cajal's reduced silver method, and Palmgren's method [30].

For the determination of the densities of the cells we used celloidin sections 10, 15, and 20 μ (Nissl preparations) in thickness. Paraffin sections also were used for calculations (for comparisons and verifications and at the same time for obtaining relative sizes), for example, in the Braché and Palmgren methods, which were convenient for calculations of the stellate cells.

2. It is convenient to carry out the measurement of the volume of the cerebellum on material after fixation in formalin before dehydration. However, all the other measurements of such sizes as the thickness of the layer, the length of the cortex, the density of the cells, etc. were conducted on sections. Therefore, for convenience in Table 1 in the Appendix there are presented the already enumerated volumes with an allowance for the compression of the brain during dehydration and embedding: the actually measurable size of the volume was multiplied by the coefficient of the volumetric compression $(k)^3 = (0.75)^3 = 0.42$, where k is the coefficient of linear compression.

3. The density of the granule cells was determined two ways: (1) under magnification 90 $\times$ 15 (oil immersion) the whole depth of the section was looked over and with the help of a drawing instrument (DI–4) the bodies of all the granule cells located in the visual field were drawn; (2) under magnification 90 $\times$ 5 with the help of an ocular mesh a visual calculation of the number of cells over the depth of the preparation was carried out.

4. The density of the spiny branches of the Purkinje cells was determined from the Golgi preparations. Cells were chosen all the spiny branches of which in the region selected for calculation appeared to be stained. The spiny branches of this region of the dendrites were drawn

with the help of a drawing instrument (magnification 40 × 15). Then from the drawings were measured the total length of these branches and the area onto which they were projected.

The density of the spinules on the branches also were determined with the help of sketches, but under greater magnification (90 × 15, immersion).

APPENDIX

Table 1. Macroscopic Parameters.

Indices	Human	Macaque	Cat	White Rat	Mouse	Frog
V_{CCb}	43,000	3,350	2,100	105	42	—
S_{Cb}	72,000	7,200	5,500	322	153	2,625
L	1,000	400	250	52	28	1.3
n	196	98	66	19	12	—
H	350	270	220	170	150	250
h	230	180	140	110	100	120

Table 2. The Densities of the Cells.

Indices	Human	Macaque	Cat	White Rat	Mouse	Frog
$p_{PC}10^6$	300	500	650	1,400	2,500	4,400
$p_{Go}10^3$	2.1	2.3	2.8	3.2	3.5	0.85
$m_{GC/PC}$	1,600	950	600	250	140	22
$a \times b$	100 × 67	75 × 65	60 × 55	50 × 30	40 × 20	—

Table 3. The Dendrites of the Purkinje Cells.

Indices	Human	Macaque	Cat	White Rat	Mouse
S_{PC}	120,000	85,000	60,000	30,000	20,000
σ	0.5	0.4	0.3	0.25	0.2
N_0	60,000	34,000	18,000	7,500	4,000

Table 4. The Parallel Fibers.

Indices	Human	Macaque	Cat	White Rat	Mouse
$2q_0$	37	36	30	30	29
λ	2.6	2.2	1.6	0.9	0.6

The symbols of basic quantities presented in Tables 1–4 are as follows:

V_{CCb} volume of the cortex of the cerebellum, mm³

S_{CCb} area of the cortex of the cerebellum, mm²

L length of the cortex measured in a sagittal section of the vermis, mm

n number of folia in the region of the vermis

H and h thickness of the molecular and granular layers, respectively, μ

p_{PC} density of Purkinje cells; $p_{PC} \cdot 10^6$: number of Purkinje cells in an area, 1 mm²

p_{GC} density of granule cell; $p_{GC} \cdot 10^9$: number of granule cells in a volume 1 mm³

$m_{GC/PC}$ number of granule cells per Purkinje cell

v eigen (characteristic) volume of the granular layer of one Purkinje cell, that is, the volume containing $m_{GC/PC}$ of granule cells, μ^3

a and b spacing between Purkinje cells in $\perp$-and $\parallel$-sections, respectively, the parameters of the network formed by the Purkinje cells: the step down the row and the step along the rank

S_{PC} area of the branching region of the Purkinje cell dendrites in a $\perp$-section, μ^2

N_0 total number of spinules of a single Purkinje cell

σ density of the spinules, the number of spinules per unit of area of the Purkinje cell branching region

λ length of a parallel fiber, mm

$2q_0$ average number of contacts formed by one parallel fiber with the Purkinje cells

M number of sublayers of the molecular layer

S_{ML} and S_{GL} area of the molecular and granular layers measured in $\perp$- or $\parallel$-sections

REFERENCES

1.
V. M. Bekhterev, *Provodyashchie puti spinnogo i golovnogo mozga* (*Conduction Pathways in the Brain and Spinal Cord*) vol. 2, St. Petersburg (Leningrad): 1896.
2.
S. M. Blinkovi and I. I. Glezer, *Mozg cheloveka v tsifrakh i tablitsakh* (*The Brain of Man in Numbers and Tables*), Moscow: Medgiz, 1965.
3.
S. T. Bok, *Histonomy of the Cerebral Cortex*, Amsterdam: Elsevier, 1959.
4.
V. Braitenberg, *Nature* 1961:6, 4775.
5.
—— and R. P. Atwood, *J. Comp. Neurol.* 1958:109, 1–30.
6.
R. Cajal, *Histologie du système nerveux de l'homme et des vertébrés*, vol. 2, Paris, 1911.
7.
V. Dahl, S. Olsen, and A. Birch-Andersen, *Acta Neurol. Scand.* 1962:38, 81–97.
8.
R. S. Dow, *Neurophysiol.* 1949:12, 245–256.
9.
—— and G. Moruzzi, *The Physiology and Pathology of the Cerebellum*, Minneapolis: University of Minnesota Press, 1958.
10.
J. C. Eccles, *The Physiology of Synapses*, Berlin: Springer, 1964.
11.
——, P. Andersen, and P. E. Voorhoeve, *Nature* 1963:199, 655.
12.
——, R. Llinas, and K. Sasaki, *Nature* 1964:203, 245 (N4942).
13.
C. Estable, in *Brain Mechanism and Learning: A Symposium*, ed. J. F. Delafresnaye, Oxford: Blackwell, 1961.

14.
C. A. Fox, *J. Comp. Neurol.* 1959:112, 39–54.
15.
——— and J. W. Barnard, *J. Anat.* 1957:91, 202–313.
16.
J. Freznik, *Acta Morph. Acad. Sci. Hung.* 1962:12, 9–14.
17.
R. Friede, *Acta Anat.* 1955:25, 65–72.
18.
———, *Proc. N.A.S.* 1963:49, 187–193.
19.
I. M. Gelfand and M. L. Tsetlin, *Dokl. AN, SSSR* 1960:131, no. 6.
20.
P. Glees, C. Pearson, and A. G. Smith, *Quart. J. Exp. Physiol.* 1958:43, N1.
21.
E. G. Gray, *J. Anat.* 1961:95, 345–356.
22.
Ya. M. Iontov, in collection *Morfologiya mezhneyronnykh svyazey* (*Morphology of Interneural Connections*), edited by N. G. L. Kolosova, AMN, SSSR (Publishing House of the Academy of Medical Sciences of the USSR) 1953, p. 20.
23.
T. Inukai, *J. Comp. Neurol.* 1928:45, 1–32.
24.
B. Jacob, *Das Kleinhirn, Handbuch der Mikroskopischen Anatomie,* vol. 4, Berlin, 1928.
25.
J. Jansen and A. Brodal, *Das Kleinhirn, Handbuch der Mikroskopischen Anatomie,* vol. 8, Berlin, 1958.
26.
A. C. U. Kappers, G. C. Huber, and E. C. Crosby, *The Comparative Anatomy of Nervous Systems of Vertebrates, Including Man,* New York. Hafner, 1936 (reprinted 1960).
27.
O. Larsell, *J. Comp. Neurol.* 1947:87, 85–129.
28.
———, *J. Comp. Neurol.* 1953:99, 135–200.
29.
F. Ovsyannikov, *Izv. Rossiysk Akad. Nauk.* 1905:20, seriya 5, 30.

325 THE CEREBELLAR CORTEX

30.
A. Palmgren, *Acta Zoologica* 1948:25, 378–392.
31.
A. Pezard, *Bull. Biol. France et Belg.* 1941:75, 1–62.
32.
H. A. Riley, *Res. Publ. Assn. Nerv. Ment. Dis.* 1929:6, 37–192.
33.
Schaper, *J. Comp. Neurol.* 1898:8, 130–141.
34.
M. Scheibel and A. Scheibel, *J. Comp. Neurol.* 1954:101, 733–764.
35.
——— and A. Scheibel, *J. Comp. Neurol.* 1954:102, 77–133.
36.
Ye. K. Sepp, *Istoriya razvitiya nervnoy sistemy pozvonochnykh* (*A History of the Development of the Nervous System of Vertebrates*), Moscow: Medgiz, 1959.
37.
D. A. Sholl, *The Organization of the Cerebral Cortex*, London: Methuen, and New York: Wiley, 1956.
38.
Ph. Stöhr, *Z. Ges. Anat.* 1923, Abt. 1.
39.
J. Szentágothai, *Information Processing in the Nervous System*, Amsterdam, 1964.
40.
——— and K. Rajkovits, *Z. Anat. u. Entwicklungs.* 1959:121, 130, 130–141.

III
The Regulation of
Movements

9

Some Problems in the Analysis
of Movements

I. M. Gelfand
V. S. Gurfinkel
M. L. Tsetlin
M. L. Shik

This article serves as a type of preface to data dealing with an experimental investigation of movements, but not in the slightest degree is it a systematic review of the literature on this question. We wanted to set forth in it some general concepts which seemed to us helpful in the pursuit of the physiology of motor activity, and which assist us in choosing the direction of further work. Many of these views are closely connected with studies on the structure of movements as developed in the remarkable works of N. A. Bernstein [2–4].

The control of movements is one of the most important functions of the nervous system. The structure and function of the nervous system undoubtedly is determined to a large extent by this problem. The physiology of movements is basically a study of the purposeful activity of the nervous system as a whole. Therefore, the control of movements seems to us one of the most natural objects for study of those integral functions of the nervous system which are connected, so to say, with *operative control*.

The final result of the work of the nervous system on the control of movements is the sending of impulses to the muscles and the basic question of the physiology of movements is the study of the mechanism for the development of expedient combinations and sequences of such signals.

Here the simplest viewpoint is the concept of the presence of some higher nerve center (situated, for example, in the cortex), where there stems the generation of commands of the sort which completely determine movement, so that the role of all the remaining nerve mechanisms is only the transmission of these commands. The inconclusiveness of this viewpoint is obvious, and in works on the physiology of movements, beginning with the discovery of Fritsch and Hitzig, and after that in the classical works of Sherrington, Magnus, Pavlov, Ukhtomskiy, and in the publications of contemporary research, the question concerning the interaction of different nerve mechanisms in the process of the realization of movement takes the center place.

The fact is that in natural movements dozens of different muscles are involved working coordinately. The system of commands necessary for movement realization cannot but be in this case very complicated. During its development calculation of a rich and varied afferentation is essential, including also ones originating in the course of the movement itself. The concrete realization of movements to a large extent depends on initial conditions—the original pose, etc.

Undoubtedly, a whole series of nerve centers are occupied in the development of commands for the muscles and processing for the purpose of afferentation. The study of their interaction led us to the attempt to describe the features of complicated systems of control from a single viewpoint which will be presented below. Some results of experimental verification of this viewpoint are written up in subsequent articles of the collection. However, many and occasionally important features of the ideas presented in this article still require experimental verification; they are examined by us as a working hypothesis for further research.

This point of view—we will call it the principle of the least interaction [see 11, and also Chapter 13] is a complicated, multilevel system of control examined as an aggregate of subsystems having relative autonomy. Each of the such subsystems has its own "individual" problem consisting of the decrease of interaction with the "outer medium"; the latter for a given subsystem is made up of the medium outside as regards the whole system and the remaining subsystems. Complex systems of control can consist of several levels each of which includes a series of such subsystems.

For subsystems of a certain level actions of the outer, as regards this level, medium include the afferentation coming from below, and the organization of their interaction is determined by the interaction of higher levels. For the lowest level afferentation is exclusively prescribed; the subsystems of this level have outputs for the effectors. In section 2 we will dwell on these ideas in greater detail.

In the organization of the control of movements, utilization of such features of the motor problems which can simplify control, reduce the number of independently controlled effector parameters, and simplify the processing of the incoming afferentation, plays an important part. The problems having such features are, so to say, organized.

In the control of movements organization becomes apparent first of all in that for each motor act it is possible to single out a relatively small number of leading effector parameters and to determine the basic afferentation necessary for the realization of this movement. The tendency for such simplification is also a manifestation of the principle of the least interaction. The decrease of the number of controlled parameters lowers the overall level essential for impulse control.

This article is made up of three sections. The first of them is devoted to synergies and certain other mechanisms simplifying control of movements. In the second we will set forth our general concepts about the arrangement at the spinal level of the structure of movements. In the last part, model representations connected with the function of a pool of motor neurons are described.

1. In order for the higher levels of the central nervous system to solve effectively the problems of the organization of motor acts in the time required, it is essential that the number of controlled parameters not be too large and the afferentation requiring analysis not be too high. The so-called synergies play an important role in the establishment of such conditions of work. It is customary to call *synergies* those classes of movements which have similar kinematic characteristics, coinciding active muscle groups and conducting types of afferentation.

For each synergy there are distinctive, specific connections imposed on certain muscle groups, a subdivision of all the muscles participating in a movement into a small number of connected groups. Because of this, for realization of a movement it is enough to control a small number of independent parameters, although the number of muscles participating in movement can be large.

Although there are but a few synergies, they make possible almost the whole variety of voluntary movements to be included. We can distinguish relatively simple synergies of postural preservation (the stabilization synergy), cyclic locomotive synergies (walking, running, swimming, etc.), synergies of throwing, blowing, jumping, and a certain (small) number of others.

The synergies listed here for an adult human appear already fully worked out; the biomechanical side of the majority of them have been studied. A more detailed research of one special synergy—the respiratory synergy of standing—is written up in Chapter 13.

We will mention this respiratory synergy here very briefly. The fact is that with respiration noticeable displacements of different parts of the body take place. However, on the position of the overall center of gravity these displacements hardly show up. The reason, as explained, is that cophasally with deflection of the torso backward (during inspiration) deflection of the pelvis forward takes place; during expiration the paired displacements occur in the opposite directions. This synergy is specific and does not take place, for example, during external disturbance (a light push in the back). It breaks down during certain neurological illnesses, and then the body's center of gravity oscillates in accordance with the phase of respiration. We can suppose that the described respiratory synergy of a vertical pose is not an exception but an example of a typical mechanism participating in the most diverse natural movements.

It is natural to assume that movement training is the development of corresponding synergies which reduce the number of parameters requiring individual control. A new synergy like that is made, of course, each time not on an empty place but on the framework of a small number of basic synergies and inherent neurophysiological mechanisms which lower the number of independent parameters of the controlled system. Some of the mechanisms of this sort, although not examined by research in this project, are already well known. We refer here to such well-studied examples of functional organization as the system of interaction of the muscle motor neurons—antagonists, acting on the same joint; the system of postural reactions, using a fixed system of the interaction of different kinds of receptions (labyrinth, otolith, proprioceptive of the neck and limbs); and also the important mechanism of development of temporary connections.

The basic synergies and the simplest neurophysiological mechanisms enumerated here form, so to say, "a dictionary of movements." Using this analogy, we can say that the efforts of the muscles are the letters of the language of movements, and the synergies combine these letters into words, the number of which are much less than simply the number of combinations of letters. Moreover, the wealth of the dictionary provides a variety of admissible movements. The majority of motor problems arising before an organism are contained in the scope of this dictionary and only in exceptional cases does there arise a need for its enrichment.

Until now the question has chiefly been the effector aspect of the synergies. Undoubtedly, however, with each synergy there are connected afferent currents in which are singled out the conduction signals and the addresses characteristic of it. The language of the synergies is in this sense not only the external language of movements but also the internal language of the nervous system during control of the movements. The synergies are able to simplify the processing of afferentation, having organized it according to the motor problem.

From this viewpoint in such a problem, as for example the identification of images, in the first place it is determined to which synergy this problem belongs, which in turn predetermines the subsequent course of identification.

We can assume that new movement training consists of the development of a simple process of movement control and leads to the search and correction of suitable synergies or groups of synergies, including here also isolation of conducting afferentation. In the work of V. I. Krinskiy and M. L. Shik [30] the role of this last factor was studied in conditions when the motor problem had to be carried out under the control of specially distorted visual afferentation. It turned out that to some transformations of the visual field the available system of utilization of visual afferentation in the problem of pose retention is resistant. We can assume that such distortions of the visual field are still able to use the available synergy. The more marked distortions made the realization of the problem impossible, and some training time was needed in order that the problem again become executable. It is interesting that during some transformations of a visual signal about deviations from an assigned pose the retraining, although allowing the problem to be executed, did not reestablish that exactness of its solution available to the examinee during utilization of undistorted visual afferentation.

Utilization of the mechanism of synergies is, of course, only one of the ways of simplifying the problem of the control of movements. Another possible approach is connected with the existing similarity of the problem of movement control with the mathematical problem concerning the search for the minimum function of many variables. The language natural for this mathematical problem [14] was found suitable also for a description of the structure of movement. The combination of local adaptations with extrapolations characteristic for nonlocal search methods of the extremum apparently is also typical for the process of admission of solutions in motor problems, and certain peculiarities of such a search were successfully detected in the experiments. Such an approach in connection with the problem of preservation of the erect posture was used in the work of Gelfand, Gurfinkel and Tsetlin [10], and stimulated further research devoted to the study of tremor.

In V. I. Krinskiy and M. L. Shik's work, an attempt was made to investigate the solution method of a simple motor problem which presented, however, high demands of exactness for the execution of the assignment. The examinee was asked to find such a position of two joints of the upper extremities so that the pointer on the galvanometer, the position of which was a certain function of the angles of the indicated joints, would come into a zero position. The experimenter could simultaneously record the trajectory of the cathode ray on the screen of the oscilloscope in coordinates of the limb angles. The change of one of the limb angles caused movement of the ray along the horizontal; the other, along the vertical line. The examinee did not see the screen. The experimenter could also change the coefficients in order that during successive executions of the problem (each took 10–60 sec) the desired position of the joints was different. The function employed had the shape of a "small boat" with a single lowest point

$$f(x;y) = |(x - a) - (y - b)| + \alpha|x - a| + d|y - b|.$$

In the initial attempts all the examinees were allowed only one method: they made successive changes of the joint angles bringing the position of the ray step by step along the horizontal and vertical to the assigned point (a change of the joint was made when the pointer, passing over the minimum, again indicated a withdrawal from the required pose). But later on, together with that sort of solution method, another was applied which utilized the organization of the function, although for the examinee it remained unknown (regardless of the level of theoretical preparedness). On the screen it was seen how the examinee going to the "bottom," descended along it to the lowest point, not climbing "on board" (this is feasible only during coordinated changes of the angles of both joints), and only in the immediate vicinity of the deepest point, made the desired pose more precise by successive changes of the angles of the joints. The tactics of behavior of the examinee in this case were similar to the so-called ravine tactics [15].

If we agree with the idea stated here that the higher levels of the nervous system, revising the conducting afferentation, control the work only of a small group of muscles determining a given movement (or synergy); if, moreover, we still consider that this control is effected not immediately but by means of reorganization of interaction at the lower levels, then it becomes clear that the basic "manual" work with respect to movement realization is carried out by precisely these lower levels, while the higher ones form only functional synergies and reorganize the interaction system of the elements of the lower levels.

In many papers on modeling of pattern recognition, the aim of the model is the formation of a generalized image in which are reflected only the most significant attributes of the real object. During control of movements the reverse problem, so to say, is solved—by abstract representation the structure of a real movement with all the essential details is reproduced.

The realization of an image as a real movement requires its translation from three-dimensional, kinematic language into the language of muscle dynamics: the motor composition, the number of motor units, the spatial and temporal order of their recruitment, etc.

2. For a further account it is convenient to single out the intermediate neural structures in order to dwell at greater length, later, on the function of the last effector link—the motorneuron pool and the interneuronal structures connected with it.

We will analyse the working principles of the intermediate neural structures in an example of the organization of movements at this spinal level.

Characteristic for the spinal level is an extremely large volume of transformed afferentation and a large number of efferent outputs. Moreover, afferentation is received first hand at the spinal level, and the efferent commands are immediately realized. On all the activity at the spinal level are acting supraspinal influences organizing its activity in order to realize the assigned movement. The presence of relatively autonomous subsystems, also divided spatially, is an essential feature of the structure at this level. The presence of a series of innate (so-called "soldered") interactions is typical for those subsystems.

The autonomy (although relative) of the individual subsystems at the spinal level allow effective control to be carried out in a short time interval—the solution should not assume a complex or long concordance. Obviously, the organization of any complex movement by no means leads to the working of only one such subsystem or several soldered subsystems. Therefore, in our ideas about the organization of movements at the spinal level, interaction among individual subsystems plays a central role.

When we talk about the possibility of the relatively autonomous working of the individual subsystems, then inevitably the question arises which particular, "individual" goal in a concrete problem one or another subsystem pursues. It seems likely to us that such a goal in the first approach is the decrease of the total of impulses received by the subsystem both from the periphery and from the sides of the other subsystems and higher levels of the nervous system (the *principle of least interaction*). This total of impulses, serving as an estimate of function of one or another subsystem, is developed apparently by special mechanisms the operation of which depends on the assigned interaction system. It is natural to assume that so-called "usual" or "expected" afferentation adds a relatively small contribution to this total afferentation in comparison with the "unusual" or "unexpected." Perhaps a "nonequilibrium," "unbalanced" afferentation plays a special role. In other words, we can assume that the goal which the subsystem pursues is the minimization of external interactions tending to disturb it from that position in which it finds itself at a given moment. The tendency to minimization of interaction leads to the coordinated working of the individual subsystems subordinating the autonomous activity of each of them in the interest of the solution of the overall problem assigned by the supraspinal afferentation.

Moreover, the basic role of the supraspinal influences is the appropriate rearrangement of the interaction organization of the individual subsystems at the spinal level [34].[1] Such a rearrangement may find expression in the motor effect; however, this result is not obligatory. Rearrangement can create a "readiness for movement" for realization of which, however, a supplementary action is required. Possibly some supraspinal influences (for example, connected with the operation of the labyrinthine apparatus) generally only in this way even affect spinal activity.

The activity of each relatively autonomous subsystem, as we assumed, is directed to the reduction of the total afferentation formed from proprioceptive afferentation and afferentation appearing external as regards the given subsystem—from the neighboring subsystems of the same level and supraspinal ones. A change of the mode of function of the subsystem leads first to a change of the characteristic afferentation and is directed to minimization of the total afferentation. If the contribution of its afferentation into the total one is relatively large then the role of the rest, and among them the supraspinal afferent actions, is reduced. We can see this in conditions of special experiments [17, 18, 37, 40] or in certain situations arising during athletic activity. Probably it is also natural to examine the phenomenon of a dominant, discovered by Ukhtomskiy, from this point of view. The transmission of the control of habitual movements to the lower levels (the automation of motor habits according to N. A. Bernstein) from this viewpoint is the result of the tendency of the higher levels to minimize their own interaction with the lower levels.

The individual details of movement are formed and completed in the process of the interaction of the subsystems forming the spinal level of

---

[1] Automata games [39] are a natural mathematical model for the study of control methods by the assignment of interaction. On the digital computer, behavior of a system consisting of a large number of automata was modeled, the interaction among which was corrected by some higher level. The problem of the higher level was to develop an interaction guaranteeing beforehand the assigned behavior of the automata. The modeling on the digital computer showed that a correction of interaction can be made without waiting for the complete correspondence of the behavior of the collection of automata with the interaction assigned from above.

regulation of movement. If we somewhat schematize the Wells classification [41], then in typical movements of a human being and higher animals we can single out the muscles which carry out the basic active part of the motor problem (relatively few of them) and the muscles which stabilize the position of the basic mass of kinematic links of the body (the majority of them).

The activity of the corresponding controlling neural mechanisms is basically autonomous, and the dominant afferentation is peripheral; the interaction of these mechanisms with the rest will be relatively small. Hence, there results the typicalness of such "stabilizing" modes of function in the movements. The relatively small role of interaction in such modes of this sort guarantees their "freedom," permitting the higher levels of the central nervous system not to be responsible for the control of the corresponding cerebrospinal subsystems. Apparently in some of the simplest cases such stabilization is already realized at the segmental level [23]; if afferentation is not minimized on this lowest level, the higher levels are included. The work of a large number of stabilizing mechanisms guarantees the smoothness, the evenness of the movement; without them the movement would become ataxic.

Of course the regulation of an active movement is by no means reduced to the operation of the stabilizing mechanisms. The system of interaction assigned to the higher levels should guarantee nonequilibrium functioning of the muscles responsible for the active part of the movement, and almost always an expedient change of the angles of the joints. Although voluntary movements have been studied relatively little, we can nevertheless state some considerations resulting from the concepts stated above. In particular, if the role of the higher levels of the nervous system is not in the sending of direct commands but in the reorganization of the interaction system (the tuning) of the neural mechanisms at the spinal level, then naturally such a tuning should take more time than for the usual transmission of commands.

We know that the latent time of a simple motor reaction is virtually constant, and that even systematic training does not lessen it more than 10 to 15 msec. Moreover, of the 120 to 180 msec of latent time the communication time is by no means more than 50 to 80 msec. Inasmuch as any of the active movements for the majority of the neural structures of the spinal level signify the inclusion of stabilization mechanisms required by the given problem, so such a tuning should for the majority of spinal mechanisms be in general outline the same. Of course, the tuning may change substantially even in the course of the movement itself. However, study of such tuning is simplest when movement has not yet begun. We will call *pretuning* the phenomena connected with the preparation of the spinal neural mechanisms for movement, and we will dwell on them in somewhat greater detail.

One of the first observations, which, as is now clear, indicates the presence of pretuning, belongs to Hufschmidt [26], who discovered that as much as 60 msec before the start of voluntary contraction of the muscle, inhibition of the muscle activity of the antagonist occurs.

Our experiment consisted of the following procedure: We instructed the examinee by a signal to make a specific movement, for example, to straighten the foot. Within a certain time after the signal (but before the start of the movement) a tendon reflex was produced and the amplitude of the electromyographic response was measured. It turned out that the size of the reflex essentially depends on what time remained before the start of the movement. This dependence has the same character for the tendon reflexes and the monosynaptic H-reflex [25] for the corresponding muscle. It is interesting to note that for relatively long intervals before the start of a movement (70–50 msec) the changes of amplitude of the spinal reflexes are approximately the same both for those muscles which are to participate in the movement and for those which in this movement are not activated. As the time to the start of movement decreases, these changes show up considerably sharper in regard to precisely those muscles which will participate in the movement. In this way, apparently, tuning of the spinal mechanisms bears at the beginning a so to say diffuse nature, and as the moment of initiation of the movement approaches, the change of the system of interaction at the spinal level is localized. True, it is not ruled out that the diffuse changes of the condition of the spinal-controlling mechanisms are connected with an orienting reaction arising in the experiments described. These experiments are described in greater detail in Chapter 11 in this collection.

3. In this section we want to talk about some ideas connected with the function of the last effector link of the control system of movements—the motorneuronal pool. At the same time the mode of function of the individual motor units will interest us, so that in the center of attention are found the working features of the pool connected with its atomicity—a common characteristic of all neural structures.

The nerve cells themselves do not have and cannot have any complex behavior: the information obtained by the individual neuron is immeasurably more meager than that which the whole organism receives and its reactions are very stereotyped. Therefore the central problem in neurophysiology is the study of how the expedient behavior of the organism in interaction with the changing environment is formed from the interaction of different nerve structures and, all things considered, from the behavior of individual neurons. In connection with this the search for such principles of interaction of the neurons which would guarantee execution of the integral physiological acts is very important.

The tendency to minimization of interaction, about which we have already spoken above, gives rise to the possibility of the nonindividualized control of the neurons of one or another nerve structure by means of the influence on their own system of interaction. Of course, action on the interaction system of the neurons does not exclude direct influences on these neurons. Nonindividualized action on the neurons of some center gives rise to the possibility of a simple description of their operation.

The determining role of the autonomous collective function indicates, in particular, that the key place in the solution of any physiological problem belongs to the so-called "horizontal" interaction of the neurons as yet studied experimentally only in the simplest examples.

These ideas are applicable, probably, to the working organization of both the motor and sensory systems.

Suppose that there is a homogeneous medium of neurons connected together so that each, being excited, exerts a facilitatory influence on its neighbors. Then such a condition of the system will be stable when all the active neurons are working synchronously. Moreover, interaction of them will be minimal. The applicability of this assertion to a real neural structure is difficult to check immediately, in the first place because of methodical difficulties, and in the second, because there is no basis to think that the interaction of neurons is only positive and symmetrical.[2]

Among the neurons of a homogeneous system—the motor neurons of one muscle—there cannot but be interaction, simply because of the proximity of their electrical fields, the overlapping of the branching zones of the dendrites, and also in view of the facts that each muscle receptor is projected not on one motor neuron but on several, and that each motor neuron receives an impulse from several muscle receptors. The different receptors exert their influence on the motor neurons either directly or by interneurons. The impulse of an individual muscle receptor depends on the activity of the many motor units of the muscle. But once interference of the motor neurons exists, then if it is facilitatory the motor neurons should become excited synchronously. If they work independently, then this indicates the existence of a special mechanism preventing synchronization.

An experimental study of the activity of individual motor units of a human muscle developing moderate tension in the postural mode showed that these motor units work virtually independently [9, 20]. This is all the more surprising as the mean frequency of impulse of all the active motor units is approximately the same 7–11 per second (activity with a different frequency is evidently not stable), and the impulse of the individual motor unit in the course of several tens of beats is very stable (the ratio of the standard deviation to the mean duration of the interval between two impulses is about 0.2–0.3).

The independent impulse of different motor units of one muscle creates the principal possibility of separate control of the units. In special conditions of artificial visual and auditory control of activity of the motor units, a human being can actually voluntarily "switch on" or "switch off" the assigned motor unit (speaking more strictly, the group of motor units containing the assigned one), without changing the impulse of the second randomly chosen motor unit of the same muscle [1, 9].

[2] A mathematical model of such a structure was examined by Gelfand and Tsetlin [13]. The pacemaker of the heart, all the elements of which work synchronously, can serve as a physiological model. Moreover, the element, the impulse frequency of which is the greatest, is the "leader" [12].

The fact of independent activity of the motor units makes it possible to understand the genesis of a physiological tremor, to predict the dependence of its amplitude on the effort built up, and the spectral composition [7, 8].

The actual arrangement of the mechanism guaranteeing asynchronous activity and contrast of the active motor units, is not known. However, available information about the characteristics of certain elements of the segmental apparatus of the spinal cord and the nature of the connections among them [6, 19, 24, 27, 28, 32, 33, 35, 36, 38, 42, 43] makes it possible to suggest a model of the arrangement of this mechanism. The arrangement of the model is examined in reference [9], and a more detailed account will be given in Kotov's paper as yet unpublished [29]. Here we will note only that important for a normal mode of function of the motor units of the muscle are, according to these model representations, besides the known characteristics of motor neurons themselves, reciprocal inhibition, in particular the dependence of the volley of the Renshaw cells on the frequency of its activation, and the hypothesis concerning the collaborative activation of the corresponding alpha and gamma motor neurons. For an explanation of certain pathological modes of function of the muscle (Parkinson's disease, post-poliomyelitis paralysis) during which the activity of the motor units differs substantially from the normal, it is enough to assume in the limits of this model wholly concrete and small changes of the characteristics of the elements of the model. The study of the behavior of a model makes it possible to distinguish certain significant parameters of the architecture of the segmental apparatus and its elements, from change of which the mode of function of the system varies substantially, and parameters variations of which little effect the working of the system. Such conclusions would undoubtedly be much more difficult to obtain in direct experiments.

REFERENCES

1.
J. V. Basmajian, *Science* 1963:141, 440.
2.
N. A. Bernstein, *Arkhiv Biol. Nauk* 1935:38, 1.
3.
————, *O postroenii dvizheniy* (*Concerning the Construction of Movements*), Moscow: Medgiz, 1947.
4.
————, *Problemy kibernetiki* 1961:6, 101.
5.
V. I. Bryzgalov, I. I. Pyatetskiy-Shapiro, and M. L. Shik, *Dokl. AN, SSSR* 1965:160, 1039.
6.
J. Eccles, *The Physiology of the Nerve Cells*, Baltimore: Johns Hopkins Press, 1957.
7.
A. G. Feldman, *Biofizika* 1964:9, 726.
8.
I. M. Gelfand, V. S. Gurfinkel, Ya. M. Kots, M. L. Krinskiy, M. L. Tsetlin, and M. L. Shik, *Biofizika* 1964:9, 710.
9.
————, V. S. Gurfinkel, Ya. M. Kots, M. L. Tsetlin, and M. L. Shik, *Biofizika* 1963:8, 475.
10.
————, V. S. Gurfinkel, and M. L. Tsetlin, *Dokl. AN, SSSR* 1961:139, 1250.
11.
————, V. S. Gurfinkel, M. L. Tsetlin, collection *Biol. aspekty kibernetiki* (*Biological Aspects of Cybernetics*), Moscow: Izdvo AN SSSR (Publishing House of the Academy of Medical Sciences of the USSR), 1962, p. 66.
12.
————, S. A. Kovalev, and L. M. Chaylakhyan, *Dokl. AN, SSSR* 1963:148, 973.
13.
———— and M. L. Tsetlin, *Dokl. AN, SSSR* 1960:131, 1242.
14.
———— and M. L. Tsetlin, *Dokl. AN, SSSR* 1961:137, 295.
15.
———— and M. L. Tsetlin, *Uspekhi Matem. Nauk* 1962:17, 3.

16.
———— and M. L. Tsetlin, Chapter 11.
17.
B. E. Gernandt and H. W. Ades, *Exp. Neurol.* 1964:10, 52.
18.
————, J. Katsuki, and R. B. Livingston, *J. Neurophysiol.* 1957:20, 453.
19.
R. Granit, *Receptors and Sensory Perception: A Discussion of Aims, Means, and Results of Electrophysiological Research and the Process of Reception*, New Haven: Yale University Press, 1955.
20.
V. S. Gurfinkel, A. N. Ivanova, Ya. M. Kots, I. I. Pyatetskiy-Shapiro, and M. L. Shik, *Biofizika* 1964:9, 636.
21.
————, Ya. M. Kots, V. I. Krinskiy, Ye. I. Paltsev, A. G. Feldman, M. L. Tsetlin, and M. L. Shik, Chapter 11.
22.
————, Ya. M. Kots, Ye. I. Paltsev, and A. G. Feldman, Chapter 13.
23.
————, Ya. M. Kots, and M. L. Shik, *Regulyatsiya pozy cheloveka*, (*The Regulation of Human Posture*), Moscow: Izd-vo Nauka, 1965.
24.
J. Haase, *Pflüg. Arch. Ges. Physiol.* 1963:276, 471.
25.
P. Hoffman, *Untersuchungen über die Eigenreflexe menschlichen Muskeln*, Berlin, 1922.
26.
H. Hufschmidt, *Pflüg. Arch. Ges. Physiol.* 1962:275, 463.
27.
C. C. Hunt, *J. Gen. Physiol.* 1955:38, 801.
28.
P. G. Kostyuk, *Fiziol. Zh. SSSR* 1961:47, 1241.
29.
Yu. B. Kotov, in press.
30.
V. I. Krinskiy, M. L. Shik, *Biofizika* 1963:8, 513.
31.
———— and M. L. Shik, *Biofizika* 1964:9, 607.
32.
D. P. C. Lloyd, *J. Neurophysiol.* 1941:4, 525.

33.
P. B. C. Matthews, *Physiol. Rev.* 1964:44, 219.
34.
I. I. Pyatetskiy-Shapiro and M. L. Shik, *Biofizika* 1964:9, 488.
35.
B. Renshaw, *J. Neurophysiol.* 1946:9, 191.
36.
A. I. Shapovalov, *Dokl. AN, SSSR* 1961:141, 1267.
37.
S. M. Sverdlov and Ye. V. Maksimova, *Biofizika* 1965:10, 161.
38.
J. Szentágothai, in *Basic Research in Paraplegia*, ed. J. D. French and R. W. Porter, Springfield, Ill.: Thomas, 1962, p. 51.
39.
M. L. Tsetlin, *Uspekhi Matem. Nauk* 1963:18, 3.
40.
N. V. Veber, L. M. Rodioriov, and M. L. Shik, *Biofizika* 1965:10, 334.
41.
J. Wells, *Kinesiologie* 1955.
42.
V. J. Wilson, *Basic Research in Paraplegia*, ed. J. D. French and R. W. Porter, 1962, Springfield, Ill.: Thomas, 1962, p. 74.
43.
G. P. Zhukova, *Arkhiv Arat.* 1958:35, 43.

10

An Analysis of Physiological
Tremor by Means of a General-
Purpose Computer

V. S. Gurfinkel
L. E. Sotnikova
O. D. Tereshkov
S. V. Fomin
M. L. Shik

1. INTRODUCTION

In 1963 we carried out work in a study of normal physiological tremor with the help of a general-purpose computer. Reasons of a twofold nature determined interest in this problem. On the one hand we proceeded from the hypothesis that certain functional features of the human motor system and those problems (for example, the retention of an assigned posture) which this system executes should find reflection in tremor. Therefore, it seemed possible to us to derive specific physiological information from tremor. In particular we considered it interesting and important to answer the following questions: (1) what is the influence of hemodynamic factors on tremor; (2) what are the characteristics of a tremogram of a two-component system; (3) how do different postural problems affect the form of tremor; (4) what is the nature of the different components of tremor, in particular, how is tremor dependent on the functional features of the motor units. On the other hand, the work in the study of tremor also pursued a purely methodical purpose, namely, to test on a rather simple and convenient object a system of analog-digital cross-coding of information and its subsequent input into the computer, about which we will speak in greater detail below. It is well known that information received directly from a physiological experiment often represents a certain aggregate of continuously changing quantities (deviations from the assigned position in the case of tremor, oscillations of the electrical potential during spreading of excitation along the nerve fiber, etc.). Translation of that sort of data into discrete numerical form is an essential step in the adaption to analysis of the physiological information of multipurpose computers.

In the processing of tremograms we used the well-known methods of spectral and correlation analysis, examining each tremogram as the realization of some stationary random process. A priori, there are not, of course, any grounds to assume that just these traditional methods are well suited to detect the features of operation of such a complicated and highly nonlinear system as the human motor apparatus. However, the experiment showed that just these simple methods make it possible to obtain certain meaningful results.

2. METHODS FOR CARRYING OUT AN EXPERIMENT AND PROCESSING EXPERIMENTAL DATA

By tremor we mean random, that is, not consciously controlled by a human being, oscillations of the angles of a joint, in one or another joint. We conducted experiments in the study of the normal tremor of a human limb under different conditions of the motor problem, namely, retention of a voluntary hand position, retention of an assigned hand position, aiming with a pistol with normal and tilted positions of the head. For recording of tremor we used strain gauge pickups [10] of angular displacement in the radiocarpal joint and an inductive pickup of the position of the barrel of a pistol on a plane perpendicular to the axis of aiming (there are four mutually connected secondary coils symmetrically located along the vertical and horizontal line on the plane of displacement, in the center among which is fastened on the barrel of the pistol the primary coil of the pickup; the latter feeds from a generator of high-frequency voltage and induces electromotive forces proportional to the corresponding shifts of the barrel, in the horizontal and vertical pairs of secondary coils—see for details [12]).

A block diagram of the arrangement for recording and analog-digital conversion of a signal is shown in Fig. 1. The signals from the pickups (1) were amplified with an "Alvar" multichannel electroencephalograph

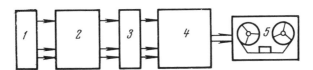

Fig. 1. Block diagram of the arrangement for the analog-digital conversion and recording of tremor. (1) pick-up; (2) Alvar electroencephalograph; (3) cathode followers; (4) UDAR-1 analog-digital converter; (5) digital tape recorder.

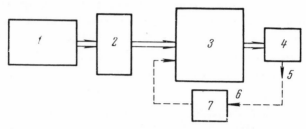

Fig. 2. The system of data input into the computer. (1) digital tape recorder; (2) input system; (3) computer; (4) tape-puncher; (5) punched tape with data; (6) punched tape with program; (7) photoinput system.

(2), and simultaneously with recording on the paper tape, the signals were fed via the corresponding cathode followers (3) to the inputs of the 10-channel analog-digital converter (4), developed in the Institute of Earth Physics, Academy of Sciences, USSR, for the digital recording of seismograms and operating according to the well-known [9] method of voltage step compensation.

The analog-digital converter carries out sequential conversion of an electrical voltage varying from 0 to 10 V fed into the inputs of each channel in an 11-bit binary code. The frequency of the conversions are 50 per second for each channel,[1] which is entirely adequate for this problem. The parallel binary code in the form of electrical impulses proceeds from the analog-digital converter into the recording heads of the multitrack digital tape recorder (5), where it is recorded on magnetic tape in a form already suitable for input into the computer. The duration of each recording in our experiments was 40 to 50 sec.

The input of data into the computer from magnetic tape was carried out with the help of the same digital tape recorder on which the entry was made, and a specially developed simple layout of the input which realizes amplification and formation of synchronizing impulses controlling selections of the channels and input. A block diagram of the input system and the processing of digital information is represented in Fig. 2. The signals computed from the magnetic tape of the digital tape recorder

[1] Later on the rate was increased to 125 conversions per second.

(1) proceed via the input system (2) into the digital computer (3). The input system is connected to the computer via a standard connector of a photoinput; therefore, no change in the construction of the machine in this case is required. The exactness of the reproduction of the initial signal (only the first 5 digits of the code were introduced) in the frequency range (0.5–8 Hz) of interest was approximately 3 percent, which is ample for this problem.

For the convenience of further processing and storage of information, the numerical data fed into the operating memory of the computer were taken out by perforator (4) on to punch tape (5). The data thus obtained together with the processing program were then fed over the photoinput device (6), and the results of the computations were printed as numbers on the printing device of the machine.

The time for complete processing of one recording (computation of a correlation function and its frequency spectrum) by the medium high-speed computer is approximately 15 min at about 7 thousand operations per second, while for similar processing of data with the help of an automatic analog correlograph of the EASP-S type, several hours are needed.

3. THE INFLUENCE OF HEMODYNAMICS ON THE TREMOR

The physiological tremor (Fig. 3A) is an irregular oscillation with an average amplitude of 1 to 2 [minutes of arc]. These oscillations according to the opinion of Brumlik [2] and Buskirk [3] are connected with hemodynamic phenomena. These authors base their conclusion on an analysis of oscillations of the tip of the finger recorded by an accelerometer. In such conditions, apparently, actual pulse oscillations show up the strongest on the record.

On the other hand, in conditions of recording of oscillations of an interlink angle (tremor of the radiocarpal joint) the form of the tremor during retention of an active posture or pose depends only to a small extent on hemodynamic phenomena. This is because the tremor of an active posture hardly changes during *tensing* (Fig. 3B). When the whole arm is resting on a support, only pulse oscillations of the size of the joint angle are visible (Fig. 3C), disappearing during tensing (Fig. 3D).

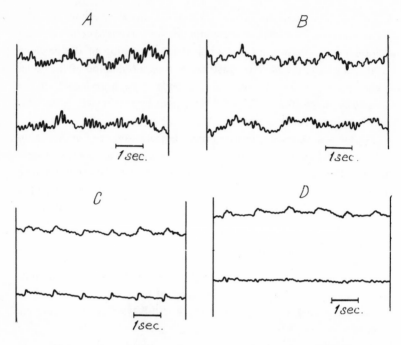

Fig. 3. Tremor of the left (upper curve) and right (lower curve) radiocarpal joints during active (A, B) dorsiflexion of the wrist, and when the wrist is resting on a support together with the forearm (C, D). (B) and (D) are for the right shoulder tensed.

Apparently, during maintenance of an active posture, tremor is mainly determined by muscle activity.

4. THE SPECTRAL AND CORRELATIVE CHARACTERISTICS OF A NORMAL TREMOR

With the help of the analog-digital conversion of the tremor and the consequent input of data thus obtained to the computer, we carried out a spectral and correlative analysis of normal tremor. In the spectrum of normal tremor it is possible to distinguish regions of low (0–3 Hz) and high frequencies (6–8 Hz). The presence of these two components is rather clearly visible as well as by the autocorrelation function of the tremor (Figs. 4–6).

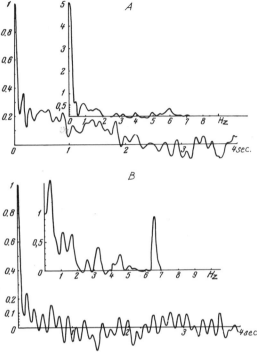

Fig. 4. Autocorrelation function of a tremor of the radiocarpal joint and its spectrum (above) during active extension of the wrist under only proprioceptive control (A) and during a precise task under visual control (B).

5. THE INFLUENCE OF DIFFERENT MOTOR TASKS ON THE CHARACTERISTICS OF TREMOR

It was shown above that the tremor is primarily determined not by hemodynamic phenomena but by characteristics of the human motor apparatus, its muscle activity. In connection with this there arises the hypothesis that a change of the motor task should be reflected in the tremor. This hypothesis was confirmed in the experiments. In the first series of tests we examined the influence of an explicit task on the tremor. With that end in view we recorded at first a tremor of the left and right radiocarpal joints of the same examinee during active extension of both wrists, with closed eyes. In this case (Fig. 5, A and B) the shape of the tremor was almost the same for both joints. Then the

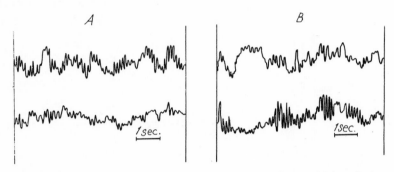

Fig. 5. Tremor of the left (top) and right (bottom) radiocarpal joints during active pose retention with eyes closed (A) and during execution of a precise task by the right hand (B).

following precise problem was put to the examinee: he took in his right hand a sharpened pencil and tried to hold the tip of it exactly against a fixed point on the support. The position of the left wrist in this case was held, as before, only under proprioceptive control. Under the conditions of execution of such a problem the form of the tremor of the left radiocarpal joint did not change (Fig. 5B), but on the tremogram of the right arm the proportion of high-frequency (6–8 Hz) oscillations increased somewhat. These changes of the tremor appear distinctly during correlative and spectral analyses of the tremor. In Fig. 4, graphs are presented of autocorrelation functions and their spectrograms corresponding to the tremograms in Fig. 5B. It is noteworthy that there is a maximum in the 6–7 Hz region in the tremor of the right arm, which had executed the precise task, whereas this maximum is missing in the tremor of the left arm, the position of which is under proprioceptive control only.

In Fig. 6, there are presented autocorrelation functions of a tremor and their spectrograms for a different examinee. Here the new spectral maximum arising in the tremor of the hand executing the precise task is wider (7–9 Hz) and there is a maximum, although weakly expressed, in the tremor of the hand retaining the posture only under proprioceptive control.

It is curious that the tremograms during retention of a pose posture

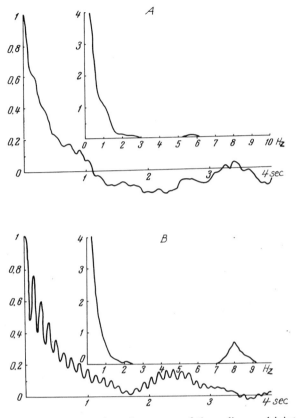

Fig. 6. Autocorrelation function of a tremor of the radiocarpal joint and its spectrum during active retention of the wrist. (A) and (B) are the same as in Fig. 4.

under only proprioceptive control and in the conditions of a precise task differ little to the eye. Oscillations with a frequency of 7 to 10 [minutes of arc] per second exist in both cases. We can assume that the absence in the first case of a high-frequency maximum in the spectrogram of the autocorrelation function of the tremor is related to the fact that the phase of these high-frequency oscillations is "floating" all the time and in the conditions of the precise problem their phase is more fixed. As a result, since the amplitude itself of the oscillations with a frequency of 7 to 10 per second changes slightly, we do not see in our estimation significant changes in tremor although they are well revealed by correlation analysis.

The form of the autocorrelation function does not depend on the nature of the position assigned, if we do not consider the appearance of

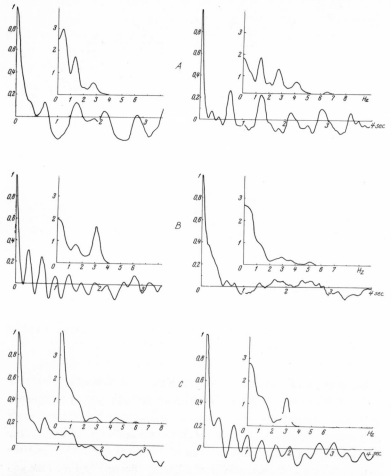

Fig. 7. Autocorrelation function and its spectrum for the vertical (left) and horizontal (right) coordinates of oscillations of the pistol muzzle. (A) without aiming; (B) ordinary aiming; (C) aiming with head tilted to the right at 90 deg.

high-frequency oscillations on the curve in the event of an exact task. On the other hand, in different people it can differ sharply. Thus, in both problems the size of the autocorrelation function of one examinee (see Fig. 4) dropped to 0.1 its value at zero (variance) in just 0.1 sec, and that of another (Fig. 6), only in 1 sec. We can assume that the sharpness of the decrement characterizes individual peculiarities of the apparatus for control of movements ("handwriting") and does not depend on the nature of the problem, while the presence or absence of a high-frequency maximum in the spectrum of the autocorrelation function is related to the nature of the posture task and does not depend on the peculiarities of the individual subject.

The relation of the tremor to the nature of the motor problem appears also in the conditions of the test with aiming. In Fig. 7, graphs are presented of the autocorrelation function and their spectra for the vertical and horizontal coordinates of oscillations of the end of the muzzle of a sport pistol (recorded with the help of a transformer pickup). The examinee in one case held the pistol with an outstretched arm, not aiming (Fig. 7A); in another, aiming (Fig. 7B); in a third he aimed tilting his head to the right at a 90-deg. angle (Fig. 7C). It is obvious that the spectral maximum in the 3 to 4 Hz zone is absent when no aiming was done; it appears in the projection of oscillations on the vertical axis during erect aiming, and on the horizontal projection during aiming with an inclination of the head.

6. THE RELATION BETWEEN TREMOR AND THE NATURE OF THE ACTIVITY OF MOTOR UNITS

The tremor of an active posture, inasmuch as it is only to a small extent determined by hemodynamic factors, as shown above, should by its direct origin have the nature of the activity of the motor units since only via them can whatever supraspinal effects there might be influence the tremor. We can define this concept concretely if we take into consideration the characteristics of the posture mode of the activity of the motor units [5, 6]. In the research mentioned it was shown that the individual motor unit functions with a 7–11 per second frequency while the ratio of the standard deviation to the average interval usually equals

0.2–0.3. The impulses of different motor units of one muscle in the range of moderate tension are practically independent. We can imagine that in the range indicated an increase of muscle tension is realized chiefly through involvement of supplementary motor units into activity. It is clear from what has been said that in an ordinary problem of retention of a posture the muscle fibers work in a mode of single contractions, but they are not tetanic. We know, finally, that the cycle duration of a single contraction of the muscle is approximately 0.08–0.15 sec (for various muscles).

Proceeding from the data presented, we could try to synthesize the spectral structure of the tremor and compare the result, theoretically obtained, with the spectrum of the tremor obtained from the experimental data. The appropriate computations were made by A. G. Feldman [4]. It turned out that we should expect a structure of the spectrogram of the tremor similar to that obtained from the experiments and presented in this paper.

7. TREMOR IN A MULTILINK SYSTEM. SYNERGISMS

When speaking about tremor, we always have meant some two-link system, for example the radiocarpal joint. It seems interesting to examine the tremor of a multilink system embracing several joints (for example, the whole arm from shoulder to fingers). This work is as yet still not complete. However, we obtained a series of results for a process very similar to a tremor of a multilink system—namely, for oscillations of the horizontal projection of the center of gravity of a human body maintaining a vertical posture. We found that here the individual links of the corresponding multilink system—of the human torso— interact in a specific way among themselves. This interaction very clearly appears in the so-called respiratory synergy. The fact is that in oscillations of the center of gravity of the body (Fig. 8, upper curve) of a healthy man, respiration hardly shows up in spite of the fact that "respiratory" displacement of masses in the organism undoubtedly occur. Parallel recording of the angles of the truncal and hip joints showed (Fig. 8) that their respiratory oscillations are antiphasic. There is a correlation also between the "respiratory" oscillations of several other

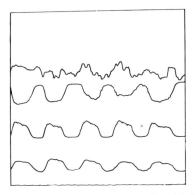

Fig. 8. Sagittal stabilogram, truncal angle, angle of the hip joints and pneumo-gram (from top to bottom, respectively), while standing upright with closed eyes, by a healthy male subject.

joints of a human subject maintaining a comfortable stance. Thus, there exists a "respiratory synergy" of the vertical posture thanks to which respiration does not lessen the stability of the vertical pose of the human being.

In patients with cerebellar disease, especially when their eyes are closed, the respiratory oscillations appear rather distinctly in the stabilogram. In these patients there are not the strict relations between the "respiratory" oscillations of the joints characteristic for a healthy person.

The synergy examined is evidently not a unique example of the union of several muscles (among them even ones distant from each other) into a single-parametric control group.

8. NORMAL AND PATHOLOGIC TREMOR

An experiment of the investigation of quantitative characteristics of normal tremor make it possible to turn to a study of the pathologic tremor of patients with neurological illnesses. Proceeding from the idea that the tremor to a considerable degree reflects the activity of the motor units, we paid primary attention to an analysis of the form of pathologic tremor.

One such attempt was undertaken by us [11] in order to control the effectiveness of the surgical treatment of Parkinson's disease. It was shown that after apparent disappearance of the sweeping motion of Parkinson's tremor with chemothalamotomy, two essentially different variants may disappear. In certain cases the normalization of the amplitude of the tremor is also accompanied by the return of the shape of the tremor to normal. In such patients the effect of the operation turns out to be stable. In other cases only the average amplitude of the tremor is normalized, but its shape remains pathologic—sinusoidal oscillations with a frequency of 3 to 5 per second. The synchronized activity of the motor units operating as *packets* [6] correspond to them. In such patients, within several weeks a relapse sets in—the typical high amplitude (a few degrees, even tens of degrees of arc) and Parkinson's tremor reestablishes itself.

Thus, in recording the tremor during the operation it is possible then and there to make a prognosis of the effectiveness of the operation for a given, specific patient. Such timely findings can be used by the neurosurgeon for choosing optimum tactics for further treatment.

9. CONCLUSION

The data presented above allow the following conclusions to be made.

1. Tremor of an interlink angle during maintenance of an active posture reflects, basically, the activity of the motor units of the muscle, although a hemodynamic component is also contained in it. If we examine the process of retention of a posture as a problem in minimization of deviation from a required position, then we can consider the tremor as the trajectory of search for the minimum of the estimator. From this viewpoint we can treat the changes of the spectrum of the tremor during a precise task, in comparison with the retention of a posture merely under proprioceptive control, as an indication of different tactics of neural regulation of a posture under these conditions.

2. The mechanism by means of which respiratory oscillations of the common center of gravity are prevented during retention of a natural vertical stance is the coordinated activity of many muscle groups related to respiration. Of a family of this sort the *functional synergy* of

the typical form of motor activity indicates a significant decrease of a number of degrees of freedom of the controlling system. This is an example of the physiological organization of an estimator of motor activity which reduces the number of parameters determining the value of the function.

3. The presence of a specific connection between the characteristics of the tremor (spectrum) and the condition of the motor apparatus makes it possible to use an analysis of the tremor for diagnostic purposes.

4. The use of general-purpose computers in conjunction with analog-digital devices was found to be very convenient in the group of questions examined by us. It is apparent that they are also expedient in many other problems related to the processing of a large volume of physiological information.

The authors are deeply grateful to O. G. Sorokhtin and V. A. Ponomarev for help in the organization of the analog-digital recording of the results of the experiments.

REFERENCES

1.
Ye. B. Babskiy, V. S. Gurfinkel, E. L. Romel, and Ya. S. Yakobson, 2-*ya nauchnaya sessiya TsNIIPP* (*The Second Scientific Session of the Central Scientific Research Institute of Application and Construction of Protheses*), 1952, pp. 37–43.
2.
D. V. Brumlik, *Neurology* 1962:12, 159–179.
3.
Van C. Buskirk and R. A. Fink, *Neurology* 1962:12, 361–370.
4.
A. G. Feldman, *Biofizika* 1964:9, 726–730.
5.
I. M. Gelfand, V. S. Gurfinkel, Ya. M. Kots, V. I. Krinskiy, M. L. Tsetlin, and M. L. Shik, *Biofizika* 1964:9, no. 6, 710.
6.
————, V. S. Gurfinkel, Ya. M. Kots, M. L. Tsetlin, and M. L. Shik, *Biofizika* 1963:8, 475–486.

7.

————, V. S. Gurfinkel, and M. L. Tsetlin, *Dokl. AN, SSSR* 1961:139, no. 5.

8.

————, V. S. Gurfinkel, and M. L. Tsetlin, in collection *Biologicheskie aspekty kibernetiki* (*Biological Aspects of Cybernetics*) Moscow, 1962, pp. 66–73.

9.

E. I. Gitis, *Preobrazovateli informatsii dlya elektronnykh tsifrovykh vychis-litel'nykh ustroystv*, (*Information Converters for Digital Electronic Computers*) Moscow-Leningrad: GEI (State Scientific Technical Publishing House of Power Engineering Literature) 1961.

10.

V. S. Gurfinkel, P. K. Isakov, V. B. Malkin, and V. I. Popov, *Byull Eksperim Biol. Med.* 1959:11, 12–18.

11.

————, E. I. Kandel, Ya. M. Kots, and M. L. Shik, *Vopr. Neyrokhirurgii* 1963:4, 1–6.

12.

V. I. Krinskiy and M. L. Shik, *Biofizika* 1963:8, 513–515.

11

Concerning Tuning before
Movement

V. S. Gurfinkel
Ya. M. Kots
V. I. Krinskiy
E. I. Paltsev
A. G. Feldman
M. L. Tsetlin
M. L. Shik

By now there has been accumulated much morphological and physiological data indicating that the spinal cord is not a passive activating
apparatus reproducing supraspinal instructions, but is a complex
system the interaction of the elements of which are no less essential for
motor result than the supraspinal impulses feeding into motor neurons.

As the basis of such a viewpoint rests the following information
[1, 3–6, 8, 12–14, 19, 21].

1. The presence in the spinal cord of a large number of interneurons,
many of which are not afferent neurons of either the second or third
order, which are situated in the ventral horn and in the interstitial zone.

2. The overwhelming part of the descending fibers from the brain
terminate not at the motor neurons, but at the interneurons of the
interstitial zone, as well as of the ventral and dorsal horn.

3. The greater part of the synapses in the spinal cord are formed by
spinal neurons on each other; only the minority are formed from axons
coming from the brain and spinal ganglion.

4. The presence of a powerful system of differentiated intracentral
interaction of the motor neurons of the different muscle groups at the
segmental level (reciprocal facilitation and inhibition).

5. The presence of a reflex system of interaction of the muscles at the
segmental level by the muscle receptors (myotactic reflex action).

6. The influence of segmental afferentation on withdrawal reflexes via
the propriospinal and spinobulbar-spinal systems.

We can assume therefore that the cerebrospinal system of interaction
of the motor neurons (being realized via the corresponding interneurons
intracentrally and reflexively) underlies the essential supraspinal
influences; on the other hand, the effect of the supraspinal actions

depends on the condition of the system of interaction among the motor neurons. The control of movements in many respects occurs due to a spinal change of this intrinsic system of interaction.

A series of experimental data about the intrinsic spinal system of interaction of the motor neurons makes it possible to assume that the stable state of silence of almost all the motor neurons (of the muscles) corresponds to it and that it exerts a stabilizing influence on the motor neurons leading to the elimination of the disturbance effect (afferentation); moreover, this process, depending on the conditions, can take place both aperiodically and with oscillations (rhythmic reflexes). However, in the conditions of integral motor acts, the segmental apparatus receives such a tuning that the interaction of the elements realized on the segmental level is subordinate to the solution of the overall motor problem.

A series of experimental data (tests with sudden discharging, simultaneous recording of the oscillations of the overall center of gravity and the electrical activity of the muscles during maintenance of a vertical pose) shows that in the conditions of integral motor acts a correction of movement is realized in the time that is characteristic for spinal reactions. We can conclude from this that in the conditions of natural motor acts the activity at the spinal level is subordinate and conforms to the overall motor task. An elaboration of the questions arising here is encountered in the initial stage; in connection with this it seems essential to us to formulate several possible courses of research: (1) experimental proof of the original hypothesis that a rearrangement of the segmental relations before movement actually takes place; (2) attempts to define concretely the physiological mechanisms guaranteeing the use of the spinal system of interaction of neural elements in the control of the movements; and (3) reasoning out (in view of the difficulty and often impossibility of direct experiments) of mathematical working models of the segmental level and the principles of control of its activity.

The present article is devoted to the first of the courses of research indicated. From methodical deliberations we chose as the simplest experimental problem a simple, unmeasured movement carried out by a command in compliance with a previous instruction.

As an indicator of the condition of the segmental apparatus of the spinal cord before movement the tendon and H-reflexes [10] were examined during single and paired stimulation. The reflexes were tested in the original state and at different times in the interval between the giving of the command to movement and the start of the movement itself. The reflexes were recorded by the electrical response of the corresponding muscles of the lower extremities.

For evoking a tendon reflex, an apparatus was employed consisting of a small electromagnetic hammer and an electronic delay circuit which made it possible to produce a tap on the tendon at a specified time after giving the movement command (Fig. 1). For evoking the H-reflex, we used a "Multistim" 2-channel stimulator which makes it possible to specify the necessary interval between the first and second stimuli and to regulate their amplitudes independently. We placed the stimulating electrode for evoking the H-reflex in the popliteal fossa over the posterior tibial nerve, the stimulation of which provokes in the muscles of the tibia a direct M-response and a reflex monosynaptic H-reflex [10, 17].

In the first series of investigations we studied the amplitude of the tendon reflex in relation to the time interval since the moment of giving the movement command. At this command the examinee was to

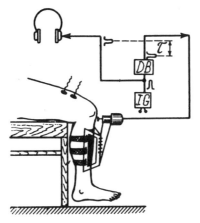

Fig. 1. Diagram for the setup of the experiment for testing the tendon reflex. IG-impulse generator; DB-delay block.

straighten his leg at the knee joint. The recording of the surface electromyogram of the quadriceps femoris muscle (tibia extensor) shows that usually within 160 to 180 msec after giving the command, tibia extension occurs.

During the first 100 msec after giving the movement command, the amplitude of the knee reflex remains fixed. If the tap on the tendon is applied within 100 msec or more after giving the command, then the amplitude of the reflex is the greater the closer to the command it is

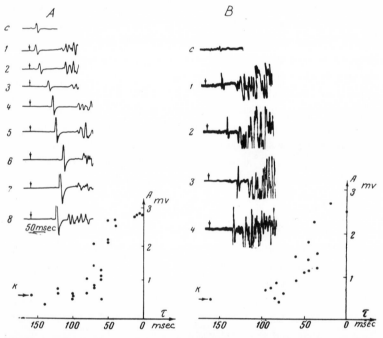

Fig. 2. Amplitude change of the tendon reflexes during the latent period before arbitrary movement. (A) electrical responses of the rectus head of the quadriceps femoris muscle to a standard intensity tap on the patellar tendon at rest (C) according to the time to the start of movement (traces 1–8). The arrow indicates the sound click, the command for the start of the arbitrary movement. (B) is the same for the gastrocnemius muscle during evoking of the Achilles reflex. The graphs are of the dependence of the amplitude of the tendon reflexes on the time to the start of movement; the ordinate is the amplitude of the electrical response of the muscle (mV).

evoked. Before the start of movement it is increased 2 to 3 times (Fig. 2A). We should note that the increase of the knee reflex occurs rather abruptly in the 100–130 msec interval, after giving the command (70–40 msec before the start of movement), while in the last 30 to 40 msec before the start of movement the amplitude of the reflex grows insignificantly. From Fig. 2A it is also apparent that the reflex evoked in a specific phase of the latent period (traces 4–6) delays the start of the motor reaction. Similar changes occur during testing with the Achilles reflex (Fig. 2B) when the examinee by command carried out plantar flexion of the foot. This can serve as an indication that a change of condition of the segmental apparatus before the start of movement does not lead just to a facilitation of the alpha-motor neurons, since in this last case we would expect that the growth of the amplitude would continue right up to the start of movement.

The data obtained indicate that only in the first part of the latent period of a simple motor reaction there do not appear noticeable changes in the segmental apparatus of the spinal cord. However, as much as 70 msec before the start of movement, the condition of the segmental apparatus changes, a fact which is revealed in the increase of the amplitude of the knee reflex. With what are these changes connected? Do they concern only the muscle group which according to a previous instruction must take part in the movement, or are they nonspecific changes which involve different links of the motor apparatus?

For an answer to this question, research was carried out in which the examinee was ordered by a command to execute two different movements: in one experiment to flex his leg from the hip joint, and in the other, to extend the knee. In both cases there was determined the size of the tendon reflex of the rectus (two-joint) head of the quadriceps femoris muscle and the lateral (one-joint) head of the same muscle, which as we know, has a common tendon. From Fig. 3A and B it is apparent that 70 to 80 msec before the start of flexure in the hip joint an increase of amplitude of the tendon reflex occurs in the rectus head of the quadriceps femoris muscle (participating in this movement) and is absent in the reflex responses of the lateral head. During straightening of the leg at the knee joint, in the realization of which both portions of

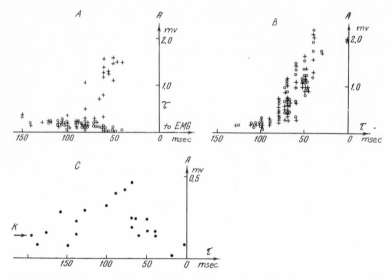

Fig. 3. Dependence of the amplitude of the tendon reflex from the time to the start of arbitrary flexure in the hip joint (A), before arbitrary straightening (B) and flexure (C) of the knee joint. The ordinate is the amplitude of the electrical response of the rectus head of the quadriceps femoris muscle, (+ and ●) and the lateral head (○); the abscissa is the time to the start of movement.

the muscle take part, the amplitude of the reflex of both the rectus and lateral heads increases.

In another form of experiment the examinee was ordered by a command to bend his leg at the knee joint. The patellar reflex began to increase 70 to 80 msec before the start of movement, but after this it returned again to the initial level (Fig. 3C). We could observe a small amplitude increase of the patellar reflex also before flexure movement of the ipsilateral arm in the elbow joint.

Thus, changes of the segmental apparatus, incipient 60 to 70 msec before movement, include specific and nonspecific components. We can relate this to the well-known data of Granit [8] and others, who discovered in acute experiments on animals that the supraspinal activity of the gamma-motor neurons has, as a rule, a diffuse nature. However, the more significant changes on that muscle which was to participate in the movement make it possible to assume that the increase of the patellar reflex before movement is related not only to diffuse activation of the gamma-motor neurons, but includes some other mechanisms. In this connection it seemed expedient to study the influence of Jendrassik's maneuver (arbitrary tension of the muscles of the arm) on the size of the patellar reflex at different time intervals before movement. We can assume that with Jendrassik's maneuver besides intracentral facilitation [20] there also occurs activation of the gamma-system which leads to the increase of the tendon reflexes.

Measurements showed that Jendrassik's maneuver increases the amplitude of the patellar reflex altogether by 30 percent, and this facilitated reflex subsequently changes before the start of movement as usual.

As noted above, it is difficult to explain the abrupt increase of the patellar reflex as only facilitation of the alpha-motor neurons. Possibly that is why before the start of movement there occurs not only activation of the systems facilitating monosynaptic reflexes, but also the suppression of the inhibitory systems tonically blocking the monosynaptic reflexes [7]. For verification of this hypothesis, Series IV of the research was conducted, in which the condition of the inhibitory systems of the segmental apparatus of the spinal cord were tested with the help of single H-reflexes of different strengths and with paired stimulation (Series V of the research).

As we know, with an increase of intensity of the stimulation, the H-reflex at first increases but then lessens [17]. In previous work [9] we showed that a decrease of amplitude of the H-reflex in proportion to an increase of the stimulation intensity of the posterior tibial nerve is related not only to the blockade of the reflex volley by the antidromic excitation of the axons (and bodies) of the motor neurons, but also with the process of central inhibition. During study of the changes of the H-reflexes before movement (Series IV of the experiments) we found that the amplitude of the H-reflex evoked by a weak stimulation (during the absence of, or with an insignificant M-response), significantly increases in just 60 msec, before the start of movement, and that the larger it is the nearer it is to the edge of the myogram (Fig. 4, Fig. 5, curve I).

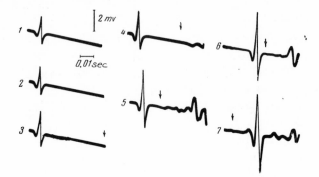

Fig. 4. Increase of amplitude of the H-reflex of the gastrocnemius muscle in the latent period of a simple motor reaction. (1) control recording in conditions of rest; (2–6) records at different moments before the start of movement; (7) record during movement. The start of the myogram (determined from another pair of electrodes with a greater amplification) is shown by the arrows. Traces 1–5 were produced with a smaller delay of the triggering of the sweep of the beam than traces 6–7. Stimulus intensity = 24.5 v.

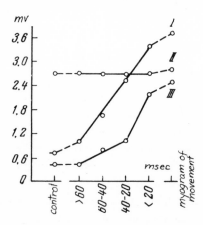

Fig. 5. Increase of amplitude of the H-reflex of the soleus muscle according to the time to the start of movement. The ordinate is the amplitude of the H-reflex (in mV) of the electrical response of the muscle; the abscissa, the intervals (in msec) to the beginning of the myogram; I–III: reflexes of the I–III zones.

As we can see from the recordings cited, the dependence of the change of amplitude of the reflex response on the time prior to the start of movement has a rather even nature. The amplitude of the H-reflex aroused by a strong stimulation (with a large M-response) significantly changes 40 msec before the start of movement (Fig. 5, curve III). Its increase, in proportion to the nearness to the start of movement, has a different nature than the amplitude increase of the H-reflex evoked by weak stimulation. The slow growth of the amplitude of the H-reflex with a greater M-response changes within a 30-msec interval before movement into an abrupt increase. This fact may indicate a blockade before the start of movement of that inhibiting system which is partially responsible for the depression of the H-reflex for a greater intensity of stimulation.

In the Series V experiments another indication was obtained that before movement the condition changes of not only the alpha-motor neurons [11] but also the interneurons of the segmental apparatus of the spinal cord. Paired stimulation with an interval 25–60 msec was applied to the posterior tibial nerve. As we know, it is easy at the same time to select such intensities of stimulations that the response to the second stimulation will be delayed. It turns out that if such paired stimulation

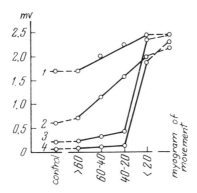

Fig. 6. Increase of amplitude of a solitary H-reflex (1 and 2) and of the amplitude of a second H-reflex during paired stimulation with the interval between stimuli 60 msec (3) and 25 msec (4). With the paired stimuli the second stimulus was the same as during the single stimulus evoking the reflex of curve I.

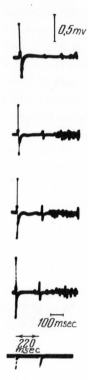

Fig. 7. Increase of amplitude of the second knee reflex before movement upon evoking of two reflexes with an interval of 220 msec.

is applied before the start of movement, then the response to the second stimulation is inhibited (Fig. 6, curves 3 and 4). This occurs only if the second reflex is 20 msec or less from the start of the arbitrary movement. A comparison of the slope of the curve of inhibition of the second reflex with the curve of the amplitude change of a solitary H-reflex evoked by stimulation of the same intensity (Fig. 6, curve 1) shows that the effect of inhibition cannot be explained simply by facilitation of the alpha-motor neurons before movement, even if we interpret curve 1 in Fig. 6 only in this way. Similar changes are observed during testing with a paired tendons reflex. Moreover, the second reflex in the pair, originally significantly weakened (in comparison with the size in the control during an isolated application) is partially restored before movement (Fig. 7). Consequently a profound reorganization of the condition of the segmental apparatus of the spinal cord precedes the beginning of movement. It can be detected before the start of a jerking physical movement. Probably changes of this type have great significance both in the problem of assumption and retention of poses, where they may be related to the basic mechanism of control of motor activity [2, 15, 16, 18].

Different changes of the size of the reflexes in different groups of muscles, discovered in connection with movement (or the preparation for movement) can be used for study of constitutional and functional motor synergies.

The authors are grateful to N. A. Bernstein, I. M. Gelfand, and I. I. Pyatetskiy-Shapiro for helpful discussion of this paper.

REFERENCES

1.
J. T. Aitken and J. E. Bridger, *J. Anat.* 1961:95, 38.
2.
N. A. Bernstein and Ya. M. Kots, *Tonus–BME* 1963:32.
3.
J. Eccles, *The Physiology of the Nerve Cells*, Baltimore: Johns Hopkins Press, 1957.
4.
K. Frank and M. J. F. Fuortes, *J. Physiol.* 1956:31, 424.
5.
S. Gelfan, *Nature* 1963:198, 162.
6.
B. Gernandt, in *Basic Research in Paraplegia*, ed. J. D. French and R. W. Porter, Springfield, Ill.: Thomas, 1962, p. 93.
7.
S. Giaguinto, O. Pompeiano, and J. Somogyi, *Experientia* 1963:19, 652.
8.
R. Granit, *Receptors and Sensory Perception: A Discussion of Aims, Means, and Results of Electrophysiological Research and the Process of Reception*, New Haven: Yale University Press, 1955.

9.
V. S. Gurfinkel, Ya. M. Kots, V. I. Krinskiy, and M. L. Shik, *Byull. Eksperim. Biol. Med.* 1965:48, no. 5, 15–18.

10.
P. Hoffman, *Untersuchungen über die Eigenreflexe menschlichen Muskeln*, Berlin, 1922.

11.
H. J. Hufschmidt, *Pflüg. Arch. Ges. Physiol.* 1962:275, 463.

12.
C. C. Hunt, *J. Gen. Physiol.* 1955:38, 801.

13.
D. P. C. Lloyd, *J. Neurophysiol.* 1941:4, 525.

14.
P. G. Kostyuk, *Fiziol. Zh. SSSR* 1961:47, 1241.

15.
V. I. Krinskiy and M. L. Shik, *Biofizika* 1963:8, 513.

16.
——— and M. L. Shik, *Biofizika* 1964:9, 607.

17.
J. W. Magladery, *Pflüg. Arch. Ges. Physiol.* 1955:261, 302.

18.
I. I. Pyatetskiy-Shapiro and M. L. Shik, *Biofizika* 1964:9, 488.

19.
A. I. Shapovalov, *Dokl. AN, SSSR* 1961:141, 1267.

20.
J. Sommer, *Dtsch. Z. Nervenheilk.* 1940:150, 249.

21.
V. J. Wilson, in *Basic Research in Paraplegia*, ed. J. D. French and R. W. Porter, Springfield, Ill.: Thomas, 1962, p. 74.

12

The Control of Movements
of the Joints and Kinesthetic
Afferentation

Ya. M. Kots
V. I. Krinskiy
V. L. Naydin
M. L. Shik

1. In certain joints changes of angles occur, as a rule, in natural movements. The problem of control of a mechanical system with a considerable (often, about ten) number of degrees of freedom arises for the nervous system. A long time ago, N. A. Bernstein [1], who defined the basic problem of the coordination of movements as "the overcoming of an abundant number of degrees of freedom," called the attention of physiologists to this situation. Cited below are some published and personal data indicating that the number of degrees of freedom of the system controlling movement are much less than the number of mechanical degrees of freedom of the controlled system. Coordination of movement is ensured, however, in spite of this situation. It is natural to suggest that this coordination is achieved because of the fact that corresponding to each characteristic form of motive activity is a subdivision of all the joints participating in the movement into a small number of connected groups of the sort that for the control of each of them one degree of freedom of the control system is enough (*functional synergies*—see also Chapters 9 and 13).

Functional synergies can be considered a unique way of decreasing the number of degrees of freedom of the controlled system. Moreover, their own subdivision of the joints (muscles) into such groups probably corresponds to each characteristic form of motor activity.[1]

The significance of kinesthetic afferentation for the development of the simplest functional synergies will be the subject of this chapter. Joints are chosen here as elements of the motor apparatus in view of the fact that forward movements of mammals are synthesized from rotational ones.

2. Everyday examples indicating that the number of degrees of

[1] A way of reducing the dimensions used essentially when seeking the minimum of the function of many variables is by the ravine method [6]. This makes it possible to curtail sharply the time of finding the minimum for well-organized functions.

freedom of the system controlling movements is very small are widely known. It is enough to recall how difficult it is to write with one's hand the letter d and at the same time to make circular movements with the foot. In Gunkel's paper [7] we are shown that when a human being makes movements of different rhythms with the two hands the amplitude of the movements of at least one hand is modulated by the frequency of the movements of the second.

In the study of a model task of locating a pose by the function of the proximity to the goal of two joints we were convinced that a human being most often changes the angle in each of the joints successively. The problem is worked out, however, in the shortest time when the examinee succeeds simultaneously and independently in changing the angles of two joints [9].

In other experiments on two joints of the arm (for example, with the elbow and the radiocarpal joint) potentiometric goniometers were fastened, the outputs of which were fed via a direct current amplifier to the horizontal and vertical plates of an oscilloscope. The examinee was instructed to follow its movements on a screen. The beam moved along the horizontal when the examinee changed the angle in one of the joints, and along the vertical when he changed the angle in the other joint. The examinee was told to make a simultaneous change of the two joint angles; at the same time he saw on the screen the moving of the focus along a sloping line. If movements in the joints were made alternately, then the point moved not in a straight line but step by step. The results obtained are seen from the table. A healthy subject without previous training can carry out simultaneous flexion or simultaneous extension in the elbow and radiocarpal joints, and in the elbow joint and the metacarpophalangeal joint of the index finger. However, to make simultaneous flexion of one of these joints and extension of the other, the examinee must previously have made 4 to 10 practice trials.

Thus, the results of all the experiments described indicate that independent control of even two joints causes a definite difficulty even for a healthy subject although it is possible. We should note that in the natural movements of a human being there occur, in spite of this circumstance, fully coordinated changes of dozens of joint angles

Table 1. The Execution of a Simultaneous Movement in Two Joints

Combination of Joint and Direction of Movements	Normal Joints	Both Joints Deafferented	Deafferented Distal Joint
Elbow joint—radiocarpal			
flexion—flexion	+	−	− +
extension—extension	+	−	− +
flexion—extension	− +	−	−
extension—flexion	− +	−	−
Elbow joint—metacarpophalangeal joint			
flexion—flexion	+	−	− +
extension—extension	+	−	− +
flexion—extension	− +	−	−
extension—flexion	− +	−	−

Note: a + task is executed stably at once; a − + task is carried out stably after several "pretrials"; a − task generally is not executed stably.

(walking, eating, dressing, sports, etc.) Our attention is also attracted by the fact that for the same pair of joints the possibility of simultaneous control depends on the direction in which the change of the angle in each of the joints is made. Possibly these differences are related by how much the combination being investigated is typical for natural human movements. In other words the combinations of movements being investigated in the experiment can have an unequal degree of training because of their varied use in everyday life.

3. The results of a series of experimental studies suggest the idea that kinesthetic afferentation is an important link of the mechanism which ensures coordinated movements of several joints. A mammal's deafferented limb usually does not participate in the natural movements of the animal [13]. However, if we immobilize the healthy paws, then the animal develops the ability to make rather varied, including locally directed, movements with the deafferented limb. Thus, Munk [14] observed that a monkey did not use the deafferented front limb. However, if the healthy paw was tied behind the back, the monkey began to use the deafferented one and could carry out with it all the movements

of which an unimpaired monkey is capable. These movements of certain animals could be carried out by a deafferented limb immediately upon emergence from narcosis. In the normal moving of an animal and in climbing about the cage the monkey does not use the deafferented limb. Munk makes the well-founded conclusion that the movements of the paw are possible even without afferentation of this paw. However, those movements of an deafferented limb which are made in connection with the movements of other limbs (Gemeinschaftbewegungen) are seriously affected. He particularly noted that in contrast to the improvement which is observed in an animal in the use of only one deafferented limb (for example in taking food), there never was seen any improvement of the associative movements of the deafferented limb. Of course, all the movements of the deafferented limb were clumsy and ataxic.

Sherrington [15] showed that electrical stimulation of the motor cortex easily evokes the usual motor effects in a deafferented limb of a monkey.

Bickel [3] observed that a dog often does not use a deafferented limb in running, although it can carry out localized scratching movements with it. This observation was subsequently confirmed by K. I. Kunstman and L. A. Orbeli [10] and others.

When running on a treadmill at full speed (8–10 kilometers per hour) the deafferented rear paw also makes active walking movements although the dog does not lean on it as before. When running at low speed (2–4 kilometers per hour) the deafferented paw not only does not support but does not even make active movements.

Twitchell [16], performing complete deafferentation of the upper limb of a macaque, showed that such a limb participates in protective movements only if there is visual control. It was the same if the dorsal root C_5 or D_2 remained uncut (both have nothing to do with innervation of the limb more distal than the shoulder joint). However, if one of the dorsal roots C_6, C_7, C_8, or D_1 remains uncut, then such a limb differs in no way from a normal one and participates in walking, climbing, and protective movements even with the eyes blindfolded.

Thus, the experiments of a series of investigators lead to the conclusion that deafferentation of a limb inhibits not the involvement of its muscles in activity but the use of such a limb in movements requiring coordinated activity between it and the other extremities. From the data presented it follows, evidently, that the use of a deafferented limb suffers not from the fact that activation of the alpha-neurons is difficult because of the interruption of the gamma-loop, but in connection with the difficulties of coordination of activity of the deafferented and un-impaired limbs.

4. We know that with disease of the parietal region of the cortex profound disturbances of kinesthetic sense arise accompanying motor disorders to the point of afferent paresis of the corresponding limb [2, 4, 11, 12]. The deafferented arm appears paretic; the patient does not use it in day-to-day living. The impression arises that this is a superfluous limb; the patient prefers to remove it or hold it fixed. In addition, the strength of voluntary contraction of the muscles of the arm is hardly reduced, and the tendon reflexes remain.

However, although the patient practically does not use the deaffer-ented arm, he can under visual control make rather precise voluntary movements with it (fasten a button, write, etc.). It is interesting that without visual control he can on command successively touch the thumb with the other fingers of the hand, although in doing this he is absolutely unable to say which fingers are touching at a given moment.

The movements enumerated are usually carried out by the deafferented arm inaccurately, with misses. However, some movements can be performed very accurately. We will cite examples. Patient X, who had had a tumor of the right parietal region removed two years before the study, did not feel pressure on the index finger. She was told to press on an inflexible strain-gauge dynamometer connected (by an amplifier) with a galvanometer needle. Watching the dial of the instrument she had to increase her effort so that the indicator would come to rest at the assigned division (the effort was far from maximal). After several repetitions she was told to reproduce the same stress with her eyes closed. Here are the results of the experiment. The assignment was 60, the reproduction by the left (affected) arm: 60, 60, 58, 50, 50, 54; by the right (normal) one: 60, 60, 58, 60, 60. Another assignment was 100. The reproduction of the left arm was 100, 105, 100, 90, 90; by the right arm: 100, 98, 100, 104, 100. Hence, the patient could in the first repetitions accurately reproduce the assigned effort with the deafferented arm.

Patient K who had had a deep tumor of the lower parietal region removed six months before the study, with loss of kinesthetic sensitivity of the interphalangeal, radiocarpal, and elbow joints and gross impairment of sensitivity in the shoulder joint of the left arm, had to make a flexure of the radiocarpal joint with an amplitude of 30 deg. from an initial position of 180 deg. A strain gauge angle gauge, connected via an amplifier with the galvanometer needle, was fastened on the joint under observation. The patient, watching the indicator, made a movement of the assigned amplitude several times. Then the patient was told to reproduce this movement with closed eyes. Completion of the task by the left (affected) arm was 30, 25, 30, 28, 20; by the right (normal) arm: 30, 30, 29, 30, 30.[2]

The examples cited indicate that with a single group of muscles or with one joint of the deafferented arm a person can make rather accurate movements.

5. Unusual impairments appear during an attempt to carry out simultaneously movement in two joints of the deafferented arm. Ten patients were studied. They were asked to carry out the same task as did the normal ones described in section 2. Simultaneous changes of the joint angles in any combination of movement directions in the joints were impossible in patients having complete absence of kinesthetic sensation in both the joints being studied. Moreover, the patient executed easily and accurately a change of the angle separately in each deafferented joint to an assigned size (with control by the oscilloscope screen).

In a study of two joints in which kinesthetic sensation was absent, only in one (the distal) did there appear an impossibility of simultaneously carrying out movement by the two joints without previous

---

[2] The examples cited, showing the feasibility of producing rather precise movements with a deafferented arm without visual control, compels us to assume that either (1) a human can remember and reproduce several times rather accurately the structure of a voluntary motor act, or (2) adjustment of the movement on the basis of proprioceptive afferentation in these conditions is possible without the help of the cortical level.

training. However, after 10 to 20 trials beforehand, the patient begins to accomplish certain movements with these two joints in a stable manner; while certain other combinations of movements he cannot stably carry out even after many days of instruction. From Table 1 we see that in a pair of joints of which one is deafferented after training, execution is possible of only those motor combinations which a normal person (or the same patient with a normal arm) makes at once. The same motor combinations the execution of which requires some training of a normal person become impossible.

Thus, deafferentation diminishes coordination of the movements not so much because of an incorrect gradation of contraction of the individual muscles as because of impairment of coordinated control of different muscle groups.

6. The syndrome of impaired control of simultaneous movements in certain joints is observed not only in diseases of the parietal region of the cortex but also in the amyostatic form of Parkinson's disease, when it has a generalized nature. Here is what the study of one such patient showed (patient A, September 1963).

The tremor in both radiocarpal joints in form and amplitude were completely normal. Rigidity was absent. The recording of action potentials of the individual motor units showed that the pattern of their activity corresponds to a normal one and not to that characteristic for the shaking form of Parkinson's disease [5]. Muscle strength was not reduced. The tendon reflexes were diminished insignificantly (A. M. Elner). Voluntary movements in the individual joint were normal. In particular, rhythmic tapping to a metronome with the fist or finger in a rhythm of 2 to 3 per second for several minutes was possible. After the metronome was shut off the examinee could continue execution of the assignment, and the amplitude of the movements did not diminish (as usually happens with other forms of Parkinson's disease). However, any complicated movements requiring simultaneous participation of several joints were extremely difficult. Apparently these also depended on the patient's overall constraint. Walking (with support on a case) was accomplished thus: slow bending from the rear of the foot and knee of the shaking leg, then bending of the hip, and only then straightening of the knee and touching down the formerly shaking leg, changing it into a supporting one. The patient usually did not succeed on his own without outside help in rising from a chair. His walking was slow, and a change of direction of movement was especially slow. Movements of the arms in which several joints participate were difficult and sharply decelerated.

It remains unclear why simultaneous control of several joints is difficult: because of the reduction of the number of degrees of freedom of the system controlling the movements, or because of a disruption of the physiological synergies.

Finally, we also observed in patients with an impairment of kinesthetic sensation in the joints of the lower extremity (disease of the dorsal columns) a difficulty of simultaneous movements of a pair of joints of the lower extremity. The task was put to the patient the same way as during examination of movements of the joints of the arms, with the possibility of control by the screen of the oscilloscope. The patient easily carried out, under control of the screen, a change of angle in the talocalcaneal joint to an assigned value. However, not all simultaneous movements of the two joints (of the deafferented talocalcaneal and of the normal knee) were possible for this patient even after training. A similar movement in the joints of the normal leg was easily carried out [8].

In conclusion, we note that both the analytical data obtained in experiments on animals and our observations on patients indicate that impairments of kinesthetic afferentation hamper coordinated movements of different muscle groups (joints) to a greater degree than control of activity of an individual group of muscles.

The authors are deeply grateful to I. M. Gelfand, M. L. Tsetlin, and V. S. Gurfinkel, to whose ideas the author of this article is indebted.

REFERENCES

1.
N. A. Bernstein, *Arkhiv Biol. Nauk* 1935:38, 1.

2.

————, *O postroenii dvezheniy* (*Concerning the Construction of Movements*), Moscow, 1947.

3.

A. Bickel, *Untersuchungen über den mechanismus der nervosen Bewegungsregulation*, Stuttgart, 1903.

4.

O. Foerster, *Handbuch der Neurologie*, Berlin: Springer, 1936.

5.

I. M. Gelfand, V. S. Gurfinkel, Ya. M. Kots, M. L. Tsetlin, and M. L. Shik, *Biofizika* 1963:8, 475.

6.

———— and M. L. Tsetlin, *Uspekhi Matem. Nauk* 1962:17.

7.

M. Gunkel, *Pflüg. Arch. Ges. Physiol.* 1962:275, 472.

8.

Ya. M. Kots and V. L. Naydin, in press.

9.

V. I. Krinskiy and M. L. Shik, *Biofizika* 1964:9, 607.

10.

K. I. Kunstan and L. A. Orbeli, *Russk. Fiziol. Zh.* 1921:4, 253.

11.

H. Liepmann, *Monatschr. f. Psychiatrie* 1900:8.

12.

A. R. Luriya, *Higher Cortical Functions in Man*, New York: Basic Books, 1966.

13.

F. W. Mott and C. S. Sherrington, *Proc. Roy. Soc.* [Biol.] 1895:57, 481.

14.

H. Munk, *Über die Funktionen von Hirn und Rückenmark*, Berlin, 1909.

15.

C. S. Sherrington, *Brain* 1931:54, 1.

16.

T. E. Twitchell, *J. Neurophysiol.* 1954:17, 239.

13

The Compensation of Respiratory   V. S. Gurfinkel
Disturbances of the Erect Posture   Ya. M. Kots
of Man as an Example   E. I. Paltsev
of the Organization of   A. G. Feldman
Interarticular Interaction

The study of movements or retention of one or another stationary posture or pose of a multilink system, which the human body is, suggests the idea that independent control of each mechanical degree of freedom (i.e., possible changes of position in the joints), when many are involved in a movement, is probably small in most cases.

There are many facts which confirm this hypothesis. For example, the difficulties which a human subject experiences during an attempt to make rhythmic movements with different frequencies simultaneously with two arms are well known. Even if he learns this, during recording of the movement of one arm it is possible to detect a modulation of frequency of the movement of the other arm [1, 5].

The paper of V. I. Krinskiy and M. L. Shik [5] is devoted to an analysis of the possibilities of voluntary, independent control of two joints. From this, in particular, it follows that independent control of a total of only two degrees of freedom is a rather difficult problem. Apparently in a series of problems the nervous system uses for control of a large number of mechanical degrees of freedom a certain type of organization of interaction among joints.

Moreover, speaking in favor of the existence of such an organization for control of the multilink system is the hypothesis that in kinematic chains of a living organism there is absent the phenomenon of the increase of the regulation error which occurs in ordinary kinematic links with an increase of the number of the movable links (the *game of billiards paradox*, I. M. Gelfand and M. L. Tsetlin).

A convenient object for clarification of such an organization is the vertical pose of man, the retention of which is linked with the active interaction of many muscle groups and movements in many joints, and cannot be provided by the simple closing of all the joints. We can assume that the most precise interrelations of movements in the joints arise in the process of compensation of systematic disturbances of the same type, among them disorders caused by the physiological processes in the organism itself (blood flow, peristaltic waves, respiration, etc.). In this connection the sort of natural physiological disturbance of the body's position which is connected with breathing is especially suitable for study. Actually, respiratory movements of the thorax, the excursion of the diaphragm, and the resultant displacements of the organs of the abdominal cavity can lead to significant changes of position of the body's overall center of gravity.

At the same time in the process of evolution the nervous system was able to develop the sort of regulation mechanism which would effectively compensate for these changes. In addition to this, the question naturally arises, is there in fact any relation between the respiratory movements of the body and the displacements of its overall center of gravity? If not, then how are the rather large (as will be shown) disturbances of the upright posture linked with respiration compensated for?

This paper is devoted to a detailed study of the question concerning the compensation of respiration caused disturbances of the vertical position of man.

METHODS

From six normal examinees in a comforable stance were recorded with the help of tensometric goniometers appropriately adapted for our purposes [2] simultaneously the truncal, hip, knee, talocalcaneal, and cervical angles and respiration (the circumference of the chest). The changes of the angle position of the thoracic section (at the level T_3-T_4) of the spine with respect to the pelvis (S_1) were recorded as the truncal angle. Similarly the cervical angle is the angle position of the head with respect to the vertebrae C_5 to C_6. At the same time a stabilogram was recorded [2]. With the help of an elongated strain gauge pickup with a small movable pilot we studied the mobility of the different sections of the spine with respect to the pelvis. Signals from the strain gauge goniometer were fed in the inputs of the TU–4M strain gauge amplifier and were recorded by means of a H 102 loop oscillograph, a VEKS-4 3-channel electronic oscillograph, or an ink-recording encephalograph. The

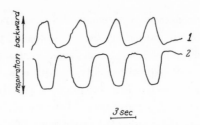

Fig. 1. Changes of the truncal angle (T) during quiet respiration. (1) TA; (2) pneumogram.

rectilinear recording of the processes on the VEKS-4 made it possible to judge beyond all doubt the phase relations of the changes of the different joint angles during the respiratory cycle. After each experiment a calibration of the goniometers and the stabilogram was made.

RESULTS

A biochemical analysis of the respiratory movements of a human subject [4] show that during inspiration simultaneously with movements of the ribs and diaphragm there is straightening out of the spine. Actually, a simultaneous recording (Fig. 1) of the truncal angle and respiration (circumference of the chest) reveal noticeable deviations, especially of the upper and middle parts of the torso backwards during inspiration and correspondingly forward during expiration. Subsequent measurements show that the spine, moreover, does not deflect as a whole, that is, during respiration displacements of the vertebrae with respect to one another. Thus, a recording (Fig. 2) of the movement of the torso at the level of different vertebrae shows that at inspiration the lumbar section of the torso (L_2–L_4) deflects a small amount forward reaching at most 5 to 6 minutes of arc with respect to the pelvis, S_1. At the L_1 level, deviations are usually absent; the thoracic (T_2–T_{12}) and cervical (C_3–C_7) sections deflect backward with maximal amplitude equal on the average to approximately 20 minutes of arc at the C_7–T_1 level.

We observe that shiftings of the torso forward during inspiration at the level of the lumbar section are the result either of a passive deflection of the spine under the action of a large mass deflecting backward, or an

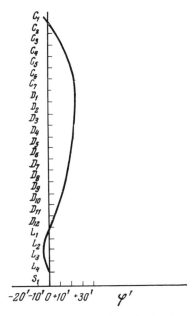

Fig. 2. Deviations of the torso recorded at the level of different vertebrae during quiet respiration. Along the vertical axis are vertebrae on the level of which were recorded angle displacements of the trunk with respect to the pelvis (S₁). Along the horizontal axis is the average amplitude of the deviations of the torso in minutes of angle. The positive angles correspond to the deviation of the torso backward in the sagittal direction at inspiration; and the negative angles correspond to the deviation forward.

active one which can be caused by the action of the iliopsoas muscle (which is attached at its upper part to the L_1–L_4 vertebrae, and at the lower part to the lesser trochanter of the thigh, and consequently which can produce bending of the spine in the lumbar section). Somehow or other the displacements forward connected with this section of the spine of the torso mass are insignificant in comparison with the deviation of the main bulk of the torso backward; consequently, because of this, the respiratory movements of the upper and middle parts of the torso are uncompensated for and can lead to a significant displacement of the

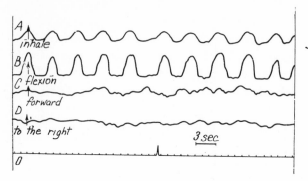

Fig. 3. Simultaneous recording of respiration (A), the hip angle (B), and the sagittal (C) and frontal (D) stabilograms. The mark on the bottom line indicates closing of the eyes.

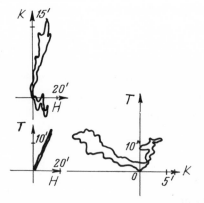

Fig. 4. Simultaneous recording of the truncal (T), hip (H) and knee (K) angles on the phase plane during two cycles of respiration. Calibration of the changes of the corresponding angles (in minutes of arc) is indicated on the axes. They correspond to the positive changes of the angles at inspiration: for the T, deviation of the torso backward; for the H, bending; for the K, straightening. While the changes in the T and H obviously are linearly related and occur simultaneously, the direction, size, and also time displacement of the changes of the K angle with respect to changes in the T and H vary from cycle to cycle of respiration.

overall center of gravity of the body which would be reflected in a recording of the stabilogram. In spite of that, synchronous recording of a stabilogram and pneumogram (Fig. 3) indicate the absence of a direct relation between the oscillations of the body and the respiratory movements of the torso. A stabilogram while holding the breath when the respiratory movements are absent hardly differs from a stabilogram during normal respiration. Only during forced respiration it is possible to record a weak effect of the respiratory movements on the stabilogram. At the same time in the stabilogram of several patients with disorders of the higher parts of the central nervous system a respiratory rhythm is distinctly detected. What has been said convinces us that normally there must exist some special mechanism as a result of which the specific shiftings of the mass are compensated for and therefore are not reflected in the position of the overall center of gravity.

In order to clarify the existence of such a compensatory mechanism, a series of recordings of the different joint angles (hip, knee, cervical, talocalcaneal), and of the respiration, and a stabilogram of a standing human subject was carried out. It turned out that at inspiration at the same time with the deviation of the torso backward, bending takes place in the hip and cervical angles (the head inclines forward), that is, antiphase changes of the truncal angle (Fig. 4). The amplitude of the respiratory oscillations of the hip angle is of the same order as the truncal, but the amplitude of the cervical angle is somewhat less than the hip.

Thus, every time respiration causes a specific displacement of the mass leading to a change of the truncal angle, a change of the hip angle occurs of approximately equal size but directed in the opposite direction. Thanks to this, respiratory-induced oscillations of the position of the overall center of gravity of the body are not observed in the horizontal plane.

The compensating role of the hip joint is supplemented by respiratory changes of the cervical angle; however, we can suppose that the basic purpose of these movements is not one of compensation but retention in space of an absolute position of the head, which apparently is especially important for normal visual and auditory perception.

Recordings of the angular changes of the talocalcaneal joints showed the absence in them, as in the stabilogram, of oscillations of a respiratory nature. The knee joint, apparently, participates in compensation, although the degree and constancy of this participation are not clear and require further study.

The suspicion can arise that the similar interrelations of the truncal,

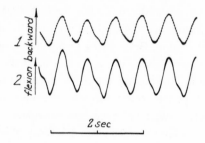

Fig. 5. Retention of interdependence of the changes of the truncal and hip angles during an increase of rate of respiration. (1) truncal angle; (2) hip angle.

hip, and cervical joints are clearly of mechanical origin. This would be so if the lower sections of the spine deflected forward at inhalation not 5 to 6 minutes of arc but by amounts comparable with the actual changes of the hip joint. Moreover, we will say, in anticipation, that under certain conditions it is possible to carry out voluntary changes of the truncal angle which generally do not lead to a change of the hip angle (see Fig. 7). We know that it is impossible to explain changes of the cervical angle during respiration as purely mechanical effects.

It is also impossible to explain respiratory changes of the hip angle as the active action of the iliopsoas muscle: according to the law of mechanics for a lower fulcrum, it can turn the torso only in the direction of the hip and not vice versa. Thus with the action of this muscle the hip remains immovable and the torso, if it moves forward at all under its effect, does so, as we have already said, by not more than 5 to 6 minutes of arc.

We should note that respiratory changes in the hip angle and the cervical section of the spine are strictly antiphase to the changes in the truncal, without any discernible time displacement: with recording of respiration (or the truncal angle) and the corresponding angles on the phase level, a figure is obtained having the appearance of a straight line (Fig. 4). Thus, changes of the angles indicated are always rigidly related. Such a relation is maintained during different changes of respiration. Thus, with an increase of the depth of respiration, proportional with the

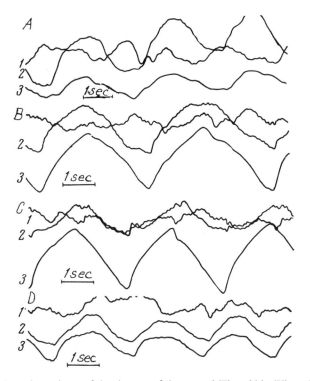

Fig. 6. Interdependence of the changes of the truncal (T) and hip (H) angles and the sagittal stabilogram (SS) during deep breathing (hypoxia with hypercapnia). (A) control recording during forced respiration; (B) initial stage of hypoxia; (C) hypoxia with hypercapnia; (D) normal respiration several minutes after (C); (1) sagittal stabilogram (upward direction on the curve corresponds to deviation of the body forward); (2) upward deviation of the curve corresponds to flexion; (3) upward deviation on the curve corresponds to deviation of the torso backward).

increase of the amplitude of the changes of the truncal angle the amplitude of the changes of the hip and cervical angles increases (Fig. 6A).

The relation of the changes in these angles to changes in the truncal angle is invariant with respect to the frequency of respiration (Fig. 5). During difficulty of respiration (the examinee in this case breathed an oxygen-depleted mixture) the amplitude of the respiratory movements of the torso vigorously increases, and the amplitude of the hip angle

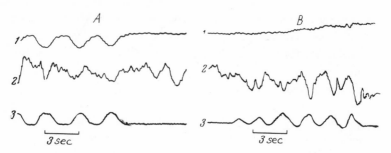

Fig. 7. (A) control recording of the T, H, and SS during untroubled respiration; (B), H, and SS during voluntary oscillations of the trunk while holding breath. Changes of the H angle related to voluntary oscillations of the torso are absent. Notations same as in Fig. 6.

(Fig. 6B) increases proportionally. Only in conditions of ensuing hypoxia with hypercapnia were signs detected of a disruption of the relation of the changes in the hip joint to the respiratory movements of the torso: the form of the oscillations of the hip angle noticeably changes, and the increase of the amplitude lags behind the increased volume of truncal movements (Fig. 6C).

The final experiment of the study made it possible to answer the question of whether this rigid connection of the joint angles develops only with respect to respiration or also with respect to other disturbances of the position of the torso. The experiment gave the following result. Under ordinary conditions of the experiment the examinee was asked to hold his breath and while holding his breath to vary arbitrarily the truncal angle roughly with the frequency and amplitude of the respiratory disturbances of the torso. The truncal and hip angles and a stabilogram were recorded. It turns out that in such conditions the hip joint generally does not reveal signs of respiratory oscillations (Fig. 7). Respiration is thus apparently the physiological process which plays the essential role in the organization of the fixed interaction of specific joints.

DISCUSSION

The data reported make it possible to form an opinion about the compensation mechanism of respiratory disturbances of the upright posture of man. The basic role in this compensation belongs to movements of the hip joint. Energetically this is justified since in comparison with the knee and talocalcaneal the musculature of the hip joint needs to develop smaller moments for compensation of truncal oscillations. Thanks to this, greater stability of the body and an economy of muscle work are achieved.

The compensation mechanism which we will call the *respiratory synergy* has a series of specific characteristics based on which we can formulate an overall definition of a synergy as follows. A synergy is a fixed and reproducible interaction of the joints or of groups of joints, developed as a result of training or innate, organized, and controlled by the central nervous system for effective solution of a specific motor problem.

There is no reason at all to assume that the synergy examined, of leveling the respiratory disturbances of the upright posture, is something exceptional. On the contrary it is very probable that there are rather many different synergies of pose (not just upright) and movement, and one of the important problems is the detection and study of the mechanism of their realization.

When there are many synergies it is possible to describe the retention of pose of man in the following way. Under the effect of a complex disturbance both random and determined, the nervous system selects at once one or another synergy or several for compensation. In such a way, control of the multilink system is simpler by far than with the independent control of each mechanical degree of freedom separately.

Apparently it is essential that the new synergy form as a result of a more or less prolonged training. Indirectly speaking in favor of this is the fact that we could not detect the appearance of a new respiratory synergy in those conditions when the hip joints of the examinee were fixed. In such cases when the hip joints were more or less fully immobilized, an increase was observed of the amplitude of the oscillations of the body's overall center of gravity (Fig. 8) in the rhythm of respiration. Apparently it turned out that the recording time was insufficient for the nervous system to be "retuned" and to develop a new synergy corresponding to the mechanically changed system. Once developed, the synergy proves to be rather stable. Thus, the infraction of the respiratory

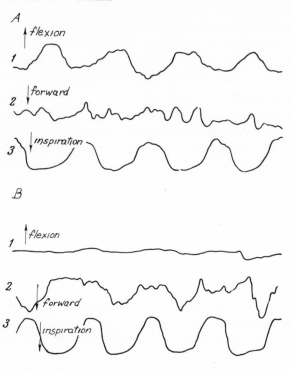

Fig. 8. Changes of a sagittal stabilogram with immobilization of the hip joints. (A) initial recording; (B) during immobilization of the hip joints by a temporary splint. (1) hip angle (flexion is the deviation of the line upward); (2) sagittal stabilogram (shifting of the line downward corresponds to the deviation of the body forward), (3) pneumogram (the shifting of the line downward indicates inspiration).

synergy can be detected only during pronounced changes of the condition of the central nervous system (for example during acute hypoxia) or local disturbances of it. Thus, for example, in a patient with a tumor in the posterior fossa there can be seen a disruption of the respiratory synergy leading to a reduction of the stability of the upright posture. In such patients, prominent slow oscillations of the stabilogram are frequently detected having the rhythm of respiration (Fig. 9). A study of the interaction of the joints in these cases shows that the phase relation of the oscillations of the position of the body's overall center of gravity with the respiratory movements of the chest are not constant. The observation presented attests to the fact that these oscillations are not simple mechanical reflections of respiratory movements.

Thus, the research cited shows that many muscles of the spine (the cervical and the lumbar sections) and the hip joints participate in the compensation of the respiratory disturbances of the position of the body's overall center of gravity. Their simultaneous involvement in compensation testifies to the presence of a fixed (within the limits of a given task) interaction of the motor centers of the corresponding muscles, which provides the basis of examining such interaction of them as a functional union (synergy) thanks to which the necessity of individual control of each mechanical degree of freedom included in the synergy becomes superfluous; that is, the dimension of the controlling system is reduced. Each synergy is organized and controlled by a definite physiological function and subserves it.

Apparently the functional synergy described is not unique. The presence of a number of synergies that reduce the number of degrees of freedom of the control system creates the opportunity of choosing one or another blocks of movements in the process of training of a complicated movement, or during retention of one or another stable posture that is disturbed randomly or determinedly.

In this chapter we have described only the phenomenology of a respiratory synergy; we did not touch upon its physiological mechanism,

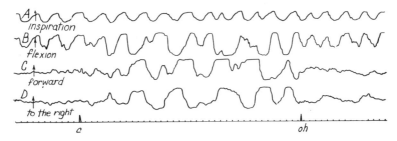

Fig. 9. Disorder of the respiratory synergy of a patient with a tumor in the posterior fossa, when the eyes are closed. (A) pneumogram; (B) truncal angle; (C) sagittal and (D) frontal stabilograms; (c) closed eyes; (o) open eyes.

which ensures a particular interaction of different muscle groups serving very different joints. In the literature there are many data about the presence of irradiation of excitation from the respiratory center evoking corresponding changes of activity of the spinal cord, the spinal reflexes, and the skeletal musculature [3, 6–9]; but obviously the interaction of the musculature ensuring respiratory synergy requires differentiated and gradual distribution of the activity of the different muscles at a time which is difficult to explain as only irradiation of activity from the respiratory center.

REFERENCES

1.
M. Gunkel, *Pflüg. Arch. Ges. Physiol.* 1962:275, 472.
2.
V. S. Gurfinkel, *Stoyanie zhorovykh lyudey i protezirovannykh posle amputatsii nizhnikh konechnostey*, dokt. diss. (*The Erect Posture of Healthy Persons and of People with Artificial Limbs After Amputation of the Lower Limbs*, doctoral dissertation), Moscow, 1960.
3.
Hildebrandt, *Pflüg. Arch. Ges. Physiol.* 1963:278, 113.
4.
A. G. Kotikova, *Biomekhanika fizicheskikh uprazhneniy* (*The Biomechanics of Physical Exercise*), Moscow, 1938.
5.
V. I. Krinskiy and M. L. Shik, *Biofizika* 1964:9, 607.
6.
L. P. Latash, *Elektricheskie yavleniya v spinnom mozgu* (*Electrical Phenomena of the Spinal Chord*), Moscow-Leningrad, 1962.
7.
O. S. Merkulova, *Interoretseptory i skeletnaya muskulatura* (*Interoceptors and the Skeletal Musculature*), Moscow-Leningrad, 1959.
8.
A. I. Roytbak, *Myshechnaya deyatel'noct' i fazy dykhaniya. I. Vsesoyuznoe soveshchanie po voprosam fiziologii vegetativnoy nervnoy sistemy i mozzhechka* (*Muscle Activity and Phases of Respiration. First All-Union Conference on Problems of the Physiology of the Vegetative Nervous System and the Cerebellum*), Yerevan, 1961.

9.
Yu. S. Yusevich, *Elektromiografiya tonusa skeletnoy muskulatury chelo-veka v norme i patologii* (*The Normal and Pathological Electromyography of the Tone of the Human Skeletal Musculature*), Moscow: Medgiz, 1963.

Abnormal rectification, phenomenon of, 114
Absolute refractory period, 155, 156
Action, definition of, xxiv
Action potential, 165
  potassium permeability and, 108
  of Purkinje fibers, 106, 107
  shortening of, 108
Adrian, R. H., 113
Afferentation
  expected, 336
  incoming, 20, 330
  kinesthetic, 373–381
  minimized, 338
  proprioceptive, 337
  segmental, 361
Agarwal, G. C., xix
Aizerman, M. A., xviii, xx
"All or nothing," phenomenon of, 119
Analog, digital converter, 348
Apnea, 204, 215, 216
Araki, T., 44
Arshavskiy, Yu. I., 16, 25–77, 161, 163, 166
Ashby's principle of homeostasis, 17
Asymptotic optimums, 9
Atrioventricular delay, lag, 103, 167–171, 177, 180
  blocking of impulses and, 179–181
Atrioventricular region, hypothetical element of, 103
Attenuation constant, 90, 91
Atwood, R. P., 251, 276, 298, 303
Autocorrelation function, 354, 355
Automata, xiii
  finite, 7
  game of, 10, 18
AV delay, See Atrioventricular delay
Axons
  of basket cells, 306, 308
  giant, of squid, 106, 107, 109, 123
  of stellate cells, 308

Babskiy, Ye. B., 170, 172, 173, 174, 175
Balakhovskiy, S. S., 186
Barnard, J. W., 251, 292, 298
Basket cells, 304, 314, 316

Basket plexuses, 305
Batsel, H., 196
Baumgarten, A., 199
Baumgarten, R., 220, 234
Berkinblit, M. B., 27–77, 79–131, 155–192
Bernstein, N. A., xviii, 329, 337, 371, 373
Bickel, A., 376
Biophysics, xvii
Bishop, G. H., 43
Blinkov, S. M., 310
Blocking of conduction, 101
  nature of, 167
  periodic, 181
Bok, S. T., 251, 253
Boroviagin, V. L., 271
Brache preparations, 320
Brady, A., 109, 122
Brain stem, synaptic switchings in, 208
Braitenberg, V., 251, 276, 298, 303, 315
Brookhart, J. M., 51
Brumlik, D. V., 349
Burns, D., 200, 203, 209, 211, 234
Buskirk, Van C., 349

Cable element, 88. See also Fibers
  dendrite section as, 30, 31
  passive electrical properties of, 103
  potential attenuation in, 83
  structure of, 87
Cable theories, 26
Cajal, R., 251, 264, 268, 278, 279, 280, 288, 289, 291, 298, 305
Carbon dioxide, levels of, 204, 218
Carmeliet, E., 117
Cells, cylindrical, 29
  densities of, 321
  of reticular formation, 51
  type A, 236, 249
  type B, 236, 241, 243, 244, 245, 247, 249
Cellular contact, model of, 83
Central nervous system (CNS)
  integral function of, 329–345
  mathematical modeling of, 1–22
Cerebellar cortex
  geometry of, 251

Cerebellar cortex (*continued*)
  granular layer of, 273
  hidden, 262–263
  length of, 262
  stellate cells of, 32, 70–71
  values for, 322
Cerebellar, disease, 357
Cerebellar islets, 265, 268
Cerebellum
  cortical functioning of, 317
  cortical layers of, 258
  granular layer of, 264–279
  orders of distortion, 261
  sagittal sections of, 256
  volume of, 254, 320
Cerebral cortex, pyramidal cells of,
    54, 61–65
Chain
  ergodic, 8
  equivalent, 145
  of excitable elements, 184–185,
    186
  expiratory, 210
  homogeneous, 140–142
  kinematic, 382
Chain method, 136
Chang, H. T., 43
Chaylakhyan, L. M., 15, 16, 25–77,
    79–131, 186
Chepurnov, S. A., 166
Chetaev, A. N., 212
Chromatolysis, 49
Climbing fibers, 309–310, 315
Closed systems, electrical properties
    of, 79–99
Close packings, principle of, 269
Cohen, M. H., xvii, 216
Cole, 78
Collaterals, of basket cell axon, 307
Compactness, principle of, 263
Complex systems, functional repre-
    sentation of, 1
  modeling of, 7
Computer, data input into, 348
Conduction, all-or-none, xiv
  atrioventricular, 103
  through membrane, 180
  retrograde, 102–103
  wave, xvii
Continuity, electrical, 82, 83, 87
  principle of, 269, 273

Continuity of layer, principle of,
    265
Continuous activity, 208
Continuous medium, 14, 104
Control, hierarchical form of, xxii
Control systems, xxii
Coordination, of movement, 373
Coraboeuf, E., 116
Correlation analysis, 346
Cortex
  macroscopic organization of,
    253–264
  responses of, 42–43
  straightened, 253, 318
  thickness of, 254
Cortical folding
  causes of, 263
  laws of, 255
  mechanical hypothesis of 263–264
Coughing, respiratory neurons and,
    205
Cranefield, P. F., 110, 118, 120
Crill, W. E., 82–98, 120
Current, postsynaptic, 29
Curtiss, 78

Daue, N. S., 102
Deafferentation, 376, 379
Deleze, J., 116
Dendrites
  of basket cell, 306, 307
  branching of, 37, 57
  of cells of reticular formation, 51
  conduction of excitation along,
    37–40, 46
  electrical properties of, xiv
  electrotonic attenuation in, 42
  excitation arising in, 45
  functional properties of, 25–77
  overlapping of, 292–295
  in Purkinje cells, 288–292 (*see
    also* Purkinje cells)
  smooth, 288, 315
  spiny, 288
  of stellate cell, 307
Dendritic tree, 31
Dense packings, 317
Depolarization, 49
  of inspiratory motor neurons, 222
  maximum rate of, 122
  supraliminal, 110

Diaphragm, 220. *See also*
    Respiration
    posture and, 383
Diastole, membrane resistance
    during, 107
Dirken, M., 198
Donskikh, Ye. A., 170, 172, 173,
    174
Dow, R. S., 298
Draper, M. H., 107
Dropouts, of impulses, 164, 166,
    168, 173

Eccles, R., 25–77, 222, 315, 316
Economy, principle of, 264
Edwards, C., 96
Electron microscope, 50, 79, 294
Elner, A. M., 379
Encephalograph, 347, 384
EPSP. *See* Potential, excitatory
    postsynaptic
Estable, C., 268
Estimator function, 2, 3
Excitable element, analytical models
    of, 155–163
    during rhythmic stimulation,
        181–183
Excitation, 303
    all-or-none, xiv
    in branched structures, 26
    conduction of, 79
    membrane potential during, 37
    rate of, 38
    rate of conduction of, 41
    spreading in heart, 16
    spread of, xvi
    transmission of, 56
    types of, 52–53
Excitation fronts, 236, 237, 238
Extremum
    blind search for, 3
    of function, 2
    local search for, 4
    nonlocal search for, 4
    ravines method of, xii, xiii, 4, 5,
        19, 335

Fadiga, E., 51
Fatt, P., 44
Feldman, A. G., 361–372, 382–395
Fibers
    auricular, 103

Fibers (*continued*)
    branching, 143
    cable properties of, 39
    compound, 136–140
    cylindrical, 133–136
    homogeneous, 163
    as lattice structures, 147
    treelike (*see* Trees)
Folium, 259
    formation of, 262
    optimum, 260
Fomin, S. V., 346–360
Fourier series, 151
Fox, C. A., 251, 292, 298
Frank, K., 97
Freide, R., 251, 265
Frequency division, 181
    principle of, 157
Fritsch, 329
Functional synergies, 358, 373
Functions
    homotopy classes of, xxiii
    interchangeability of, xxiv
    piecewise realization of, xxiii
    structure and, xviii
Fuortes, M. G. F., 97
Fusiform cells, 277

Game, in distribution, 11, 13, 18
    on periphery, 13
Game theory, language of, 9
Gamma-drum machine, 246
Ganzen, V. A., 184
Gelfand, I. M., xx, xxi, 1–22, 104,
        105, 161, 166, 235, 252,
        329–345, 371, 380, 382
Gemeinschaftbewegungen, 376
George, E. P., 88, 89, 132
Gesell, R., 198
Giant cell, dendrites of, 51
Glomeruli, density of, 271–274
    parenchymatous, 268
Golgi cells, 32
    "axon cloud" of, 318
    dendrites of, 298
    electrotonic attenuation of
        potential in, 69
    input resistance in, 69
    types of, 278, 279
Golgi preparations, 32, 269, 275–
        277, 294, 298, 305, 308, 320

Goniometer, 374, 383, 384
Goodwin, B. C., xvii
Gottlieb, G. L., xix
Granit, R., 366
Granovskaya, R. M., 184
Grant, R. P., 174
Granular layer, 317
Granule cells, 265
    axons of, 298
    dendritic processes of, 276
    density of, 270–271, 320
    phylogenetic tendencies of, 299
    soma-somatic contacting of, 267,
        268–270, 318
Gray, E. G., 268, 301, 305
Green's function, 133
Grundfest, H., 43, 44
Gunkel, M., 374
Gurfinkel, V. S., 329–345, 346–360,
    361–372, 380, 382–395
Gur game, 11, 13, 18

Hall, A. E., 112
Harmon, L. D., 161, 182
Healing, phenomenon of, 81
Heart. *See also* Myocardium
    cellular membrane of, 79–131
    intracellular stimulation of, 105
    muscle fibers of, 85
Hemodynamics, 180, 346
    influence on tremor of, 349
Heterogeneity, 101
    of excitable tissue, 180
Hild, W., 44
Histograms, 284, 285
Hitzig, 329
Hodgkin, A. L., 40, 106, 109–111,
    131, 182, 183
Hoff, H., 211
Hoffman, B. E., 110, 118, 120
Homeostasis, Ashby's principle of,
    17
H-reflexes, 363, 367, 369, 370
    of gastrocnemius muscle, 368
Huftschmidt, H., 339
Hukuchara, T., 198
Hutter, O. F., 115, 117
Huxley, A. F., 40, 106, 109, 182,
    183
Huyghens' construction, xvii
Hypercapnia, 204, 389, 390

Hyperpolarizing electrode, 112
Hyperventilation, 204, 218
Hypoxia, 389, 390

Impulses, auricular, 175
    dropout of, 165
    electrical transmission of, 81
    periodic blocking of, 163–179
    periodic dropout of, 166
    volley and, 315
Inhibition, lateral, 278
    of neuron, 208
    presynaptic, 55
Inhomogeneities, distribution of
    potential and, 99–106
Input conductivity, determination of,
    146
Input resistance, of closed hexag-
        onal synctium, 88
    determination of, 95, 134, 137,
        143
    of functioning myocardium, 120
    in Golgi cell, 69
    in motor neuron of cat, 59–60
    of neuron, 30–32, 33, 36
    nucleus interpositus and, 66–68
    of pyramidal cell, 62
    of rabbit ventricular fibers, 96
    in stellate cells, 70–71
    of syncytia, 90, 154
Inspiration, 219
    postural changes and, 387
    protracted, 207
    spinal cord and, 384
Institute of Automation and Remote
        Control, xx
Institute of Earth Physics, of USSR,
        348
Institute of Problems of Information
        Transmission, xi
Interaction, interarticular, 382–395
    principles of, 340
Interneurons, in spinal cord, 361
Intersectedness, degree of, 248
Ivanov, A. F., 161

Jacobian matrix, 138, 150
Jacobian structure, 139, 153
Jendrassik's maneuver, 367
Jenik, F., 182, 184
Johnson, E. A., 96, 97, 111, 116,
    117, 118, 120, 123, 124

Joints, interaction of, 392

Kandel, E., 49
Kanzow, E., 199
Katz, B., 54, 79, 106
Keder-Stepanova, I. A., 16, 193–233, 234–250
Kirchhoff's law, 148
Kirpichnikov, E. S., 82
Kirzon, M. V., 166
Kitamura, S., 181, 183
Kositsyn, N. S., 50
Kotov, Yu. B., 342
Kots, Ya. M., 361–372, 373–381, 382–395
Kovalev, S. A., 15, 25–77, 79–131, 161, 163, 166, 170, 174, 186
Krinskiy, V. I., 186, 333, 334, 361–372, 373–381, 382
Kronecker, N., 215
Kuffler, 96
Kunstman, K. I., 376
Kupfmuller, K., 182

Λ-model, 158, 161, 177, 178
Latmanizova, L. V., 166
Least interaction, principle of, xiv, 16, 19, 330, 336
"Length constant," 99, 100
Lewis, E. R., 183
Limb, deafferented, 375, 376, 377
    tremor of, 347 (see also Tremor)
Llinás, R., xv
Local currents, theory of, 37
Lungs, ventilation of, 225
Lykashevich, L. P., 16

Macko, D., xiv
Magnus, 329
Marckwald, M., 215
Markov chain, 8
Matrix method, 138, 147
McCulloch, 14
Mechanism, soldered, 316
Media, activity in, xvii
    continuous, 252
    excitable, xvi
Medulla oblongata, inspiratory neurons of, 207
    respiratory neurons of, 194, 197–206, 211, 214, 224

Medulla oblongata, inspiratory reticular formation of, 235
Membrane
    of dendrites, 37, 41–57
    disk, 83, 84
    of heart cells, 106–124
    of myocardial cells, 94
    postsynaptic, 43
    somatodendritic, 45–46
Membrane potential, during re-polarization, 111
    ion conductivity and, 118
    of respiratory neurons, 201
Membrane resistance, 32, 34, 38
    during diastole, 107
    of Purkinje fibers, 107
Mesarovic, M. D., xiv
Miledi, R., 54
Minimum function, search for, 334
Mislavskiy, N. A., 194
Model
    behavior of, 342
    electrophysiological, 314–320
    mathematical, xi
    with several inputs, 184
    of temporary displacement, 170
Mollusks, axons of, 41
Monochotomous tree, 136, 145
Mossy fibers, in granular layer of cerebellar cortex, 274–279
Motor centers, xviii
Mortor cortex, pyramidal cell of, 32
Motor disorders, 377
Motor neurons
    diaphragmatic, 217, 219
    electronic attenuation of potential in, 59–60
    expiratory, 224
    input resistance in, 59–60
    inspiratory, 222
    spinal system, 58, 362
    synaptic excitation of, 48
Motor units, tremor and, 355–356
Movements
    control of, 6–7, 33
    coordination of, 373
    experiments on, 339
    of joints, 373–381
    natural, 330
    operation control of, 329–345
    simultaneous, 375

Movements (*continued*)
  tuning before, 361–372
  visual control of, 377
M-response, 363, 369
Multilink system, 382–395
  tremor in, 356–357
"Multistim," 363
Munk, H., 375, 376
Muscle, contraction of, xxi
  inspiratory, 207, 215
  intercostal, 223
Muscular tension, xx
Mutual complementation, principle
  of, 294
Myocardial fibers, action potential
  of, 119
  electrical continuity of, 87
Myocardium
  action potential of, 123
  cells of, 85, 115, 116
  electrical behavior of, 78–131
  excitation along, 101
  functioning, 86
  as functional syncytium, 79–85
  input resistance of, 120
  local heterogeneities of, 100
  passive electrical properties of,
    80, 89–99
  spread of excitation along,
    99–106
  structure of, 87, 88
  Wenckebach period and, 172
Myotactic reflex action, 361

Nakayama, G., 220
Naydin, V. L., 373–381
Nelson, P., 44
Nembutal, 51
Nerve fiber, branching of, 40
Nervous system, xxi, xxv. *See also*
    Central nervous system
  minimal principle in, xiii
  regulation mechanism of, 383
  traditional models of, 14
Nesland, R., 199, 200, 211, 212
Network
  cubic, 154
  definition of, 147
  input current of, 149
  Jacobian, 153
  multidimensional, 152–154

Network
  two-dimensional, 148–152
  unidimensional, 136, 149
Neurons
  activity mechanism of, 54
  chromatolyzed, 45
  continuous, 214
  excitation of, 25, 53
  expiratory, 195–215, 223
  functional features of, 52–57
  of homogeneous system, 341
  inhibition of, 55, 208
  input resistance of, 30–32
  inspiratory, 195–215
  noncontinuous, 214
  of nucleus interpositus, 66
  spike activity of, 52
  respiratory, 193, 196, 200, 201,
    203, 214
    continuous, 215
    functional properties of,
      197–206
    intercalary, 221
    location of, 196
    on spinal cord, 217–225
    of spinal level, 223
    in thoracic region, 221
    transitional, 216, 217
Ngai, S., 216
Nicholson, C., xv
Nissl preparations, 265, 266, 272,
    284, 320
Noble, D., 109, 112, 114, 116, 117,
    118, 119, 120, 122, 123
Node of branching, 56, 87
  conduction in, 101
  diagram of, 143
  spikes at, xv
Node of Ranvier, 181, 183
Nodes, excitation conduction and,
    54

Oberholzer, R., 196
Obex, 195
O'Brien, J. M., 110
Ohm's law, 28
Okudzhava, V. M., 43, 44
Orbeli, L. A., 376
Organic compounds, crystal struc-
    tures of, 6

Organization, solution methods
and, 3
Otsuka, M., 116
Overlap, phenomenon of, 108

Pacemaker, of heart, 15
latent, 105
Packets, motor units as, 358
Paintal, J., 186
Paltsev, E. I., 361–372, 382–395
Parallel fibers, 322
sublayer of, 311, 313
system of, 297–304, 319
Parameters, essential, 5
intrinsic, 2
macroscopic, 321
nonessential, 5
working, 2
Parkinson's disease, 358
amyostatic form of, 379
"Partial spikes," 47, 48
Patellar reflex, 366, 367
Pavlov, I. P., 329
Perkel, D. H., 184
Permeability
character of change of, 111
membrane potential and, 109
Pezard, A., 287
P fibers, 124
Phase, active tissue and, 14
Phylogenesis, Purkinje cells and, 300
sublayers in, 313
Pitts, T., 14, 195, 197, 220
Plum, F., 199, 200, 211, 212
Pneumogram, 386, 392
Polarization, point, 119
Ponomarev, V. A., 359
Pons, respiratory neurons of, 213
rhythmic respiration and, 216
Posture
active, 355
erect, respiratory disturbances
and, 382–395
posed, 352–353, 371
Potassium, conductivity of, 114
Potential, applied to syncytium, 92
"dendritic," 42, 43
distribution of, 99–104
electronic, 56
electronic attenuation of, 35, 36,
53

Potential, applied to syncytium
in Golgi cell, 69
in motor neuron of cat, 59–60
nucleus interpositus and, 66
of pyramidal cell, 62–65
in stellate cell, 70–71
excitatory postsynaptic (EPSP),
27, 28, 29, 47, 51, 219
inhibiting postsynaptic (IPSP),
219
"pacemaker," 109
postsynaptic (PSP), 27–37, 45
prespike synaptic, 48
prethreshold growth of, 39, 40
somatic, 45
"Prepotentials," 80
PSP. See Potential, postsynaptic
Pumps, ionic, 180
Purkinje cells, 45, 82, 255, 279–297
activity of, 315–316
axons of, 281
basket cells and, 307
bodies of, 280, 297
contacts of, 299–301
dendrites of, 288–295, 300, 301,
304, 310–316, 321
density of, 282–283
distribution of, 263, 283–287
eigenvolumes of, 296, 302
examples of, 289
functioning of, 316
intervals between, 284, 285
network of, 283–287, 295
phylogenetic tendencies of, 299
resting potential, 106
rows, 313
spacing in, 257
spiny branches of, 320
volley of, 316
Purkinje fibers, 95, 106–115
action potential of, 107
repolarization phase of, 113
sodium conductivity in, 109
Purpura, D., 43, 44
Pyatetskiy-Shapiro, I. I., 371

Rall, W., 31, 132
Rank, 286
elementary, 301
hypothesis of, 294, 313–314

Rashton, 78
Ravines method, xii, xiii, 4–5, 19, 135
Recursion method, 138
Refractoriness, 14
  gradient of, 185
Refractory period, 172, 173
  absolute, 155, 156
  relative, 156
Renshaw cells, 342
Repolarization, 108, 121
  "all-or-nothing," 112, 114
  hypotheses of, 109
  ion permeability in, 111
  mechanism of, 118
Respiration
  action potentials and, 193
  apneic, 216
  difficulty of, 389
  hip angle and, 388
  postural changes and, 389 (see also Posture)
  rhythm of, 206, 207–213, 215, 234
  spine and, 384
Respiratory oscillations, 358
Responses, of cortex, 42, 43
Rhythm, frequency division type, 156
  transformation of, 182
Rhythmic volley, working behavior of, 242
Rikko, N. N., 16, 213, 234–250
R-model, of excitable element, 158, 159, 177, 178. See also Rosenblueth
Robertson, A. D. J., xvii
Robertson, P. A., 116
Rosenblueth, A., xvii, 157, 169, 170, 177, 178
Rows, hypothesis of, 294
Rumery, R. E., 82

Safety factor, 165
  AV delay and, 179
  in heart, 180
Salmanovich, V. S., 170, 172, 173, 174
Salmoiraghi, G., 200, 203, 209, 211, 234
Samojloff, A. F., 168, 170, 172

Schaper, 288, 289
Sears, T., 221, 222
Sherrington, C. S., 329, 376
Shidlovskiy, V. A., 102
Shik, M. L., 329–345, 346–360, 361–372, 373–381, 382
Sholl, D. A., 251
Sinus node, synchronization of, 104–105
Smolyaninov, V. V., 25–77, 79–131, 132–154, 162, 185, 251–352
Sneezing, respiratory neurons and, 205
Sodium, permeability, 117
Soma, dendritic synapses and, xiv
Sorokhtin, O. G., 359
Sotnikova, L. E., 346–360
Specificity, of connections, 235
Spectral analysis, 346
Spencer, W., 49
Sperelakis, N., 79
Spikes, generation of, xv
Spinal cord, 361
  motor neurons of, 32
  respiratory neurons of, 217
  segmental apparatus of, 371
Spinules, 307
  of Purkinje cell dendrites, 291
Stabilograms, 357, 384, 386, 390, 392, 393
Stark, L., xix
Stationary random environment, 8
Stellate cells, 304–314
  contrast mechanism of, 314
  of cortex, 70–71
  dendrites of, 306
  electrotonic attenuation of potential in, 70–71
  input resistance in, 70–71
  types of, 307
Stimulation, frequency of, 161
  paired, 369
Stimuli, successive responses to, 159
Stöhr, Ph., 294
Strain gauge pickups, 347
"Successive steps" method, 31
Sumi, T., 224
Summation, 181
Summator, neuron as, 25
Swallowing, respiratory neurons and, 205

Synapse
  activation of, 53
  axodendritic, 50
  axosomatic, 50
  cholinergic, 43
  dendritic, 49–52, 53, 54
    local effectiveness of, 34
  excitability of, xix
  location of, 27–37
  in spinal cord, 361
Synaptic action, excitatory, 27
Synchronism, principle of, 264
Synchronization, of B type cells, 245
  mechanism of, 15
  phenomena of, 249
Syncytium
  close-mesh, 101
  as closed network, 132
  electrical properties of, 132–154
  hexagonal, input resistance in, 122
  open, 146
  spread of excitation in, 99
  as treelike structures, 132
  types of, 87
Syndrome, of impaired control of
    simultaneous movements,
    379
Synergy, 331
  functional, 358, 373
  respiratory, 332, 356, 391, 392,
    393, 394
Systems, expedient, homogeneous,
    18
Szentágothai, J., 251, 268, 298, 305,
    307

Tachycardia, auricular, 174
Takahara, Y., xiv
Tasaki, J., 44
Telesnin, V. R., 161
Tendon reflex, testing of, 363–367
Tensing, 349
Tereshkov, O. D., 346–360
Terzuolo, K., 44
Threshold, dropping, 159
  of excitable element, 155–163,
    165, 170
  excitation, 236
  level of excitation and, 137
Tille, J., 96, 97, 111, 116, 117, 118,
    120, 123, 124

Tissue, active, excitation in, 15
  model of, 18
  properties of, 14
  flat excitable, 15
  inhomogeneity of, 163–166
  myocardial, 98
  periodic blocking of impulses in,
    155
Tofani, W., 196
Transmission, synaptic, 81
Transverse fibers, 311
Trautwein, W., 96
Trees, 142–147
  arbitrary, 144–146
  homogeneous, 147
  monochotomous, 136, 145
  normal, 142
Tremograms, 346, 352
Tremor, characteristics of, 351–355
  definition of, 347, 349
  normal, 350
  pathologic, 357–358
  physiological, 346–360
  of radiocarpal joint, 350, 351,
    352, 353
Trigger zone, 25, 52, 53
Truncal angle, 384
Tsetlin, M. L., xx, xxi, 1–22, 104,
    105, 161, 166, 235, 252, 329–
    345, 361–372, 380, 382
Tubercles, auditory, 195
Tubocurarine, 43
Twitchell, T. E., 376

Udelnov, M. G., 103
Ukhtomskiy, 329
Ulyaninskiy, L. S., 175, 337

Vagal system, 198
Variables. See Parameters
Ventricular fibers, action potential
    of, 122
Volley activity, 210
  in Purkinje cells, 315
  respiratory, 220
  rhythmic, 249
von Neumann, 11

Wang, S., 216
Wang, 211
Waves, phase-shift between, xvii

Waves (*continued*)
  propagation of, xvi
Wedensky inhibition, 183
Weidmann, S., 82, 107, 108, 111,
    112, 113
Weiner, N., xvii
Wells, J., 338
Wenckebach period, 168–179
Wilde, W. S., 110
Wilson, D. M., 161, 163, 168
Woldring, S., 198
Woodbury, J. W., 82, 83, 85, 86, 87,
    88, 89, 97, 98, 109, 112, 113,
    116, 117, 120

X-ray structure, analysis of, 6

Zeevi, Y. Y., xv
Zhabotinsky, A. M., xvii